T0260304

Image Processing Technologies

Signal Processing and Communications

1. Digital Signal Processing for Multimedia Systems, *edited by Keshab K. Parhi and Takao Nishitani*
2. Multimedia Systems, Standards, and Networks, *edited by Atul Puri and Tsuhan Chen*
3. Embedded Multiprocessors: Scheduling and Synchronization, *Sundararajan Sriram and Shuvra S. Bhattacharyya*
4. Signal Processing for Intelligent Sensor Systems, *David C. Swanson*
5. Compressed Video over Networks, *edited by Ming-Ting Sun and Amy R. Reibman*
6. Modulated Coding for Intersymbol Interference Channels, *Xiang-Gen Xia*
7. Digital Speech Processing, Synthesis, and Recognition: Second Edition, Revised and Expanded, *Sadaoki Furui*
8. Modern Digital Halftoning, *Daniel L. Lau and Gonzalo R. Arce*
9. Blind Equalization and Identification, *Zhi Ding and Ye (Geoffrey) Li*
10. Video Coding for Wireless Communication Systems, *King N. Ngan, Chi W. Yap, and Keng T. Tan*
11. Adaptive Digital Filters: Second Edition, Revised and Expanded, *Maurice G. Bellanger*
12. Design of Digital Video Coding Systems, *Jie Chen, Ut-Va Koc, and K. J. Ray Liu*

13. Programmable Digital Signal Processors: Architecture, Programming, and Applications, *edited by Yu Hen Hu*
14. Pattern Recognition and Image Preprocessing: Second Edition, Revised and Expanded, *Sing-Tze Bow*
15. Signal Processing for Magnetic Resonance Imaging and Spectroscopy, *edited by Hong Yan*
16. Satellite Communication Engineering, *Michael O. Kolawole*
17. Speech Processing: A Dynamic and Optimization-Oriented Approach, *Li Deng and Douglas O'Shaughnessy*
18. Multidimensional Discrete Unitary Transforms: Representation, Partitioning, and Algorithms, *Artyom M. Grigoryan and Sos A. Agaian*
19. High-Resolution and Robust Signal Processing, *Yingbo Hua, Alex B. Gershman, and Qi Cheng*
20. Domain-Specific Embedded Multiprocessors: Systems, Architectures, Modeling, and Simulation, *Shuvra Bhattacharyya, Ed Deprettere, and Jürgen Teich*
21. Watermarking Systems Engineering: Enabling Digital Assets Security and Other Applications, *Mauro Barni and Franco Bartolini*
22. Biosignal and Biomedical Image Processing: MATLAB-Based Applications, *John L. Semmlow*
23. Image Processing Technologies: Algorithms, Sensors, and Applications, *Kiyoharu Aizawa, Katsuhiko Sakaue, and Yasuhito Suenaga*

Additional Volumes in Preparation

Image Processing Technologies

Algorithms, Sensors, and Applications

EDITED BY

KIYOHARU AIZAWA
University of Tokyo
Tokyo, Japan

KATSUHIKO SAKAUE
Intelligent Systems Institute
Tsukuba, Japan

YASUHITO SUENAGA
Nagoya University
Nagoya, Japan

CRC Press is an imprint of the
Taylor & Francis Group, an **informa** business

Library of Congress Cataloging-in-Publication Data
A catalog record for this book is available from the Library of Congress.

ISBN 13: 978-0-8247-5057-2

Headquarters
Marcel Dekker, Inc., 270 Madison Avenue, New York, NY 10016, U.S.A.
tel: 212-696-9000; fax: 212-685-4540

Distribution and Customer Service
Marcel Dekker, Inc., Cimarron Road, Monticello, New York 12701, U.S.A.
tel: 800-228-1160; fax: 845-796-1772

Eastern Hemisphere Distribution
Marcel Dekker AG, Hutgasse 4, Postfach 812, CH-4001 Basel, Switzerland
tel: 41-61-260-6300; fax: 41-61-260-6333

World Wide Web
http://www.dekker.com

The publisher offers discounts on this book when ordered in bulk quantities. For more information, write to Special Sales/Professional Marketing at the headquarters address above.

Series Introduction

Over the past 50 years, digital signal processing has evolved as a major engineering discipline. The fields of signal processing have grown from the origin of fast Fourier transform and digital filter design to statistical spectral analysis and array processing, image, audio, and multimedia processing, and shaped developments in high-performance VLSI signal processor design. Indeed, there are few fields that enjoy so many applications—signal processing is everywhere in our lives.

When one uses a cellular phone, the voice is compressed, coded, and modulated using signal processing techniques. As a cruise missile winds along hillsides searching for the target, the signal processor is busy processing the images taken along the way. When we are watching a movie in HDTV, millions of audio and video data are being sent to our homes and received with unbelievable fidelity. When scientists compare DNA samples, fast pattern recognition techniques are being used. On and on, one can see the impact of signal processing in almost every engineering and scientific discipline.

Because of the immense importance of signal processing and the fast-growing demands of business and industry, this series on signal processing serves to report up-to-date developments and advances in the field. The topics of interest include but are not limited to the following:

- Signal theory and analysis
- Statistical signal processing
- Speech and audio processing
- Image and video processing
- Multimedia signal processing and technology
- Signal processing for communications
- Signal processing architectures and VLSI design

We hope this series will provide the interested audience with high-quality, state-of-the-art signal processing literature through research monographs, edited books, and rigorously written textbooks by experts in their fields.

Preface

Research and development of image processing technologies has advanced very rapidly in the past decade. Thanks to recent progress in VLSI technologies, low-cost computers, huge storage, and broadband and wireless technologies, handling image and video became very familiar to us. Digital still cameras, digital video camcorders and internet video are already around us in daily life. These technologies enable us to easily manipulate image and video on desktop PCs. Research and development of applications of image processing have been also accelerated and widened by the rapid growth of hardware development. A number of image processing and computer vision techniques in current use, that used to be in a topic of academic discussion, have now become reality as working real-time systems.

In this book we describe the state of the art of image processing technologies from various points of view. We focus on sensing, algorithms and applications of the growing image processing fields. In order to provide accurate coverage, we invited world renowned scholars and experts from both universities and companies to contribute to this book.

All 10 chapters were originally published in 1999 as a special issue of survey papers on image processing in the *IEICE Transactions on Information and Systems*. It was a unique special issue because it covered a wide spectrum of image processing from sensing to applications and it presented the state of the art of the fields in detail. The issue was well-received, which encouraged us to publish the articles as a book by updating the contents.

Chapter 1 to 4 describe image processing methodologies mainly related to three-dimensional imaging. Chapter 1 is on depth computation from images. Chapter 2 describes optimization techniques that are related not only to three-dimensional recovery but also to segmentation, edge detection, etc. Chapter 3 focuses on reconstruction of motion and structure from multiple images. Chapter 4 covers the emerging research field of three-dimensional image communication from the point of view of representation and compression.

Chapter 5 and 6 examine sensors for advanced image processing systems. Chapter 5 describes various omnidirectional sensors and their applications. Chapter 6 is on computational sensors, in other words, smart sensors that integrate image acquisition and processing on a single VLSI.

Chapter 7 to 10 describe four different application fields of image processing. Chapter 7 covers a wide spectrum of facial image processing, which includes detection, recognition, practical recognition systems, synthesis of facial images and useful databases. Various application systems for human–machine interaction, agents, robots, and facial image cartoons are described. Chapter 8 discusses image processing techniques used in document analysis and recognition. Chapter 9, on medical image processing, focuses on computer-aided diagnosis (CAD) and computer-aided surgery (CAS). CAD and CAS are in the application areas of pattern recognition and visualization, respectively. Chapter 10 covers Intelligent Transport Systems (ITS). Image processing for controlling vehicular traffic is described.

We thank all the contributors, as without them this book could not exist. Indeed, it is their efforts that make this text valuable. Image processing is continuously expanding in many different directions. In this single book, we could not travel everywhere, but hope that the reader will use the book as a roadmap for future exploration. Finally, we acknowledge the patience and guidance of Mr. B. Black, Mr. B. J. Clark, and others at Marcel Dekker, Inc., who helped us to produce this book.

Kiyoharu Aizawa
Katsuhiko Sakaue
Yasuhito Suenaga

Contents

Contributors

Kiyoharu Aizawa
University of Tokyo, Tokyo, Japan

Akira Amano
Graduate School of Informatics, Kyoto University, Kyoto, Japan

Jun Fujiki
National Institute of Advanced Industrial Science and Technology, Tsukuba, Japan

Hiroshi Harashima
The University of Tokyo, Tokyo, Japan

Jun-ichi Hasegawa
Chukyo University, Toyota, Japan

Osamu Hasegawa
Imaging Science and Engineering Laboratory, Tokyo Institute of Technology, Nagatsuta, Yokohama, Japan
Neuroscience Research Institute, National Institute of Advanced Industrial Science and Technology, Tsukuba, Ibaraki, Japan
PRESTO, Japan Science and Technology Corp. (JST), Kawaguchi, Saitama, Japan

Masayuki Kanbara
Nara Institute of Science and Technology, Nara, Japan

Masahide Kaneko
Department of Electronic Engineering, The University of Electro-Communications, Chofu, Tokyo, Japan

Kensaku Mori
Nagoya University, Nagoya, Japan

Takeshi Naemura
Stanford University, Stanford, CA

Shinji Ozawa
Keio University, Tokyo, Japan

Katsuhiko Sakaue
National Institute of Advanced Industrial Science and Technology, Tsukuba, Japan

Takeshi Shakunaga
Okayama University, Okayama, Japan

Junichiro Toriwaki
Nagoya University, Nagoya, Japan

Toyohide Watanabe
Nagoya University, Nagoya, Japan

Yasushi Yagi
Osaka University, Osaka, Japan

Naokazu Yokoya
Nara Institute of Science and Technology, Nara, Japan

1

Passive Range Sensing Techniques: Depth from Images

Naokazu Yokoya and Masayuki Kanbara
Nara Institute of Science and Technology, Nara, Japan

Takeshi Shakunaga
Okayama University, Okayama, Japan

SUMMARY

Acquisition of three-dimensional (3-D) information of a real-world scene from two-dimensional images has been one of the most important issues in computer vision and image understanding in the last two decades. Non-contact range acquisition techniques are essentially classified into two classes: passive and active methods. This chapter concentrates on passive depth extraction techniques that have the advantage that 3-D information can be obtained without affecting the scene. Passive range sensing techniques are often referred to as *shape from x*, where *x* is one of the visual cues such as shading, texture, contour, focus, stereo, motion, and so on. These techniques produce 2.5-D representations of visible surfaces. This chapter discusses several aspects of this research field and reviews the literature including some recent advances such as video-rate range imaging sensors, emerging themes, and applications.

1. INTRODUCTION

One of the most important tasks in computer vision and image understanding is to three-dimensionally interpret two-dimensional (2-D) images of a real-world scene. Thus, the acquisition of three-dimensional (3-D) range or depth information of a scene from 2-D images has attracted much attention in the last two decades. Non-contact range acquisition techniques are essentially classified into two categories: passive and active methods. The former is generally based on solving an inverse problem of the process of projecting a 3-D scene onto a 2-D image plane and has an advantage that 3-D information can be obtained without affecting the scene. The latter is accomplished by emitting some radio or light energy to a target scene and receiving their reflections. Previous reviews of passive methods are found in [1]-[4] and those of active sensors in [5], [6]. This chapter concentrates on reviewing a wide range of passive techniques for extracting depth information from images.

Passive range sensing techniques are often referred to as *shape from x*, where *x* is one of the visual cues such as shading, texture, contour, focus, stereo, motion, and so on. As mentioned above, this type of technique requires solving inverse problems of image formation processes (see Fig. 1): Some may be referred to as *optical* or *photometric* inverse problems and others as *geometric* inverse problems. These passive techniques are roughly classified into two approaches: One is based on solving photometric inverse problems using one or more images taken mainly from a single viewpoint and the other is based on solving geometric inverse problems using multiple images taken from different viewpoints. Both techniques produce 2.5-D representations of visible surfaces. We discuss several aspects of this research field and review the literature including some recent advances.

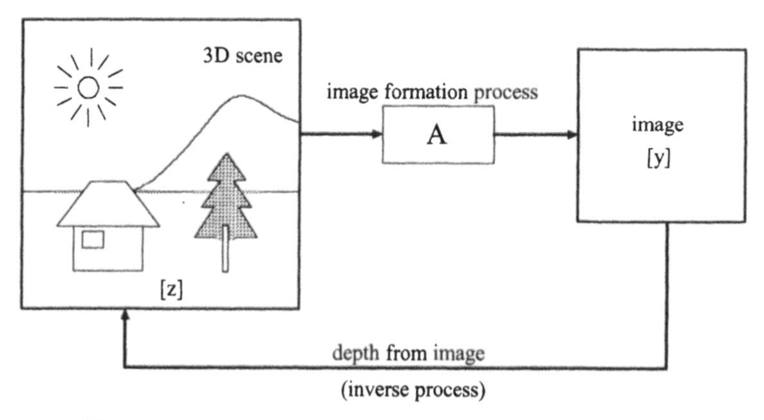

Figure 1 Image formation of a 3-D scene and its inverse process.

This chapter is structured as follows. First, Section 2 describes several methods for acquiring depth mainly from single-view images; that is, shape from shading, shape from texture, shape from contour, and shape from focus/defocus. We then address the problem of acquiring depth from multiple-view images, which includes binocular stereo, multi-ocular stereo, and motion stereo, in Section

3. After reviewing the principles of a variety of passive range sensing techniques, we finally describe recent advances including video-rate passive range sensors as well as some emerging themes and applications in Sections 4 and 5.

2. DEPTH FROM SINGLE-VIEW IMAGES

2.1 Shape from Shading

Shape from shading has been discussed in a wide variety of contexts since the Horn's pioneering work [7]. In the ideal case, the shape from shading problem is formulated as:

$$I(x) = \rho I_s \cos\phi(x),$$

where x denotes a point in the image, $I(x)$ the reflected light intensity observed at x, I_s the illuminant intensity, ρ the constant albedo on the surface, and $\phi(x)$ the angle between the light source direction and the surface normal at the 3-D point on the object surface corresponding to x (see Fig. 2). It is known that $\phi(x)$ can be calculated for a given ρI_s. Thus, the shape of the object surface can be reconstructed from $\phi(x)$ by combining additional constraints such as those obtained from photometric stereo [8] and smoothness assumption on the surface [9] and so on.

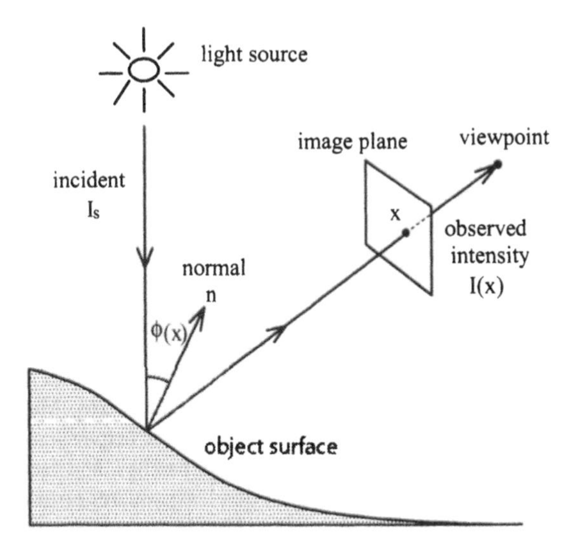

Figure 2 A process of determining brightness of a Lambertian surface.

Shape recovery in a real situation is not easy because real scenes include a lot of complex photometric phenomena affected by interreflections, non-constant albedo distribution, specular reflections, and so on. These factors make it difficult to solve real problems. However, some real problems can be solved when they are controllable and carefully formalized. Wada et al. [10] addressed the problem of recovering the 3-D shape of an unfolded book surface from the shading information in a scanner image. Taking into account several factors included in a real environment, they formulated the problem as an iterative, non-linear optimization problem. Cho et al. [11] showed another approach to the same problem using a divide-and-conquer strategy.

Integration of shape from shading and stereo has many advantages to reconstruct both the shape and reflectance properties of surfaces. In the 1990's, several algorithms were proposed for the integration of shape from shading and stereo modules [12]-[15].

Shape from shading has a lot of variations, some of which focus on particular optical phenomena such as interreflection [16], [17], shadows [18], specular reflection [19], and highlights [20]. Among them, specular reflection is often discussed for its integration with other depth cues. Bhat and Nayar [21] discussed the problem of accurate depth estimation using stereo in the presence of specular reflection based on the analysis of specular reflections. Recently, Samaras and Metaxas [22] proposed another integration scheme that uses non-linear holonomic constraints within a deformable model framework for the recovery of shape and illuminant direction from shading. This is rather general as it does not require knowledge of illuminant direction.

2.2 Shape from Texture

Shape from texture is a cue to 3-D shape, which is very closely related to binocular stereo and shape from motion. All of them are based on the information available in multiple perspective views of the same surface in the scene. However, shape from texture is different from the others in that correspondence should be solved in a single image.

In typical approaches, texture distortion is first calculated from the image, then 3-D orientation and shape of the scene surface are inferred from the texture distortion. In this scheme, identification of texture elements (texels) is an important problem. For texel identification, several techniques are proposed such as a multiscale region detector in a Laplacian-Gaussian scale space [23], [24], scale selection based on a windowed second moment matrix [25], and measurement of local spatial frequency moments using Gabor functions [26]. Malik and Rosenholtz [27], [28] proposed a technique for estimating the texture distortion as a 2-D affine transformation between neighboring image patches, and recovered the surface orientation by non-linear minimization of a least-squares error criterion. Other investigations of this problem are included in [29]-[31].

2.3 Shape from Contour

Shape from contour is another strong cue to 3-D shape reconstruction. In the case of curved objects, rich and robust information on the shape is provided by occluding contours. The corresponding contours on the surface, called rims, are viewpoint dependent, and the viewing directions at their points are tangential to the surface. That is, shape from contour techniques are incomplete by itself for 3-D reconstruction except for a limited class of objects, but it may provide a strong cue when it is used with other 3-D reconstruction methods. With using controlled object rotation, Zheng [32] explored a shape from contour technique for acquiring 3-D graphics models from a sequence of object contours. This is actually a multiple-view approach. 3-D surface reconstruction techniques using occluding contours are also discussed in [33]-[35].

There exist different but similar techniques called *shape from silhouette.* Some of these techniques are sometimes referred to as space carving [36], [37], and are also used to reconstruct 3-D graphics models from multiple-view images obtained from a controlled moving camera or multiple fixed cameras of known postures. These may be classified into multi-ocular or motion stereo and will be described later.

2.4 Shape from Focus/Defocus

Blurring phenomena due to defocusing are also among the important cues for depth recovery. There are two approaches to the utilization of the blurring phenomena for shape reconstruction. The first approach, called *depth from focus,* is a search-based method. A sequence of images are taken with a changing focus in small steps, and the setting that optimizes image focus is determined [5], [38]-[41]. Considerable work has been done to improve focus criterion functions and search algorithms.

On the other hand, the second approach, called d*epth from defocus*, is a model-based method [42]-[49]. A couple of images are taken with different optical settings. The depth map is reconstructed by solving the inverse problem of the blurring process based on camera and edge models. Xiong and Shafer [50] and Asada et al. [48] proposed integrated methods.

Recently, several attempts were reported concerning the integration of the focus cue and other cues. Ahuja and Abott [51] proposed an integration of stereo and focus as well as vergence cues for the surface recovery. Hiura and Matsuyama [52] proposed using coded apertures with depth from defocus. Schechner and Kiryati [53] analyzed the effect of noise in different spatial frequencies and discussed performance comparison between depth from focus/defocus and stereo algorithms.

3. DEPTH FROM MULTIPLE-VIEW IMAGES

3.1 Binocular Stereo

Binocular stereo imitates the human stereo vision and is typical of non-contact passive range sensing methods [1]. A pair of images are obtained from two different viewpoints under perspective projection as illustrated in Fig. 3. Provided that the corresponding points are determined in both image planes, a distance measurement (depth) from a known reference coordinate system to the surface point is computed based on triangulation. The main problem in binocular stereo is to find correspondences in a stereo pair. This problem, which is called the *stereo correspondence problem*, has been attacked by many researchers over a quarter century [54]-[56], following the Julesz's pioneering work [57] on random-dot stereograms in the early 1960's.

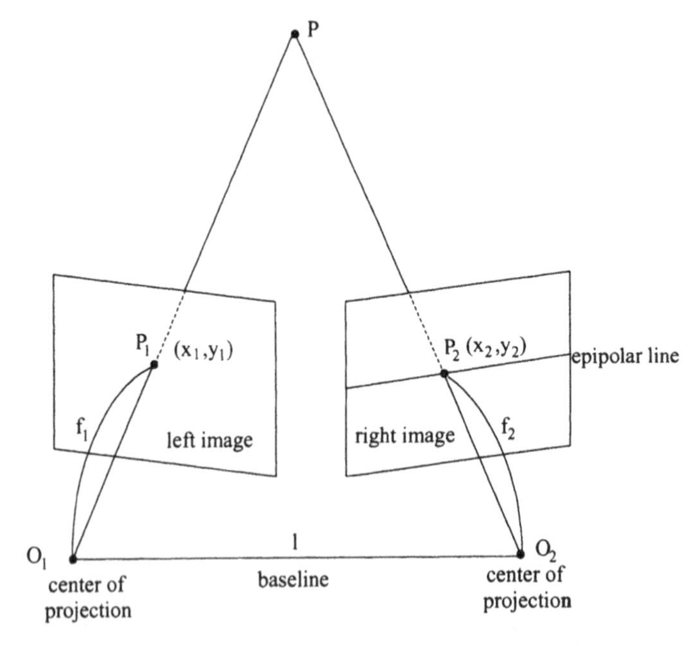

Figure 3 Geometry of binocular stereoscopic imaging.

The stereo correspondence search is restricted to a line called an *epipolar line*, which is determined as an intersection of an image plane and an *epipolar plane,* containing an image point and two projection centers of cameras (see Fig. 3). In the standard stereo imaging geometry using two equivalent cameras with parallel viewing directions, the epipolar line is a scanline. In general, the stereo correspondence problem is formulated as a global optimization problem; see Sakaue's paper [58] for a detailed survey on computer vision algorithms from the viewpoint of optimization. Different features are chosen for stereo matching: Some are computed for image points (pixels) [54], while others are segments (lines or contours) [59]. These are called area-based and segment-based methods, respectively. Area-based stereo matching methods usually produce dense depth maps, while segment-based matching methods compute depth values at sparse segments.

The first step in designing a stereo matching algorithm is to define similarity or difference measures between the left and right images. Typical measures for area-based matching are the sum of squared differences (SSD), the sum of absolute differences (SAD), and the normalized cross correlation (NCC). These measures of standard stereo geometry are defined as follows.

$$SSD(d) = \sum_i \sum_j \{I_L(x+d+i, y+j) - I_R(x+i, y+j)\}^2,$$

where I_L and I_R represent the left and right images of a stereo pair and d denotes the disparity at a point (x, y) in the right image.

$$SAD(d) = \sum_i \sum_j \left| I_L(x+d+i, y+j) - I_R(x+i, y+j) \right|.$$

$$NCC(d) = \frac{C(I_L, I_R) - \sum_i \sum_j \mu_L \mu_R}{\sum_i \sum_j \sigma_L \sigma_R},$$

where μ_L and μ_R represent intensity averages in the corresponding windows of the left and right images, respectively, and σ_L and σ_R are standard deviations in the windows, respectively. $C(I_L, I_R)$ is a cross correlation between the corresponding windows:

$$C(I_L, I_R) = \sum_i \sum_j I_L(x+d+i, y+j) I_R(x+i, y+j).$$

The disparity d is found by minimizing the difference measure or by maximizing the similarity measure. However simple minimization or maximization procedures cannot always determine the correspondences uniquely because of the ill-posed nature of the problem [60]. Resolving the ambiguities in stereo matching requires additional constraints (local supports) such as continuity constraint [61]-[63], disparity gradient constraint [64], and disparity similarity function [65]. Poggio et al. [66] clearly pointed out that the regularization theory provides a unified approach for integrating similarity measures and continuity constraints. Using the framework of standard regularization theory, the stereo matching problem is formulated as that of finding a disparity function $d(x, y)$ that minimizes the following energy functional.

$$E(d) = P(d) + \lambda S(d),$$

where $P(d)$ is called a penalty functional, which measures the differences between corresponding points, and $S(d)$ is a stabilizing functional, by which a continuity or smoothness constraint is imposed upon the solution. Note that the regularization parameter λ controls the strength of the continuity or smoothness constraint. For example, Yokoya [67] used the following penalty and stabilizing functionals in continuous forms.

$$P(d) = \iint \{ I_L(x + d(x,y), y) - I_R(x,y) \}^2 dxdy.$$

$$S(d) = \iint \{ d_x(x,y)^2 + d_y(x,y)^2 \} dxdy.$$

As mentioned earlier, the stereo matching for computing a depth map is a global optimization problem. A number of techniques are used to obtain the global minimum or an approximation of the global minimum; for example, correlation-based template matching [54], [68], relaxation method [69], gradient descent [67], [70], continuation method [71], Kalman filter-based estimation [72], dynamic programming [55], [73], and stochastic optimization algorithms such as simulated annealing [56] and genetic algorithms [74]. Some deterministic algorithms are integrated with multi-scale or coarse-to-fine strategies for avoiding local minima [75], which also can reduce computation time [76]. The regularization-based formulas above can result in a variational approach with multi-scale techniques. Figures 4 and 5 show the result of disparity surface reconstruction from binocular stereo images by a multi-scale regularization approach [67].

Most of the existing approaches assume the continuity or smoothness of visible surfaces. This is usually true for a single surface, but it is not true for an entire scene because a scene usually consists of different surfaces between which there is a discontinuity. If discontinuities are known in the scene, we can easily estimate the depth map preserving discontinuities, however, it is difficult to localize discontinuities before obtaining the depth map. Thus an important problem is how to estimate a depth map and discontinuities simultaneously. The controlled continuity constraint [77], [78] gives a promising approach to the problem. Discontinuities in depth produce occlusions in binocular images; that is, there may exist a region in one image that has no correspondence in the other image. Discontinuities and occlusions often cause mismatches. This problem in binocular stereo can be relaxed to some extent by employing disparity gradient-based iterative depth updating techniques [67], [79], non-linear diffusion techniques [80], [81], and an adaptive window approach [82], [83].

Another important problem with respect to stereo vision is stereo camera calibration. In almost all of the existing approaches, it is assumed that internal and external parameters (or their equivalents) of stereo cameras are known. Moreover, the left and right images are assumed to be rectified before stereo matching in many cases. The stereo calibration problem is discussed in [84]-[89]. The most important issue in stereo calibration is to find solutions that are not significantly affected by noises such as digitization errors and feature localization errors.

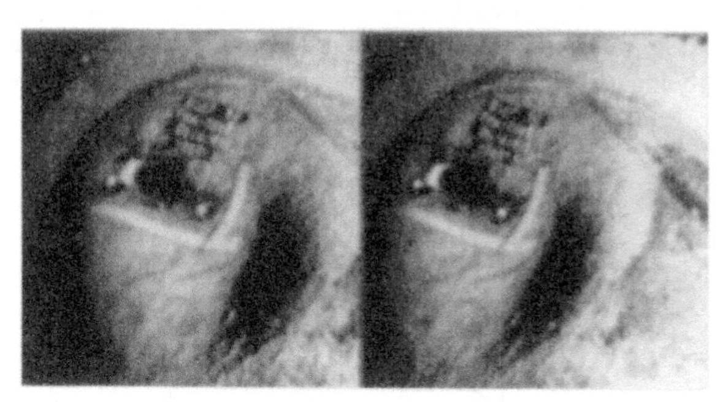

Figure 4 Binocular stereo images of a damaged underwater construct.

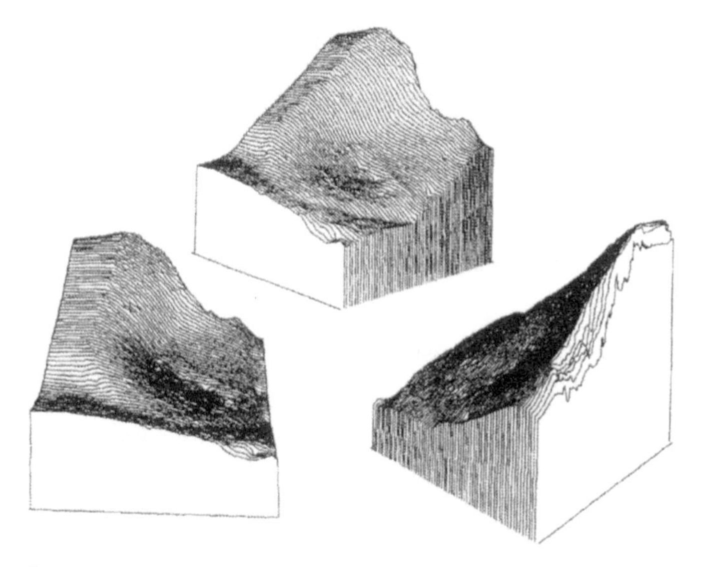

Figure 5 3-D relief of a reconstructed disparity surface viewed from three different positions.

3.2 Multi-ocular Stereo

Some difficulties in binocular stereo such as mismatches and occlusions can be eliminated by adding a third camera [90]-[96]. Stereo with three cameras is generally called *trinocular stereo.* There are typically two types of trinocular camera settings in the literature. The trinocular stereo with three, linearly located cameras can reduce stereo matching errors and also makes it easy to detect occlusions. On the other hand, the trinocular stereo with three, two-dimensionally located cameras can potentially find disparities for linear features parallel to an epipolar line in one stereo pair. Figure 6 illustrates a standard trinocular stereo using three, two-dimensionally located cameras, which combines horizontal and vertical stereopsis.

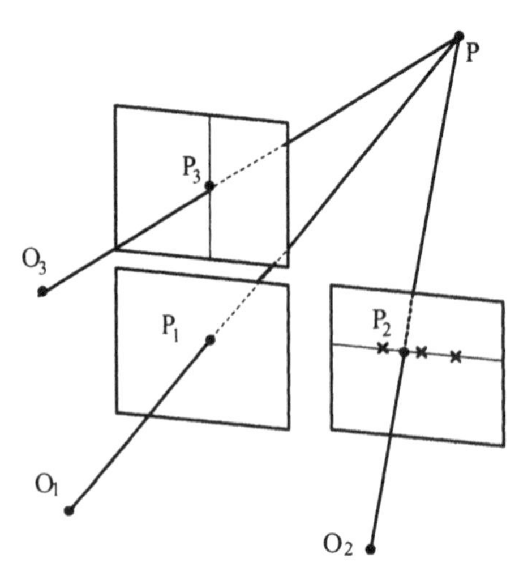

Figure 6 Geometry of trinocular stereo vision with three, two-dimensionally located cameras.

The idea of trinocular stereo vision can be easily extended to the multi-ocular stereo vision with more than three cameras as shown in the Moravec's early work [97]. Recently an increasing number of multi-ocular stereo vision systems with 1-D, 2-D, or arbitrary 3-D camera placements have been developed [94], [95], [98]-[102]. Some of these systems employ the *multiple-baseline stereo* technique originally proposed by Okutomi and Kanade [98], which integrates information from all the binocular stereo pairs of combination and can improve the reliability of stereo matching by reducing mismatches in each stereo pair. Figure 7 shows that summing up the SSDs with different baselines reduces the ambiguity in finding the minimum of an objective function. The objective function employed in multiple-baseline stereo is the sum of SSDs (SSSD), which is simply defined in terms of depth d as $SSSD(d) = \sum_i SSD_i(d)$ at each pixel in the reference image.

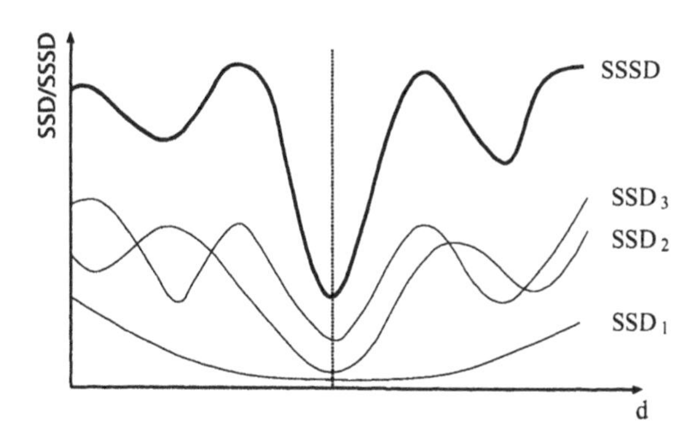

Figure 7 Similarity measures used in multiple-baseline stereo: SSDs and their summation SSSD.

A single moving camera also acquires multi-ocular stereo images of a static scene [103]. Especially, a camera moving linearly with a constant velocity produces a dense sequence of images that is represented as a 3-D volume of data, which is sometimes referred to as a *spatio-temporal image*. A section of the spatio-temporal image consisting of epipolar lines, which is referred to as an *epipolar plane image (EPI)* [104], drastically eases the stereo matching problem; that is, stereo matching actually results in detecting line segments in the epipolar plane image [103]-[105]. Discontinuities and occlusions can also be easily detected in the epipolar plane image. The EPI approach was originated by Yamamoto as the visualized locus method [106] for motion analysis.

There is another type of technique based on controlling the motion of a camera or an object, which can be regarded as shape from contour (described in Section 2.3). Zheng [32] proposed a method of acquiring a 3-D model of a rotating object on a turntable from sequences of its occluding contours (silhouettes) taken by a fixed camera. This approach also employs an EPI representation of image sequences. It should be noted that the same situation as in [32] can be realized by rotating a camera around a fixed object. Approaches using an uncontrolled moving camera will be discussed in detail in the next section.

3.3 Motion Stereo

It is known that a rigid shape can be reconstructed from an image sequence taken from a camera when the camera moves relatively around the object. The simplest class of this problem is equivalent to multi-ocular stereo discussed in the previous section. That is, when camera motion is controlled and the camera is well calibrated, the problem is regarded as multi-ocular stereo. Several navigation systems have been built using a well-calibrated camera [107], [108]. However, the recovery of shape with a well-calibrated camera is not effective in many real applications. Shape recovery is often adversely affected by small errors in the supposed calibration, or is sensitive to small changes that occur due to mechanical vibrations, lens distortion, and so on. These drawbacks cannot be effectively solved in the framework of naïve multi-ocular stereo. A lot of algorithms have been proposed to make robust and reliable reconstruction in a variety of real environments. These attempts are classified into three approaches.

The first approach is posed in the simultaneous recovery of motion and shape from an image sequence. Shape recovery can be made even if the camera motion is unknown. The origin of this approach was founded in the 1970's [109], [110], and a lot of algorithms were proposed in the 1980's [111] in the field of motion analysis. After several batch algorithms were discussed by many researchers, Tomasi and Kanade [112] proposed a *factorization method* to directly decompose shape and motion from an image sequence under orthographic projection. This elegant (and elephant) approach has been modified into several directions, to cope with paraperspective projection [113], multi-body problems [114], line-based estimation [115], and sequential methods [116]. Extension to perspective projection was also attempted in the framework of factorization methods [117]-[121]. The simultaneous recovery of shape and motion is also discussed in other formulations [122]-[124]. Most shape from motion techniques cannot determine the scale factor and suffer from occlusions. Sato et al. [125] used a limited number of 3-D known points in a scene as markers to determine the scale factor. In their approach, the tracking of markers and natural features determines extrinsic camera parameters and then the shape recovery problem results in multi-ocular stereo described in the previous section. They applied an extended multiple-baseline stereo to compute dense depth maps from multiple long sequences of images obtained by a hand-held video camera. Figure 8 shows an example of a textured 3-D model and camera paths recovered by this approach.

Figure 8 Textured 3-D model and camera paths recovered from two image sequences obtained by a hand-held video camera.

In the second approach, an uncalibrated camera was used instead of a calibrated one. Therefore, self-calibration is needed along with either Euclidean or projective shape recovery. The recovery of Euclidean structure is discussed in [126]-[128]. For example, Azarbayejani and Pentland [126] proposed a recursive algorithm for estimation of motion, structure, and focal length. The recovery of projective structure, that is, structure modulo a projective transformation, is generally discussed in [129]-[132]. Beardsley et al. [133] proposed a sequential algorithm for projective and affine structure recovery. Quan et al. [134]-[136] discusses self-calibration of an affine camera, and its application to affine structure recovery and to Euclidean motion/structure recovery. This research is based on the investigation of epipolar geometry. Related topics are discussed in [137]-[143].

The integration of binocular stereo and motion stereo, the third approach, is one of the most practical approaches to the stereo navigation [79], [144]-[151]. Matthies and Shafer [144] modeled triangulation errors using the three-dimensional Gaussian distribution and showed an integration scheme of the stereo and motion approaches. For real-time implementation, some strong cues are often effectively used such as spatio-temporal consistency [146] and focus of expansion [145].

4. REAL-TIME PASSIVE RANGE SENSORS

It has been recognized over a long period that one of the most significant problems in the practical application of passive range sensing is its high computational cost. Due to recent advances in algorithms and hardware, however, a number of real-time passive sensors have appeared since the 1990's [46], [95], [99], [101], [152]-[156]. The term "real-time" means the acquisition of a dense depth map or a range image at video-rate or nearly at video-rate. Roughly speaking there are two types of real-time passive range sensors: One is multi-ocular stereo based on the multiple-baseline stereo theory [95], [99], [152], and the other is based on the principle of depth from focus or depth from defocus [41], [46], [52], [153].

4.1 Real-Time Stereo

Among various real-time stereo sensors, the pioneering CMU stereo machine developed by Kanade et al. [95], which was constructed of special-purpose hardware including digital signal processors, has accomplished the performance of acquiring a 200×200×5bit range image at the speed of 30 frames per second (actually at video-rate) using five two-dimensionally located cameras. Table 1 summarizes the performance of the CMU video-rate stereo machine [95].

An increasing number of real-time stereo machines, most of which are based on trinocular or multi-ocular stereo, are now available as commercial products. Moreover, simple binocular stereo algorithms can run in real-time even on standard PCs.

Table 1 Performance of CMU Stereo Machine

Number of cameras	2 to 6
Processing time/pixel	33ns×(disparity range + 2)
Frame rate	Up to 30 frames/sec
Depth map size	Up to 256×240
Disparity search range	Up to 60 pixels

4.2 Real-Time Depth from Focus/Defocus

Another example of a real-time passive range sensor was developed by Nayar et al. [46]. This system is based on the shape from defocus theory and produces a 512×480 depth map at video-rate using commercially available pipeline processors. This sensor has been shown to have accuracy better than 0.3%. Hiura and Matsuyama [52], [153] have also developed a real-time focus range sensor using a newly designed multi-focus camera. They reported that the implementation of the method on a SIMD parallel image processor could obtain a 200×200 range image at the speed of 25 frames per second [153].

Though most of the existing video-rate passive range sensors were demonstrated only in laboratories and at exhibitions in the early stage, some commercial versions of these sensors are now available as mentioned above. They are creating new applications requiring real-time depth sensing and interaction between users and the machines.

5. EMERGING THEMES AND APPLICATIONS

Improvements in the speed, accuracy, reliability, and cost of passive range sensors have made a number of conventional applications such as inspection and robot vision [157] more realistic and practical, and have opened up new vistas to applications. Especially video-rate sensors are now playing an important role in a new class of applications: Seamlessly merging the real and virtual worlds. This new field is characterized by the term *augmented reality* [158], *virtualized reality* [159], or *mixed reality* [160], [161]. This field will cooperate with a wearable computing framework [162] in a ubiquitous network environment.

One main issue in the research field is to properly align 3-D coordinates of the real and virtual environments with respect to each other. This problem is sometimes referred to as *geometric registration* of the real and virtual. This is equivalent to acquiring a 3-D position and orientation of user's eyes. Several vision-based approaches to geometric registration can be found in the literature, for example, see [163]-[165]. Vision-based geometric registration has the advantage that there are potentially no limitations in measuring area so that the workspace is not restricted. Moreover, such a technique can easily realize a video see-through augmented reality system, which mainly consists of a head-mounted display (HMD) with a small video camera. The real-time stereoscopic tracking of three unknown feature points in the real environment makes it possible to realize a stereoscopic video see-through system which provides us with a rich 3-D sensation of a mixed environment consisting of real and virtual objects (see Fig. 9 for an early prototype of stereo vision-based video see-through augmented reality system with two cameras [166], [167]).

Figure 9 Appearance of a stereoscopic video see-through HMD.

Another problem in merging the real and virtual worlds is to resolve occlusions among the real and virtual objects; that is, virtual objects should be merged into a real scene image without occlusion conflicts. Video-rate stereo sensors, which produce dense depth maps, can be applied to such applications as generating composite stereoscopic images of the real and virtual in real-time [95], [167], [168].

A recent important topic in stereo vision is an omnidirectional stereo [169]-[177] using catadioptric omnidirectional image sensors for robotics, video surveillance, and virtual/mixed reality applications. Yagi [178] comprehensively reviewed various types of omnidirectional image sensors including omnidirectional stereo vision systems.

A recent trend in computer vision and image processing research is toward media-oriented applications including real-time human-computer interaction [179], while most conventional applications mainly concentrate on the automation of machines in restricted environments. In this stream, passive range sensing techniques will play an important role in acquiring human behavior from captured images and in augmenting real and virtual worlds. Key issues are real-time and robustness of vision-based range sensing algorithms. The robustness may be achieved by integrating partial 3-D cues from different vision modules. This should be investigated further.

6. CONCLUSIONS

This chapter reviewed a number of passive range sensing techniques in computer vision that are sometimes referred to as *shape from x*, and briefly described the emerging media-oriented applications such as mixed reality. Depth extraction from images has attracted much attention in two aspects: Scientific interest in human depth perception and development of practical 3-D computer vision algorithms and systems. This survey mainly focused on, but was not limited to, the latter aspect.

REFERENCES

[1] S.T. Barnard, M.A. Fischler. Computational stereo. ACM Computing Surveys 14(4):553-572, 1982.
[2] B.K.P. Horn. Robot Vision. New York: McGraw-Hill, 1986.
[3] J. Aloimonos. Visual shape computation. Proc. the IEEE 76(8):899-916, 1988.
[4] M. Okutomi. Stereo vision. In: T. Matsuyama, Y. Kuno, A. Imiya, eds. Computer Vision: Reviews and Perspectives. Tokyo: Shin-Gijutsu Communications, 1998, pp.123-137 (in Japanese).
[5] R.A. Jarvis. A perspective on range finding techniques for computer vision. IEEE Trans. Pattern Anal. Mach. Intell. PAMI-5(2):122-139, 1983.
[6] P.J. Besl. Active, optical range imaging sensors. Machine Vision and Applications 1:127-152, 1988.
[7] B.K.P. Horn. Obtaining Shape from Shading Information. New York: McGraw-Hill, 1975, pp.115-155.
[8] R.J. Woodham. Photometric method for determining surface orientation from multiple images. Opt. Eng. 19(1):139-144, 1981.
[9] K. Ikeuchi. Determining surface orientation of specular surfaces by using the photometric stereo method. IEEE Trans. Pattern Anal. Mach. Intell. PAMI-3(6):661-669, 1981.
[10] T. Wada, H. Ukida, T. Matsuyama. Shape from shading with interreflections under a proximal light source: Distortion-free copying of an unfolded book. Int. Journal Computer Vision 24(2):125-135, 1997.
[11] S.I. Cho, H. Saito, S. Ozawa. A divide-and-conquer strategy in shape from shading problem. Proc. IEEE Conf. on Computer Vision and Pattern Recognition, Puerto Rico, 1997, pp.413-419.
[12] Y.G. Leclerc, A.F. Bobick. The direct computation of height from shading. Proc. IEEE Conf. on Computer Vision and Pattern Recognition, Lahaina, Maui, HI, 1991, pp.552-558.
[13] D.R. Hougen, N. Ahuja. Estimation of the light source distribution and its use in integrated shape recovery from stereo and shading. Proc. 4th Int. Conf. on Computer Vision, Berlin, Germany, 1993, pp.148-155.

[14] P. Fua, Y.G. Leclerc. Object-centered surface reconstruction: Combining multi-image stereo and shading. Int. Journal Computer Vision 16(1):35-56, 1995.

[15] S. Pankanti, A.K. Jain. Integrating vision modules: Stereo, shading, grouping, and line labeling. IEEE Trans. Pattern Anal. Mach. Intell. 17(9):831-842, 1995.

[16] S.K. Nayar, K. Ikeuchi, T. Kanade. Shape from interreflections. Proc. 3rd Int. Conf. on Computer Vision, Osaka, Japan, 1990, pp.2-11.

[17] J. Yang, D. Zhang, N. Ohnishi, N. Sugie. Determining a polyhedral shape using interreflections. Proc. IEEE Conf. on Computer Vision and Pattern Recognition, Puerto Rico, 1997, pp.110-115.

[18] M. Daum, G. Dudek. On 3-D surface reconstruction using shape from shadows. Proc. IEEE Conf. on Computer Vision and Pattern Recognition, Santa Barbara, CA, 1998, pp.461-467.

[19] M. Oren, S.K. Nayar. A theory of specular surface geometry. Int. Journal Computer Vision 24(2):105-124, 1996.

[20] X. Yi, O.I. Camps. 3D object depth recovery from highlights using active sensor and illumination control. Proc. IEEE Conf. on Computer Vision and Pattern Recognition, Santa Barbara, CA, 1998, pp.253-259.

[21] D.N. Bhat, S.K. Nayar. Stereo and specular reflection. Int. Journal Computer Vision 26(2):91-106, 1998.

[22] D. Samaras, D. Metaxas. Incorporating illumination constraints in deformable models for shape from shading and light direction estimation. IEEE Trans. Pattern Anal. Mach. Intell. 25(2):247-264, 2003.

[23] D. Blostein, N. Ahuja. Shape from texture: Integrating texture-element extraction and surface estimation. IEEE Trans. Pattern Anal. Mach. Intell. 11(12):1233-1251, 1989.

[24] K.M. Lee, C.C.J. Kuo. Direct shape from texture using a parametric surface model and an adaptive filtering technique. Proc. IEEE Conf. on Computer Vision and Pattern Recognition, Santa Barbara, CA, 1998, pp.402-407.

[25] T. Lindeberg, J. Garding. Shape from texture from a multi-scale perspective. Proc. 4th Int. Conf. on Computer Vision, Berlin, Germany, 1993, pp.683-691.

[26] B.J. Super, A.C. Bovik. Shape from texture using local spectral moments. IEEE Trans. Pattern Anal. Mach. Intell. 17(4):333-343, 1995.

[27] J. Malik, R. Rosenholtz. A differential method for computing local shape-from-texture for planar and curved surfaces. Proc. IEEE Conf. on Computer Vision and Pattern Recognition, New York, NY, 1993, pp.267-273.

[28] J. Malik, R. Rosenholtz. Computing local surface orientation and shape from texture for curved surfaces. Int. Journal Computer Vision 23(2):149-168, 1997.

[29] L.S. Davis, L. Janos, S.M. Dunn. Efficient recovery of shape from texture. IEEE Trans. Pattern Anal. Mach. Intell. PAMI-5(5):485-492, 1983.

[30] L.G. Brown, H. Shvaytser. Surface orientation from projective foreshortening of isotropic texture autocorrelation. IEEE Trans. Pattern Anal. Mach. Intell. 12(6):584-588, 1990.

[31] C. Marinos, A. Blake. Shape from texture: The homogeneity hypothesis. Proc. 3rd Int. Conf. on Computer Vision, Osaka, Japan, 1990, pp.350-353.

[32] J.Y. Zheng. Acquiring 3-D models from sequences of contours. IEEE Trans. Pattern Anal. Mach. Intell. 16(2):163-178, 1994.

[33] E. Boyer. Object models from contour sequences. Proc. 4th European Conf. on Computer Vision, Cambridge, UK, 1996, Volume 2, pp.109-118.

[34] E. Boyer, M.O. Berger. 3D surface reconstruction using occluding contours. Int. Journal Computer Vision 22(3):219-233, 1997.

[35] L.X. Zhou, W.K. Gu. 3D model reconstruction by fusing multiple visual cues. Proc. 14th IAPR Int. Conf. on Pattern Recognition, Brisbane, Australia, 1998, pp.640-642.

[36] K.N. Kutulakos, S.M. Seitz. The theory of shape by space carving. Proc. 7th Int. Conf. on Computer Vision, Kerkyra, Greece, 1999, I, pp.307-314.

[37] A. Broadhurst, T.W. Drummond, R. Cipolla. A probabilistic framework for space carving. Proc. 8th Int. Conf. on Computer Vision, Vancouver, BC, 2001, pp.388-393.

[38] E. Krotkov. Focusing. Int. Journal Computer Vision 1(3):223-237, 1987.

[39] H.N. Nair, C.V. Stewart. Robust focus ranging. Proc. IEEE Conf. on Computer Vision and Pattern Recognition, Champaign, IL, 1992, pp.309-314.

[40] S.K. Nayar. Shape from focus system. Proc. IEEE Conf. on Computer Vision and Pattern Recognition, Champaign, IL, 1992, pp.302-308.

[41] S.K. Nayar, Y. Nakagawa. Shape from focus. IEEE Trans. Pattern Anal. Mach. Intell. 16(8):824-831, 1994.

[42] A.P. Pentland. A new sense for depth of field. IEEE Trans. Pattern Anal. Mach. Intell. PAMI-9(4):523-531, 1987.

[43] P. Grossmann. Depth from focus. Pattern Recognition Letters 5:63-69, 1987.

[44] M. Subbarao. Parallel depth recovery by changing camera parameters. Proc. 2nd Int. Conf. on Computer Vision, Tarpon Springs, FL, 1988, pp.149-155.

[45] J. Ens, Z.N. Li. Real-time motion stereo. Proc. IEEE Conf. on Computer Vision and Pattern Recognition, New York, NY, 1993, pp.130-135.

[46] S.K. Nayar, M. Watanabe, M. Noguchi. Real-time focus range sensor. Proc. 5th Int. Conf. on Computer Vision, Cambridge, MA, 1995, pp.995-1001.

[47] A.N. Rajagopalan, S. Chaudhuri. Optimal selection of camera parameters for recovery of depth from defocused images. Proc. IEEE Conf. on Computer Vision and Pattern Recognition, Puerto Rico, 1997, pp.219-224.

[48] N. Asada, H. Fujiwara, T. Matsuyama. Edge and depth from focus. Int. Journal Computer Vision 26(2):153-163, 1998.

[49] M. Watanabe, S.K. Nayar. Rational filters for passive depth from defocus. Int. Journal Computer Vision 27(3):203-225, 1998.

[50] Y. Xiong, S.A. Shafer. Depth from focusing and defocusing. Proc. IEEE Conf. on Computer Vision and Pattern Recognition, New York, NY, 1993, pp.68-73.

[51] N. Ahuja, A.L. Abbott. Active vision: Integrating disparity, vergence, focus, aperture, and calibration for surface estimation. IEEE Trans. Pattern Anal. Mach. Intell. 15(10): 1007-1029, 1993.

[52] S. Hiura, T. Matsuyama. Depth measurement by the multi-focus camera. Proc. IEEE Conf. on Computer Vision and Pattern Recognition, Santa Barbara, CA, 1998, pp.953-959.

[53] Y.Y. Schechner, N. Kiryati. Depth from defocus vs. stereo: How different really are they?. Int. Journal Computer Vision 39(2):141-162, 2000.

[54] M.D. Levine, D.A. O'Handley, G.M. Yagi. Computer determination of depth maps. Computer Graphics and Image Processing 2(2):131-150, 1973.

[55] Y. Ohta, T. Kanade. Stereo by intra- and inter-scanline search using dynamic programming. IEEE Trans. Pattern Anal. Mach. Intell. PAMI-7(2):139-154, 1985.

[56] S.T. Barnard. Stochastic stereo matching over scale. Int. Journal Computer Vision 3(1):17-32, 1989.

[57] B. Julesz. Towards the automation of binocular depth perception. Proc. IFIP Congress '62, 1962, pp.439-444.

[58] K. Sakaue, A. Amano, N. Yokoya. Optimization approaches in computer vision and image processing. IEICE Trans. Information and Systems E82-D(3):534-547, 1999.

[59] G. Medioni, R. Nevatia. Segment-based stereo matching. Computer Vision, Graphics and Image Processing 31(1):2-18, 1985.

[60] M. Bertero, T. Poggio, V. Torre. Ill-posed problems in early vision. Proc. the IEEE 76(8):869-889, 1988.

[61] D. Marr, T. Poggio. Cooperative computation of stereo disparity. Science 194:283-287, 1976.

[62] J. Mayhew, J.P. Frisby. Psychophysical and computational studies toward a theory of human stereopsis. Artificial Intelligence 17:349-387, 1981.

[63] W.E.L. Grimson. Computational experiments with a feature based stereo algorithm. IEEE Trans. Pattern Anal. Mach. Intell. PAMI-7(1):17-34, 1985.

[64] S.B. Pollard, J.E.W. Mayhew, J.P. Frisby. Pmf: A stereo correspondence algorithm using a disparity gradient limit. Perception 14:449-470, 1985.

[65] K. Prazdny. Detection of binocular disparities. Biol. Cybern. 52:93-99, 1985.

[66] T. Poggio, V. Torre, C. Koch. Computational vision and regularization theory. Nature 317(6035):314-319, 1985.

[67] N. Yokoya. Surface reconstruction directly from binocular stereo images by multiscale-multistage regularization. Proc. 11th IAPR Int. Conf. on Pattern Recognition, The Hague, The Netherlands, 1992, I, pp.642-646.

[68] K. Mori, M. Kidode, H. Asada. An iterative prediction and correction method for automatic stereo comparison. Computer Graphics and Image Processing 2(3/4):393-401, 1973.

[69] S.T. Barnard, W.B. Thompson. Disparity analysis of images. IEEE Trans. Pattern Anal. Mach. Intell. PAMI-2(4):333-340, 1980.
[70] B.D. Lucas, T. Kanade. An iterative image registration technique with an application to stereo vision. Proc. 7th Int. Joint Conf. on Artificial Intelligence, Vancouver, BC, 1981, pp.674-679.
[71] A. Witkin, D. Terzopoulos, M. Kass. Signal matching through scale space. Int. Journal Computer Vision 1(2):133-144, 1987.
[72] L. Matthies, R. Szeliski, T. Kanade. Kalman filter-based algorithms for estimating depth from image sequences. Int. Journal Computer Vision 3:209-236, 1989.
[73] G.V. Meerbergen, M. Vergauwen, M. Pollefeys, L.V. Gool. A hierarchical symmetric stereo algorithm using dynamic programming. Int. Journal Computer Vision 47(1-3):275-285, 2002.
[74] K. Sakaue. Stereo matching by the combination of genetic algorithm and active net. Systems and Computers in Japan 27(1):40-48, 1996.
[75] D. Marr, T. Poggio. A theory of human stereo vision. Proc. Royal Soc. London B204:301-328, 1979.
[76] C. Sun. Fast stereo matching using rectangular subregioning and 3D maximum-surface techniques. Int. Journal Computer Vision 47(1-3):99-117, 2002.
[77] D. Terzopoulos. Regularization of inverse problems involving discontinuities. IEEE Trans. Pattern Anal. Mach. Intell. PAMI-8(4):413-424, 1986.
[78] R. March. A regularization model for stereo vision with controlled continuity. Pattern Recognition Letters 10(4):259-263, 1989.
[79] G. Sudhir, S. Banerjee, K.K. Biswas, R. Bahl. A cooperative integration of stereopsis and optic flow computation. Proc. 12th IAPR Int. Conf. on Pattern Recognition, Jerusalem, Israel, 1994, pp.356-360.
[80] J. Shah. A nonlinear diffusion model for discontinuous disparity and half-occlusions in stereo. Proc. IEEE Conf. on Computer Vision and Pattern Recognition, New York, NY, 1993, pp.34-40.
[81] D. Scharstein, R. Szeliski. Stereo matching with non-linear diffusion. Proc. IEEE Conf. on Computer Vision and Pattern Recognition, San Francisco, CA, 1996, pp.343-350.
[82] M. Okutomi, T. Kanade. A locally adaptive window for signal matching. Proc. 3rd Int. Conf. on Computer Vision, Osaka, Japan, 1990, pp.190-199.
[83] T. Kanade, M. Okutomi. A stereo matching algorithm with an adaptive window: Theory and experiment. IEEE Trans. on Pattern Anal. Mach. Intell. 16(9):920-932, 1994.
[84] O.D. Faugeras, G. Toscani. The calibration problem for stereo. Proc. IEEE Conf. on Computer Vision and Pattern Recognition, Miami, FL, 1986, pp.15-20.
[85] N. Ayache, C. Hansen. Rectification of images for binocular and trinocular stereovision. Proc. 9th IAPR Int. Conf. on Pattern Recognition, Rome, Italy, 1988, pp.11-16.

[86] H. Takahashi, F. Tomita. Self-calibration of stereo cameras. Proc. 2nd Int. Conf. on Computer Vision, Tarpon Springs, FL, 1988, pp.123-128.
[87] B. Boufama, R. Mohr. Epipole and fundamental matrix estimation using virtual parallax. Proc. 5th Int. Conf. on Computer Vision, Cambridge, MA, 1995, pp.1030-1036.
[88] Q.T. Luong, O.D. Faugeras. The fundamental matrix: Theory, algorithms, and stability analysis. Int. Journal Computer Vision 17(1):43-75, 1996.
[89] R.I. Hartley. In defense of the eight-point algorithm. IEEE Trans. Pattern Anal. Mach. Intell. 19(6):580-593, 1997.
[90] V.J. Milenkovic, T. Kanade. Trinocular vision using photometric and edge orientation constraints. Proc. Image Understanding Workshop, 1985, pp.163-175.
[91] Y. Ohta, M. Watanabe, K. Ikeda. Improving depth map by right-angled trinocular stereo. Proc. 8th IAPR Int. Conf. on Pattern Recognition, Paris, France, 1986, I, pp.519-521.
[92] M. Yachida, Y. Kitamura, M. Kimachi. Trinocular vision: New approach for correspondence problem. Proc. 8th IAPR Int. Conf. on Pattern Recognition, Paris, France, 1986, pp.1041-1044.
[93] N. Ayache, F. Lustman. Fast and reliable passive trinocular stereo vision. Proc. 1st Int. Conf. on Computer Vision, London, UK, 1987, p.422-426.
[94] Y. Nakamura, T. Matsuura, K. Satoh, Y. Ohta. Occlusion detectable stereo – Occlusion patterns in camera matrix. Proc. IEEE Conf. on Computer Vision and Pattern Recognition, San Francisco, CA, 1996, pp.371-378.
[95] T. Kanade, A. Yoshida, K. Oda, H. Kano, M. Tanaka. A stereo machine for video-rate dense depth mapping and its new applications. Proc. IEEE Conf. on Computer Vision and Pattern Recognition, San Francisco, CA, 1996, pp.196-202.
[96] M. Agrawal, L.S. Davis. Trinocular stereo using shortest paths and ordering constraint. Int. Journal Computer Vision 47(1-3):43-50, 2002.
[97] H.P. Moravec. Visual mapping by a robot rover. Proc. 6th Int. Joint Conf. on Artificial Intelligence, Tokyo, Japan, 1979, Volume 1, pp.598-600.
[98] M. Okutomi, T. Kanade. A multiple-baseline stereo. IEEE Trans. Pattern Anal. Mach. Intell. 15(4):353-363, 1993.
[99] B. Ross. A practical stereo vision system. Proc. IEEE Conf. on Computer Vision and Pattern Recognition, New York, NY, 1993, pp.148-153.
[100] K. Satoh, Y. Ohta. Passive depth acquisition for 3D image displays. IEICE Trans. Information and Systems E77-D(9):949-957, 1994.
[101] J. Mulligan, V. Isler, K. Daniilidas. Trinocular stereo: A real-time algorithm and its evaluation. Int. Journal Computer Vision 47(1-3):51-61, 2002.
[102] J. Neumann, Y. Aloimonos. Spatio-temporal stereo using multi-resolution subdivision surfaces. Int. Journal Computer Vision 47(1-3):181-193, 2002.
[103] M. Yamamoto. Determining three-dimensional structure from image sequences given by horizontal and vertical moving camera. Trans. of IEICE J69-D(11):1631-1638, 1986 (in Japanese).

[104] R.C. Bolles, H.H. Baker, D.H. Marimont. Epipolar-plane image analysis: An approach to determining structure from motion. Int. Journal Computer Vision 1(1):7-55, 1987.

[105] R. Cipolla, M. Yamamoto. Stereoscopic tracking of bodies in motion. Image and Vision Computing 8(1):85-90, 1990.

[106] M. Yamamoto. Motion analysis using the visualized locus method. Trans. of IPSJ 22(5):442-449, 1981 (in Japanese).

[107] N. Ayache. Artificial Vision for Mobile Robots. Cambridge, MA:MIT Press, 1991.

[108] Z. Zhang, O. Faugeras. 3D Dynamic Scene Analysis. Springer-Verlag, 1992.

[109] J.J. Koenderink, A.J. van Doorn. Local structure of movement parallax of the plane. Journal of the Optical Society of America 66(7):717-723, 1976.

[110] S. Ullman. Relaxation and constrained optimization by local processes. Computer Graphics and Image Processing 10(2):115-125, 1979.

[111] T.S. Huang, ed. Image Sequence Processing and Dynamic Scene Analysis, New York: Springer-Verlag, 1983.

[112] C. Tomasi, T. Kanade. Shape and motion from image streams under orthography: A factorization method. Int. Journal Computer Vision 9(2):137-154, 1992.

[113] C.J. Poelman, T. Kanade. A paraperspective factorization method for shape and motion recovery. IEEE Trans. Pattern Anal. Mach. Intell. 19(3):206-218, 1997.

[114] J. Costeira, T. Kanade. A multi-body factorization method for motion analysis. Proc. 5th Int. Conf. on Computer Vision, Cambridge, MA, 1995, pp.1071-1076.

[115] L. Quan, T. Kanade. A factorization method for affine structure from line correspondences. Proc. IEEE Conf. on Computer Vision and Pattern Recognition, San Francisco, CA, 1996, pp.803-808.

[116] T. Morita, T. Kanade. A sequential factorization method for recovering shape and motion from image streams. IEEE Trans. Pattern Anal. Mach. Intell. 19(8):858-867, 1997.

[117] P. Sturm, B. Triggs. A factorization based algorithm for multi-image projective structure and motion. Proc. 4th European Conf. on Computer Vision, Cambridge, UK, 1996, II, pp.709-720.

[118] T. Ueshiba, F. Tomita. A factorization method for multiple perspective views using affine projection as the initial camera model. IPSJ SIG Notes 97(83):1-8, 1997 (in Japanese).

[119] K. Deguchi. Factorization method for structure from perspective multi-view images. IEICE Trans. Information and Systems E81-D(11):1281-1289, 1998.

[120] N. Ukita, T. Shakunaga. 3D reconstruction from wide-range image sequences using perspective factorization method. IEICE Technical Report 97(596):81-88, 1998 (in Japanese).

[121] J. Oliensis. A multi-frame structure-from-motion algorithm under perspective projection. Int. Journal Computer Vision 34(2-3):163-192, 1999.
[122] J. Weng, N. Ahuja, T.S. Huang. Optimal motion and structure estimation. IEEE Trans. Pattern Anal. Mach. Intell. 15(9):864-884, 1993.
[123] Y. Cui, J. Weng, P. Cohen. Extended structure and motion analysis from monocular image sequences. CVGIP: Image Understanding 59(2):154-170, 1994.
[124] J. Weng, Y. Cui, N. Ahuja. Transitory image sequences, asymptotic properties, and estimation of motion and structure. IEEE Trans. Pattern Anal. Mach. Intell. 19(5):.451-464, 1997.
[125] T. Sato, M. Kanbara, N. Yokoya, H. Takemura. Dense 3-D reconstruction of an outdoor scene by hundreds-baseline stereo using a hand-held video camera. Int. Journal Computer Vision 47(1-3):119-129, 2002.
[126] A. Azarbayejani, A.P. Pentland. Recursive estimation of motion, structure, and focal length. IEEE Trans. Pattern Anal. Mach. Intell. 17(6):562-575, 995.
[127] M. Pollefeys, L.V. Gool. A stratified approach to metric self-calibration. Proc. IEEE Conf. on Computer Vision and Pattern Recognition, Puerto Rico, 1997, pp.407-412.
[128] P. Sturm. Critical motion sequences for monocular self-calibration and uncalibrated Euclidean reconstruction. Proc. IEEE Conf. on Computer Vision and Pattern Recognition, Puerto Rico, 1997, pp.1100-1105.
[129] R. Mohr, F. Veillon, L. Quan. Relative 3D reconstruction using multiple uncalibrated images. Proc. IEEE Conf. on Computer Vision and Pattern Recognition, New York, NY, 1993, pp.543-548.
[130] R.I. Hartley. Projective reconstruction and invariants from multiple images. IEEE Trans. Pattern Anal. Mach. Intell. 16(10):1036-1041, 1994.
[131] R. Szeliski, S.B. Kang. Shape ambiguities in structure from motion. IEEE Trans. Pattern Anal. Mach. Intell. 19(5):506-512, 1997.
[132] R. Kaucic, R. Hartley, N. Dano. Plane-based projective reconstruction. Proc. 8th Int. Conf. on Computer Vision, Vancouver, BC, 2001, pp.420-427.
[133] P.A. Beardsley, A. Zisserman, D.W. Murray. Sequential updating of projective and affine structure from motion. Int. Journal Computer Vision 23(3):235-259, 1997.
[134] L. Quan. Self-calibration of an affine camera from multiple views. Int. Journal Computer Vision 19(1):93-105, 1996.
[135] L. Quan, T. Kanade. Affine structure from line correspondences with uncalibrated affine cameras. IEEE Trans. Pattern Anal. Mach. Intell. 19(8):834-845, 1997.
[136] L. Quan, Y. Ohta. A new linear method for Euclidean motion/structure from three calibrated affine views. Proc. IEEE Conf. on Computer Vision and Pattern Recognition, Santa Barbara, CA, 1998, pp.172-177.
[137] A. Shashua. Projective structure from uncalibrated images: Structure from motion and recognition. IEEE Trans. Pattern Anal. Mach. Intell. 16(8):778-790, 1994.

[138] L.S. Shapiro, A. Zisserman, M. Brady. 3D motion recovery via affine epipolar geometry. Int. Journal Computer Vision 16(2):147-182, 1995.

[139] A. Shashua. Relative affine structure: Canonical model for 3D from 2D geometry and applications. IEEE Trans. Pattern Anal. Mach. Intell. 18(9):873-883, 1996.

[140] Q.T. Luong, O.D. Faugeras. Self-calibration of a moving camera from point correspondences and fundamental matrices. Int. Journal Computer Vision 22(3):261-289, 1997.

[141] M. Trajkovic, M. Hedley. An algorithm for recursive structure and motion recovery under affine projection. Proc. 3rd Asian Conf. on Computer Vision, Hong Kong, 1998, Volume 2, pp.376-383.

[142] F. Kahl, A. Heyden. Affine structure and motion from points, lines and conics. Int. Journal Computer Vision 33(3):163-180, 1999.

[143] K. Astrom, A. Heyden, F. Kahl, M. Oskarsson. Structure and motion from lines under affine projection. Proc. 7th Int. Conf. on Computer Vision, Kerkyra, Greece, 1999, Volume 1, pp.285-292.

[144] L. Matthies, S.A. Shafer. Error modeling in stereo navigation. IEEE Journal Robotics and Automation RA-3(3):239-248, 1987.

[145] M. Abdel-Mottaleb, R. Chellappa, A. Rosenfeld. Binocular motion stereo using MAP estimation. Proc. IEEE Conf. on Computer Vision and Pattern Recognition, New York, NY, 1993, pp.321-327.

[146] H. Baker, R. Bolles. Realtime stereo and motion integration for navigation. Proc. Image Understanding Workshop, Monterey, CA, 1994, II, pp.1295-1304.

[147] R. Koch. 3-D surface reconstruction from stereoscopic image sequences. Proc. 5th Int. Conf. on Computer Vision, Cambridge, MA, 1995, pp.109-114.

[148] P.K. Ho, R. Chung. Stereo-motion that complements stereo and motion analyses. Proc. IEEE Conf. on Computer Vision and Pattern Recognition, Puerto Rico, 1997, pp.213-218.

[149] G.P. Stein, A. Shashua. Direct estimation of motion and extended scene structure from a moving stereo rig. Proc. IEEE Conf. on Computer Vision and Pattern Recognition, Santa Barbara, CA, 1998, pp.211-218.

[150] R. Szeliski. A multi-view approach to motion and stereo. Proc. IEEE Conf. on Computer Vision and Pattern Recognition, 1999, Volume 1, pp.157-163.

[151] F. Dornaika, R. Chung. Cooperative stereo-motion: Matching and reconstruction. Computer Vision and Image Understanding 79(3):408-427, 2000.

[152] S.B. Kang, J.A. Webb, C.L. Zitnick, T. Kanade. A multibaseline stereo system with active illumination and real-time image acquisition. Proc. 5th Int. Conf. on Computer Vision, Cambridge, MA, 1995, pp.88-93.

[153] S. Hiura, T. Matsuyama. Multi-focus camera with coded aperture: Real-time depth measurement and its applications. Proc. 2nd Int. Workshop on Cooperative Distributed Vision, Kyoto, Japan, 1998, pp.101-118.

[154] M. Bjorman, J.O. Eklundh. Real-time epipolar geometry estimation and disparity. Proc. 7th Int. Conf. on Computer Vision, Kerkyra, Greece, 1999, I, pp.234-241.
[155] H. Tao, H.S. Sawhney, R. Kumar. Dynamic depth recovery from multiple synchronized video streams. Proc. IEEE Conf. on Computer Vision and Pattern Recognition, Kauai, HI, 2001, Volume 1, pp.118-124.
[156] H. Hirschmuller, P.R. Innocent, J. Garibaldi. Real-time correlation-based stereo vision with reduced border errors. Int. Journal Computer Vision 47(1-3):229-246, 2002.
[157] Y. Sumi, Y. Kawai, T. Yoshimi, F. Tomita. 3D object recognition in cluttered environments by segment-based stereo vision. Int. Journal Computer Vision 46(1):5-23, 2002.
[158] R.T. Azuma. A survey of augmented reality. Presence: Teleoperators and Virtual Environments 6(4):355-385, 1997.
[159] T. Kanade, P. Rander, P.J. Narayanan. Virtualized reality: Constructing virtual worlds from real scenes. IEEE Multimedia 4(1):34-47, 1997.
[160] P. Milgram, F. Kishino. A taxonomy of mixed reality visual display. IEICE Trans. Information and Systems E77-D(12):1321-1329, 1994.
[161] Y. Ohta, H. Tamura, eds. Mixed Reality – Merging Real and Virtual Worlds. Tokyo:Ohmsha/Springer-Verlag, 1999.
[162] T. Starner, S. Mann, B. Rhodes, J. Levine, J. Healey, D. Kirsch, R.W. Picard, A. Pentland. Augmented reality through wearable computing. Presence: Teleoperators and Virtual Environments 6(4):386-398, 1997.
[163] M. Uenohara, T. Kanade. Vision-based object registration for real-time image overlay. Proc. IEEE Conf. on Computer Vision, Virtual Reality and Robotics in Medicine, Nice, France, 1995, pp.13-22.
[164] K.N. Kutulakos, J.R. Vallino. Calibration-free augmented reality. IEEE Trans. Visualization and Computer Graphics 4(1):1-20, 1998.
[165] T. Okuma, K. Kiyokawa, H. Takemura, N. Yokoya. An augmented reality system using a real-time vision-based registration. Proc. 14th IAPR Int. Conf. on Pattern Recognition, Brisbane, Australia, 1998, Volume 2, pp.1226-1229.
[166] N. Yokoya, H. Takemura, T. Okuma, M. Kanbara. Stereo vision based video see-through mixed reality. In: Y. Ohta, H. Tamura, eds. Mixed Reality – Merging Real and Virtual Worlds. Tokyo:Ohmsha/Springer-Verlag, 1999, pp.131-145.
[167] M. Kanbara, T. Okuma, H. Takemura, N. Yokoya. A stereoscopic video see-through augmented reality system. Proc. IEEE Virtual Reality 2000, New Brunswick, NJ, 2000, pp.255-262.
[168] M.M. Wloka, B.G. Anderson. Resolving occlusion in augmented reality. Proc. 1995 ACM Symposium on Interactive 3D Graphics, Monterey, CA, 1995, pp.5-12.
[169] H. Ishiguro, M. Yamamoto, S. Tsuji. Omni-directional stereo. IEEE Trans. Pattern Anal. Mach. Intell. 14(2):257-262, 1992.

[170] D. Southwell, A. Basu, M. Fiala, J. Reyda. Panoramic stereo. Proc. 13th IAPR Int. Conf. on Pattern Recognition, Vienna, Austria, 1996, I, pp.378-382.
[171] A. Chaen, K. Yamazawa, N. Yokoya, H. Takemura. Acquisition of three-dimensional information using omnidirectional stereo vision. Proc. 3rd Asian Conf. on Computer Vision, Hong Kong, 1998, I, pp.288-295.
[172] T. Kawanishi, K. Yamazawa, H. Iwasa, H. Takemura, N. Yokoya. Generation of high-resolution stereo panoramic images by omnidirectional imaging sensor using hexagonal pyramidal mirrors. Proc. 14th IAPR Int. Conf. on Pattern Recognition, Brisbane, Australia, 1998, Volume 1, pp.485-489.
[173] R. Benosman, J. Devars. Panoramic stereovision sensor. Proc. 14th IAPR Int. Conf. on Pattern Recognition, Brisbane, Australia, 1998, Volume 1, pp.767-769.
[174] S.A. Nene, S.K. Nayar. Stereo with mirrors. Proc. 6th Int. Conf. on Computer Vision, Bombay, India, 1998, pp.1087-1094.
[175] S. Peleg, M. Ben-Ezra, Y. Pritch. Omnistereo: Panoramic stereo imaging. IEEE Trans. Pattern Anal. Mach. Intell. 23(3):279-290, 2001.
[176] H. Tanahashi, C. Wang, Y. Niwa, K. Yamamoto. Polyhedral description of panoramic range data by stable plane extraction. Proc. 5th Asian Conf. on Computer Vision, Melbourne, Australia, 2002, Volume 2, pp.574-579.
[177] K.C. Ng, M. Trivedi, H. Ishiguro. Generalized multiple baseline stereo and direct virtual view synthesis using range-space search, match, and render. Int. Journal Computer Vision 47(1-3):131-147, 2002.
[178] Y. Yagi. Omnidirectional sensing and its applications. IEICE Trans. Information and Systems E82-D(3):569-579, 1999.
[179] Y. Ohta. Pattern recognition and understanding for visual information media. Proc. 16th IAPR Int. Conf. on Pattern Recognition, Quebec City, QC, Canada, 2002, I, pp.536-545.

2

Optimization Approaches in Computer Vision and Image Processing

Katsuhiko Sakaue
National Institute of Advanced Industrial Science and Technology, Tsukuba, Japan

Akira Amano
Graduate School of Informatics, Kyoto University, Kyoto, Japan

Naokazu Yokoya
Nara Institute of Science and Technology, Nara, Japan

SUMMARY

In this chapter, the authors present general views of computer vision and image processing based on optimization. Relaxation and regularization in both broad and narrow senses are used in various fields and problems of computer vision and image processing, and they are currently being combined with general purpose optimization algorithms. The principle and case examples of relaxation and regularization are discussed; the application of optimization to shape description that is a particularly important problem in the field is described; and the use of a genetic algorithm (GA) as a method of optimization is introduced.

1. INTRODUCTION

The field of applied research in computer vision and image processing (hereafter abbreviated as CVIP) is expanding from industrial applications under controlled lighting to vision systems that people use in their environment. What is required in such an expansion is a robust method that is resistant to noise and environmental changes. Application of optimization may provide a breakthrough for CVIP. Specifically, this is an approach to find out the most probable solution from images.

Conventionally, CVIP problems, such as extraction of curved lines, have been resolved by applying a method to determine parameters by the least squares method or some other means by giving appropriate functions, such as polynomials and spline functions. The method of Otsu [1], which was presented in 1979, contemplated optimal function fitting problems, including the simplicity of the functions to be fitted. With this method, a square norm of a quadratic differential of a function was used as the "appropriateness" of the function to be fitted. This method was considered to have made a great step forward from the previous stage of problem setting, where the functions were selected by humans and only the parameter setting was resolved numerically.

A similar means of problem setting was employed in other fields of engineering at that time [2]. The method of Isomichi [3],[4], which was presented in the same period, formulated problems to fit more general functions of one-dimensional signals in quantified empirical data. This method defined the problem explicitly, i.e., the solution is a function with which the sum of the error between the fitting function and the observed data, and the function that evaluates the appropriateness of the fitting function, becomes the minimum.

Methods to minimize the linear sum of indices with different dimensions in this manner are widely used in the processing of pattern data, where input and output are ambiguous. However, CVIP problems are often incompatible with computer calculations, since the function to be fitted is not obtained by analysis or is too complicated. Thus, attempts have been made in the field of CVIP to obtain quasi-optimal solutions, which are coherent overall, in the framework of relaxation.

To conduct numerical analysis of large-scale simultaneous equations, a series of methods called either the iteration method or the relaxation method is widely used. With these methods, the optimal solution is converged in the final stage by iterating calculations to correct an approximate solution that was obtained at the start. Although the methodology of relaxation has not been defined explicitly, its concept is applied to various problems. The definition of relaxation that is most widely used in the field of CVIP is "the general name of methods with which a local constraint is transmitted to the totality by the iteration of parallel operations to minimize (or maximize) a certain index defined to the totality."

Examples of the application of relaxation in a broad sense include Waltz's filtering [5] and the relaxation labeling [6] of Rosenfeld, et al., both of which were presented in the 1970's. Thereafter, attempts were made to solve simultaneous equations by the (original) relaxation method in some fields of computer vision dealing with early vision. In all of these attempts, simultaneous equations were resolved in a calculus of variations framework by formulating the early vision as optimization problems. This trend became more pronounced when Poggio et al. [7] explained early vision by the regularization theory and proposed an integrated framework.

In the field of neural computing, a neural network called the Boltzmann machine, which is based on the principle of analogy to statistical mechanics, was proven effective as an optimization machine for solving problems based on certain principles of energy minimization [8], [9]. Expanding on this approach, Geman et al. [10] established stochastic relaxation for applications of image restoration, which greatly affected subsequent research in this field. Relaxation and regularization in both broad and narrow senses are used in various fields and problems of CVIP, and they are currently being combined with general purpose optimization algorithms.

In this chapter, the principle and case examples of relaxation and regularization are discussed in Section 2. The application of optimization to shape description that is a particularly important problem in CVIP is described in Section 3; and the use of a genetic algorithm (GA) as a method of optimization is introduced in Section 4.

2. UTILIZATION OF RELAXATION AND REGULARIZATION IN CVIP

2.1 Relaxation and Regularization

2.1.1 Deterministic Relaxation

Relaxation as a means of numerical solution of simultaneous equations includes, for example, the Gauss-Seidel iteration method, the Jacobian method, and the successive over-relaxation (SOR) method, all of which are used broadly in various fields. In the case of CVIP, relaxation is used as a means of resolving calculus of variations, which deals with the extremal problem of a function that represents a certain index (e.g., a kind of energy) that needs to be minimized (or maximized) in the entire image. Such a problem can be resolved using either the direct method or the Euler equation. When the Euler equation, which is a necessary condition, is used, it becomes a problem of resolving simultaneous equations obtained by discretizing this partial differential equation per pixel (or datum point). Since the coefficient matrix of such simultaneous equations often becomes large-scale and non-dense, relaxation is generally used as a method for numerically solving the problems of CVIP. In this chapter, the framework of "calculus of variations +

simultaneous equations + relaxation" in the field of CVIP is called "deterministic relaxation."

Deterministic relaxation is described below using some concrete examples (Fig. 1). Now, a problem to interpolate sporadic altitudinal data obtained by observation of a curved surface (e.g., a ground surface) based on the principle of minimum curvature is assumed [11]. Here, the curved surface, $u(x,y)$, to be obtained is considered to be a thin elastic plate fixed by the observed datum points, and interpolation is conducted by obtaining a curved surface that minimizes the energy of deflection expressed by the equation below.

$$C(u) = \iint_{\Omega} \left(\frac{\partial^2 u}{\partial x^2} + \frac{\partial^2 u}{\partial y^2} \right)^2 dxdy \tag{1}$$

This is an extremal problem of a functional with observed data as the boundary condition, and is also a variational problem. The Euler equation for equation (1) is biharmonic as shown below.

$$\frac{\partial^4 u}{\partial x^4} + 2\frac{\partial^4 u}{\partial x^2 \partial y^2} + \frac{\partial^4 u}{\partial y^4} = 0 \tag{2}$$

Incidentally, Laplacian $c_{i,j} = \partial^2 u / \partial x^2 + \partial^2 u / \partial y^2$ of the coordinate (i, j) can be expressed discretely with h as a grid space as follows:

$$c_{i,j} = \frac{u_{i+1,j} + u_{i-1,j} + u_{i,j+1} + u_{i,j-1} - 4u_{i,j}}{h^2} \tag{3}$$

By substitution of the summation of all grid points, the following equation stands.

$$C = \sum_{j=1}^{n} \sum_{i=1}^{m} (c_{i,j})^2 \tag{4}$$

When C is the minimum, the following is obtained.

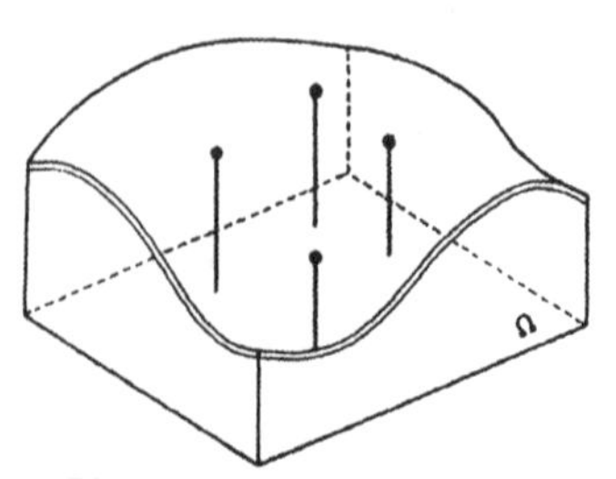

Fig. 1 Surface interpolation.

$$\frac{\partial C}{\partial u_{i,j}} = 0 \ (i = 1, \cdots; m; \ j = 1, \cdots; n) \tag{5}$$

From equations (4) and (5), C is the quadratic form of $u_{i,j}$; thus, equation (5) is a simultaneous linear equation with $m \times n$ unknowns. Furthermore, since $u_{i,j}$ appears only in terms $c_{i,j}$, $c_{i+1,j}$, $c_{i-1,j}$, $c_{i,j+1}$, $c_{i,j-1}$ in C, equation (5) can be modified as follows.

$$\begin{aligned} &u_{i+2,j}+u_{i,j+2}+u_{i-2,j}+u_{i,j-2} \\ &+ 2\,(u_{i+1,j+1}+u_{i-1,j+1}+u_{i+1,j-1}+u_{i-1,j-1} \\ &- 8\,(u_{i+1,j}+u_{i-1,j}+u_{i,j-1}+u_{i,j+1}) + 20u_{i,j} \\ &= 0 \end{aligned} \tag{6}$$

Briggs [11] modified the equation as shown below, and solved it by the Jacobian method.

$$\begin{aligned} &u_{i,j}^{(k+1)} \\ &= \frac{1}{20}\Big\{8\Big(u_{i+1,j}^{(k)} + u_{i-1,j}^{(k)} + u_{i,j-1}^{(k)} + u_{i,j+1}^{(k)}\Big) \\ &-2\Big(u_{i+1,j+1}^{(k)} + u_{i-1,j+1}^{(k)} + u_{i+1,j-1}^{(k)} + u_{i-1,j-1}^{(k)}\Big) \\ &-\Big(u_{i+2,j}^{(k)} + u_{i,j+2}^{(k)} + u_{i-2,j}^{(k)} + u_{i,j-2}^{(k)}\Big)\Big\} \end{aligned} \tag{7}$$

Here, $u_{i,j}$ at the observation points are fixed. The early value is obtained by linear interpolation from the observation points. Another method of solution is employed in the former half of reference [12]. A solution can be obtained by both of these methods by iterating local parallel operations. Equation (7) is in fact a discrete expression of the Euler equation in equation (2). After all, the Euler equation was solved by these methods with empirical data as the boundary condition.

2.1.2 Stochastic Relaxation

When the energy function is not in the quadratic form, the problem becomes nonlinear simultaneous equations. Although a formula for iterative operations can be derived in this case, the convergence to the optimal solution that minimizes the energy is not guaranteed by iterative operations, starting with an arbitrary initial value when several solutions exist (i.e., the original energy function is non-convex and has local minima). The same applies to the process of energy minimization by the Hopfield-type neural network. Thus, the means of selecting an appropriate initial value and avoiding local minima needs to be contrived. To counter such non-convex problems, a method of using multiple scales[13] is employed in the framework of deterministic relaxation, with which processing is conducted gradually from coarse scales to fine scales.

On the other hand, simulated annealing[14], [15] utilizes analogies from statistical thermodynamics to search for the global minimum of non-convex energy functions. This algorithm was proposed as a discrete combinatory optimization method, which makes the state of a system undergo a stochastic transition by establishing a stochastic process, with which the state $\boldsymbol{x}$ appears by a probability expressed by equation (8) when $\boldsymbol{x}$ (which represents the combination of labels; i.e., the state of the system) has energy $E(\boldsymbol{x})$.

$$p(\boldsymbol{x}) = \frac{1}{Z}\exp\{-E(\boldsymbol{x})/T\} \tag{8}$$

Here, Z is a constant for normalization, while T is a parameter corresponding to the "temperature" in the statistical mechanics. With simulated annealing, the system converges to one of the states of energy minimum by the probability of 1, when the initial value of the temperature parameter T is sufficiently high, and when $T \rightarrow 0$ by decreasing the temperature continuously.

The basic steps of annealing based on the probability process of Metropolis, which is often used, are as follows:

1. A sufficiently high initial temperature T_0 is determined, as is the initial state of the system. This state is considered as $\boldsymbol{x}$ (its energy is $E(\boldsymbol{x})$).
2. One freedom of the system is selected at random; its value is varied at random; and this temporary state is considered as $\boldsymbol{x}'$ (its energy is $E\boldsymbol{x}'$)).
3. The state is shifted from $\boldsymbol{x}$ to $\boldsymbol{x}'$ at a probability of min $\{1, \exp\{-(E(\boldsymbol{x}')-E(\boldsymbol{x}) / T_k\}\}$, and the new state is considered as $\boldsymbol{x}$.
4. Steps 2 and 3 are iterated until reaching the equilibrium state.
5. Steps 2 to 4 are iterated while gradually decreasing the temperature ($T_{k} \rightarrow T_{k+1}$) until the value of the energy becomes almost invariable in the equilibrium state.

Many stages of CVIP require some kind of labeling operations of pixels, such as edge detection, segmentation, or stereo matching. When the Markov random field (MRF) can be hypothesized as a probability model of image, i.e., when the state of the respective pixels depends only upon their neighborhood, the energy of the entire image is the sum of local energies. Accordingly, the energy variations of the totality $\Delta E = E(\boldsymbol{x}) - E(\boldsymbol{x}')$, which determine the state variations of the totality in Step 3 above, are equivalent to the energy variations in the neighborhood of a pixel in attention; thus, an annealing process is established by local parallel operations.

A CVIP method in which the state variations are iterated stochastically by conducting local parallel operations, is called stochastic relaxation[10], [16]. Stochastic relaxation has hitherto been used for problems of image restoration, segmentation, binocular stereo, and so on.

Geman et al.[10] applied stochastic relaxation to the problems of image restoration. They prevented the sections of discontinuity in the region boundary from becoming excessively smooth by preparing a stochastic variable for the existence of region boundary, which is called a line process, in addition to the stochastic variable for the gray levels. This contrivance is used for other applied problems of CVIP [17]-[20].

Despite their capacity for dealing with various types of problems, stochastic methods have one disadvantage in that a huge amount of operations is required to obtain solutions. It is thus desirable to use deterministic relaxation, if possible. However, there are not as many classes (types) of problems that can be solved by deterministic methods at present, and stochastic methods are the only alternative means. Thus, a massive parallel computer that can adequately handle such operations must be developed.

2.1.3 Regularization Method

In the field of CVIP, many problems of early vision are formulated as optimization problems [21]-[27].

Problems of early vision can be considered as problems to restore the descriptions of an original three-dimensional world from images created by projecting the data of the three-dimensional world onto a two-dimensional world. However, problems are often essentially indeterminable, because the data of the three-dimensional world that need to be restored are contained in the two-dimensional image in a degenerated shape. Such problems are called "ill-posed problems," as mentioned by Hadamard.

A well-posed problem satisfies the following conditions: (1) its solution exists; (2) the solution is unique; and (3) the solution depends continuously on the initial data. Should it fail to meet any of these conditions, it is called an ill-posed problem. One method available for giving a mathematically stable solution to an ill-posed problem is regularization [28]. Poggio et al. [7] proposed that the problems of early vision be discussed uniformly in terms of the theory of regularization.

Problems that can be solved by regularization are those with a possible solution space that is much greater than input. Specifically, regularization is applicable when possible solutions of problems exist infinitely, as in the case of a problem to restore a three-dimensional structure from image data that are projected two-dimensionally. Regularization is a method for obtaining a mathematically "stable" solution from such a solution space.

With standard regularization, when data that receive a linear operation A from unknown z are observed, as expressed below,

$$y = Az \tag{9}$$

an inverse problem that estimates z from data y is formulated as an optimization problem that minimizes the objective function (called energy) as follows:

$$E = \|Az - y\|^2 + \lambda \| Pz\|^2 \tag{10}$$

The first term in equation (10), $\|Az - y\|^2$, expresses the difference from the observation data, and is called a penalty functional. The second term, $\|Pz\|^2$, is called a stabilizing functional, which expresses general constraint conditions (e.g., smoothness) for solutions in the real world. λ is a regularization parameter that determines the proportions of these two functionals. This formulation is designed "to seek a solution that satisfies the constraint conditions fairly well with few contradictions to observation data." Equation (10) can be minimized in the framework of calculus of variations. With standard regularization, the norm of a quadratic form is adopted, and P is a linear operator.

A matter of great concern is how to determine the regularization parameter λ. One method to estimate a regularization parameter, taking advantage of the error from the amount obtained by observation, was proposed by Amano et al. [29], in which a function that minimizes the difference from the sample is presented as a stabilizing function. Although some samples need to be prepared in advance, this method enables the utilization of errors in observed values by presenting several samples, and estimates the regularization parameter based on these errors.

There are several other methods proposed for estimating a regularization parameter, in which a parameter is estimated as a distributed constant based on the hypothesis that the errors between the observed values and estimated values are dispersed uniformly [30]-[33]. Such methods are effective for processing the neighborhood of the boundary of discontinuity, particularly when dealing with a solution space that contains discontinuity.

2.1.4 Countermeasures for Discontinuity

Methods using a stabilizing functional may not be appropriate when the solution to be obtained is not a smooth function. To counter such problems, various studies have been conducted to establish methods to use functions in the quadratic form as the base [1]-[4]. Other methods being studied include methods to use the spline function [34], [35], GNC and other methods to realize weak continuity constraint [34], [36]-[39], and methods to use a multi-valued function [40]. Despite its great disadvantage of being time-consuming, the stochastic relaxation with the line process (discussed in Section 2.1.2), may also be effective because it is applicable to problems that cannot be solved by deterministic methods.

Terzopoulos introduced a controlled continuity constraint [34], [41]. This constraint is expressed by a generalized spline with several degrees based on continuity control functions. To solve problems of 3D surface reconstruction, a functional that expresses a controlled continuity constraint, was adopted as follows:

$$S_{pr}(u)=\frac{1}{2}\iint_{\Omega} p(x,y)\left[r(x,y)\left(u_{xx}^2+2u_{xy}^2+u_{yy}^2\right)\right. \\ \left.+\{1-r(x,y)\}\left(u_x^2+u_y^2\right)\right]dxdy \tag{11}$$

Here, $p(x,y)$ and $r(x,y)$ are continuity control functions ($0 \leqq p \leqq 1$, $0 \leqq r \leqq 1$). This functional is a weighted linear combination of an energy functional of a thin plate and an energy functional of a thin membrane, and can control the smoothness properties of the constraint by adjusting the control functions at arbitrary points.

The operations of basic control are as follows: i) $p(x,y)$ and $r(x,y)$ are set to be non-zero in all non-continuous points (x,y); ii) the value of $r(x,y)$ is set to be close to 0 when the discontinuity of direction needs to be established; and iii) the value of $p(x,y)$ is set to be close to 0 at the discontinuous points of depth.

If the values of $p(x,y)$ and $r(x,y)$ are known in advance, equation (11) is of quadratic form, and can be solved in the framework of standard regularization. To detect discontinuity in advance, the so-called edge-detection method can be used. However, during the operation of 3D surface reconstruction, Terzopoulos [41] found that the discontinuity of depth at a point where the moment of deflection of a thin plate was 0, and also discovered the discontinuity of direction at the maximum point. A similar approach is described in reference [35]. If $p(x,y)$ and $r(x,y)$ are incorporated to calculus of variations as unknown values, it is quite troublesome because dual minimization processes appear.

Grimson et al. [42] proposed a method for estimating the discontinuity for use in problems to restore a curved surface based on discrete depth data. According to this restoration method, the boundary of discontinuity is estimated by making the error distribution uniform, by using the distribution of errors between the given depth data and the fitted curved surface. This concept is the same as that in the guideline for estimating a regularization parameter.

Based on the idea that partial discontinuity is plausible but imposes a penalty, Blake [43], [44] gave the name "weak constraint" to a constraint condition that minimizes the summation of variations in continuous sections and penalties in discontinuous sections. This problem is non-convex. Graduated non-convexity algorithm (GNC) was proposed as a method to solve this problem. With this method, a sequence of functions that falls in-between the original objective function F (non-convex) and the convex envelope F^* (convex) of F is created, and local optimal solutions are obtained sequentially from F^* by the hill-climbing method to gradually approach the solution of F.

Shizawa [40] succeeded in using two functions as a constraint by expanding standard regularization to multi-valued functions. This method is applicable to a problem of restoring several overlapped surfaces called transparency. Regularization that constrains a curved surface $y = f(x)$ is expressed as a minimization problem of the following function.

$$E^{(1)}[f] = \sum_{i=1}^{N} \left(y_{(i)} - f\left(x_{(i)}\right)\right)^2 + \lambda \left\| Sf(x) \right\|^2 \tag{12}$$

When $f_1(x)$ and $f_2(x)$ are constraint functions, the constraint condition can be expressed as follows:

$$\begin{aligned} &\left(y - f_1(x)\right)\left(y - f_2(x)\right) = \\ &y^2 - \left(f_1(x) + f_2(x)\right)y + f_1(x)f_2(x) = 0 \end{aligned} \tag{13}$$

If $F(x) = f_1(x) f_2(x)$, $G(x) = -(f_1(x) + f_2(x))$, regularization with the two functions as constraint conditions becomes a minimization problem of the following function.

$$E^{(2)}[F,G]=$$
$$\sum_{i=1}^{N}\left\{F\left(x_{(i)}\right)+G\left(x_{(i)}\right)y_{(i)}+\left(y_{(i)}\right)^{2}\right\}^{2}$$
$$+\lambda_F\left\|S_F F(x)\right\|^2+\lambda_G\left\|S_G G(x)\right\|^2 \tag{14}$$

The advantage of this method is that there is no need to explicitly state or estimate which one of the two curved surfaces the empirical data belongs.

2.2 Examples of the Application of Regularization

2.2.1 Reconstruction of a Three-Dimensional Curved Surface

Here, a curved surface reconstruction problem that estimates the original curved surface is established [12], [41], assuming that data d is given as a result of applying a sampling operation S to a given curved surface u. This is identical to the problem of obtaining dense elevations from sporadic data in a digital terrain map (DTM). Although the values of empirical data were stored in the case of the problem of interpolation (introduced in Section 2.1.1), the tolerance obtained from the data of observation points needs to be given in actual operations as shown in Fig.2, due to the inaccuracy of actual empirical data. With the number of pixels of $m \times n$, the so-called quadratic variation is often used as the constraint of smoothness. As with equation (1), this is a thin plate spline, with which continuity is given to the normal vector. Equation (10) can be modified as follows:

$$E=\sum_{j=1}^{n}\sum_{i=1}^{m}S_{i,j}\left(u_{i,j}-d_{i,j}\right)^2+\lambda\iint_{\Omega}\left\{\left(\frac{\partial^2 u}{\partial x^2}\right)^2\right.$$
$$\left.+2\left(\frac{\partial^2 u}{\partial x\partial y}\right)^2+\left(\frac{\partial^2 u}{\partial y^2}\right)\right\}dxdy \tag{15}$$

To obtain a curved surface $u_{i,j}$ ($1 \leqq i \leqq m$, $1 \leqq j \leqq n$) that minimizes equation (15), simultaneous equations with $m \times n$ unknowns (as shown below) are solved by calculus of variations:

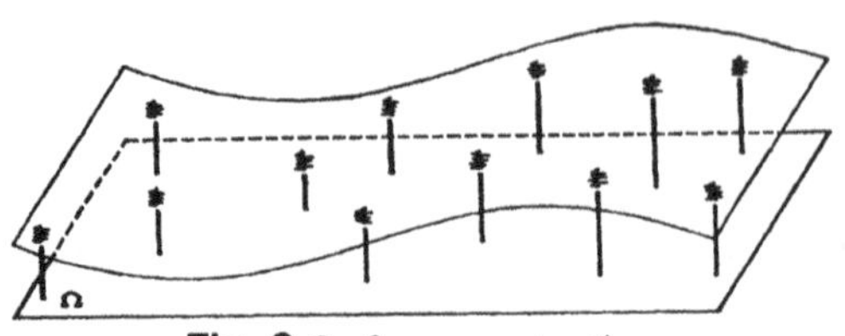

Fig. 2 Surface reconstruction.

$$\frac{\partial E}{\partial u_{i,j}} = 0 \ (1 \leqq i \leqq m, 1 \leqq j \leqq n) \tag{16}$$

If the partial differentiations, such as $\partial^2 / \partial x^2$, in equation (15) are substituted by a difference as follows (h is the grid space):

$$\frac{\partial^2 u}{\partial x^2} = \frac{u_{i+1,j} - 2u_{i,j} + u_{i-1,j}}{h^2} \tag{17}$$

Equation (16) becomes linear simultaneous equations of the values of elevations at all points in image Ω as follows, which can easily be solved [12].

$$\begin{aligned} \frac{\partial E}{\partial u_{i,j}} = {} & 2S_{i,j}\left(u_{i,j} - d_{i,j}\right) + \lambda \left\{ \frac{25}{h^4} u_{i,j} \right. \\ & - \frac{8}{h^4}\left(u_{i+1,j} + u_{i-1,j} + u_{i,j+1} + u_{i,j-1}\right) \\ & + \frac{3}{2h^4}\left(u_{u+2,j} + u_{i-2,j} + u_{i,j+2} + u_{i,j-2}\right) \\ & + \frac{1}{4h^4}\left(u_{i+2,j+2} + u_{i+2,j-2} \right. \\ & \left. \left. + u_{i-2,j+2} + u_{i-2,j-2}\right) \right\} \\ = {} & 0 \end{aligned} \tag{18}$$

Although it is nothing more than obtaining a spline curved surface, this method is greatly advantageous in that the perspective is quite favorable when adding other constraint conditions because the energy minimization is explicitly expressed by an equation using the framework of regularization. Muraki et al. [45] obtained a favorable result (Fig. 3) by adding the constraint that the variations of elevation become 0 on the contour lines of a map, as expressed below:

$$\iint_{\Omega} \left\{ \frac{\partial u}{\partial x} \Delta x(x,y) + \frac{\partial u}{\partial y} \Delta y(x,y) \right\}^2 dxdy \tag{19}$$

to equation (15), assuming that contour lines are partially obtained from a map. Here, $(\Delta x(x,y), \Delta y(x,y))$ is the direction cosine of contour lines at a point (x,y) in the image.

2.2.2 Binocular Stereo

The standard stereo camera system that consists of cameras with parallel lines of sight is examined. With this system, the points within the three-dimensional space infallibly appear on the same horizontal scanning line (epipolar line) in the images on both left and right. It is assumed that the brightness of corresponding points is similar, and that the depth of the target scene varies continuously. When the

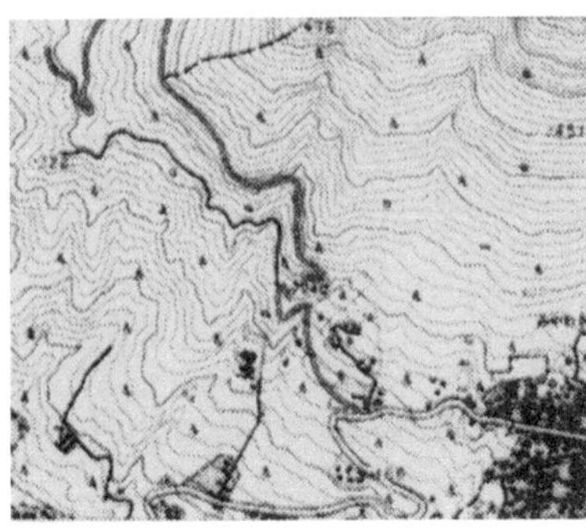

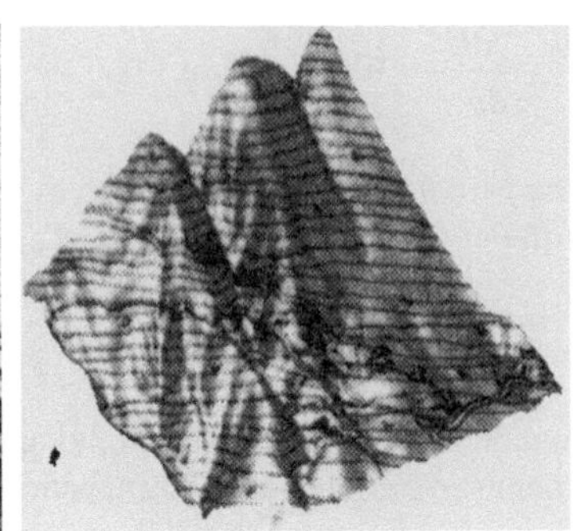

Topographical map. Reconstructed surface.

Fig. 3 3D surface from contours [45].

brightness of the left and right images is expressed as $L(x, y)$ and $R(x, y)$, respectively, a stereo matching problem, which obtains the points $(x+d\,(x, y), y)$ in the left image (that correspond to the respective points (x, y) in the right image) is formulated as a problem to obtain a disparity function $d\,(x, y)$ that minimizes the following functional in the framework of regularization.

$$E = \iint_{\Omega} \left[\{L(x + d(x,y), y) - R\,(x,y)\}^2 + \lambda \{d_x(x,y)^2 + d_y(x,y)^2\} \right] dxdy \tag{20}$$

Since this functional is not in a quadratic form in regard to the function $d(x,y)$ to be obtained, the problem is how to avoid local minima. There are two methods available to counter this problem, i.e., a method to use deterministic relaxation by multiple scales based on an Euler equation [46],

$$\{L(x+d(x,y),y) - R(x,y)\} \cdot L_x(x+d(x,y),y) - \lambda\{d_{xx}(x,y) + d_{yy}(x,y)\} = 0 \tag{21}$$

and a method to use stochastic relaxation by annealing [20].

When discontinuous regularization (proposed by Terzopoulos [41]) is applied simply to an image with high texture, a boundary of discontinuity can easily be detected as in references [47] and [48]. Boult et al. [49] proposed a method to roughly fit a curved surface to the extracted characteristic points by using a function fitting, and to interpolate their intervals. The scale of the function to be fitted here is discussed in reference [50].

2.2.3 Optical Flow

When the apparent field of velocity in the image is U, its relationship to image gray level I is expressed as follows:

$$I_t + U^T \nabla I = 0 \tag{22}$$

With the optical flow estimation by optimization operations, the solution is U, which minimizes the sum of the above formula, and is a function that expresses the smoothness of U. This method is called the gradient-based method [21].

Regarding stabilizing functionals, with which the terms of either linear or quadratic differentials of brightness are quadratic functions, in estimation problems of optical flow based on the gradient-based method, Snyder [51], [52] showed that functions that meet physical conditions (i.e., 1) they are rotation invariant and 2) the two components of the optical flow are independent), are expressed by a linear combination of two basic functions, $\nabla I \nabla^T I$ and $\nabla \nabla^T I$.

Research on problems that deal with the optical flow of discontinuous distribution has been conducted. Black [53] proposed a method to consecutively estimate the discontinuity simultaneously with the flow itself by introducing a process that is similar to the line process. There has also been an attempt to formulate the shape of the region, where the optical flow changes continuously, in the line process, using the shape and length of the boundary of discontinuity [54].

Considering that the optical flow in general scenes, such as outdoor views, is piecewise continuous, Schnörr fitted piecewise smooth functions, by dividing the region where the distribution of the optical flow obtained was initially continuous [55], [56].

2.2.4 Segmentation

Geiger et al. [57] proposed a framework for dealing with segmentation problems uniformly when the stochastic properties of discontinuity are identified. They proved that the methods to deal with region boundaries using the line process have generality, and adopted a method to use a single-line process. This is more advantageous than a method to use the line process of both the x-axis and y-axis, since the single-line process annuls both the anisotropy and the property that the boundary of discontinuity is prone to follow the axis. Then, they adopted an energy function based on the Markov field as a penalty function.

Boult et al. [58] proposed a method to fit a function that describes data of a set of pixels that are allowed to be overlapped without conducting segmentation explicitly. This is based on the idea that, with conventional methods for segmentation, the estimation of the regional boundary and the estimation problems of data within the region inevitably become a "which came first the chicken or the egg" question. This method proves that segmentation is essentially restored to a function fitting problem.

2.2.5 Edge Detection

Chen et al. [59] conducted edge detection by regularization using the cubic b-spline. With this method, problems are solved as uni-dimensional regularization problems by taking the projection to a given line in a small region. On the other hand, Gökmen et al. [60] conducted formulation in a similar manner to the line process by using a distributed constant as a regularization parameter. They also employed a method to consecutively estimate regularization parameters, so that the distribution of errors between the observed values and estimated values become uniform.

3. EXTRACTION OF SHAPE DESCRIPTIONS BY OPTIMIZATION

3.1 Parametric Model Fitting

Operations to fit functions to a set of datum points are conducted in the various stages of CVIP, such as extraction of target objects from gray-scale images and/or depth images. Basically, the fitting error between the datum points and the function is defined, and then the parameter of the function that can minimize the error is determined. The easiest means is the least squares method for the linear fitting of a function to a two-dimensional datum points.

Flexible shape models, such as superquadrics, which are created by expanding quadrics, and a Blobby model, have recently been used as shape expression models for computer graphics and object recognition. Several reports on attempts to automatically fit such function models to depth images obtained by a laser range finder have already been presented [61], [62]. When the functions are complicated, additional constraint conditions need some contrivances because the fitting errors often have numerous points of local minima.

3.2 Fitting of Non-Parametric Models

3.2.1 Snakes

There is an approach to formulate the shape extraction from images as the analogy of an energy minimization problem in dynamics. The behavior (deformation) of a deformable shape model is expressed as an energy that is a linear combination of the trend of the model itself and the constraints from the outside, and the target object, is extracted from the image by finding the state of stability of the energy local minima.

Kass et al. [63] proposed a method called the active contour model ("snakes"), with which the contour lines are detected by dynamically moving the contour lines themselves. This is a contour extraction method that takes advantage of the global properties of contour lines, and is still used widely [64]-[66].

Snakes is formulated as a minimization problem of an energy function that is a linear sum of the image energy and the internal energy, which is defined on a given contour $v(s)$ $(=(x(s),y(s));\ 0\leqq s\leqq 1)$ in the image.

$$E_{snakes}(v(s))= \int_0^1 \left\{E_{int}(v(s))+w_{image}E_{image}(v(s))\right\} ds \tag{23}$$

Here, E_{int} is the internal energy that expresses the smoothness of contours (called the energy that is concerned in the shape) to which the sum of energy that is concerned in the linear differential $v_s(s)$ of the contour $v(s)$ and the quadratic differential $v_{ss}(s)$, is often used.

$$E_{int}(v(s)) = \frac{1}{2}\left\{\alpha\left|v_s(s)\right|^2 + \beta\left|v_{ss}(s)\right|^2\right\} \tag{24}$$

α and β are weight coefficients. The value of E_{int} decreases as the contours become more circular and shorter. E_{image} is the image energy, which is expressed by the linear differential of image gray level $I(v(s))$ as follows:

$$E_{image}(v(s)) = -\left|\nabla I(v(s))\right|^2 \tag{25}$$

The value of this function decreases as the sum of the linear differential of image gray levels in the pixels on the contours increases. Accordingly, snakes converge to a smooth contour with great gray scale variations in the image. In general, an initial contour is given as the initial value by some means, and operations for obtaining the local minima are conducted, starting with the initial value *by*, for example, the steepest descent method.

Aggressive attempts are being made to expand both the "active net," which is an extended two-dimensional net model [67], and the split-and-merge type snakes [68] (Fig. 4).

3.2.2 Improvement of Snakes

Although snakes can provide satisfactory results when the quality of the target image is good, it has the following disadvantage:

- The solution varies depending on the position of the initial contour.
- Introduction of a priori knowledge is difficult.
- Operations are time-consuming.
- The solution varies depending on parameters α and β.
- It cannot follow the variations of the topology.

Several approaches have been taken to counter these problems. As for the problem that the solution depends on the position of the initial contour, Coughlan et al. [94] proposed a novel deformable template that detects and localizes shapes in grayscale images. The template is formulated as a Bayesian graphical model of a two-dimensional shape contour, and it is matched to the image using a variant of the belief propagation algorithm. A target shape contour can be localized in a cluttered image without initializing the template near the target.

Fig. 4 Split-and-merge contour model [68]. Tracking of temporarily overlapping persons; (a) initialization at 1st frame, (b) result at 1st frame, (c) 19th, (d) 32nd, (e) 35th, (f) 43rd, (g) 47th, and (h) 60th frames.

(1) Analysis of the solution space

With snakes and other methods that use optimization operations, the local minima are usually obtained by conducting iterative operations starting with an appropriate initial value. In this case, the monotonousness of the solution space is important. When the solution space has complicated non-linearity, the local minima obtained by calculations depend greatly on the initial value.

To counter this problem, Davatzikos et al.[69] identified by analysis how the properties of the functions used for minimization should be alike for them to become convex. This method is currently not practical, since it requires the values and partial differentiation of the functions, which conduct minimization, at all the positions. However, continued research on this approach should clarify the scope of applicable problems.

(2) Introduction of a priori knowledge
It is often the case that the shape of the target object for detection is already known a priori in the processing of contour detection in a given closed region from the image. Based on such a priori knowledge, research has been conducted to develop a stable contour detection method that is resistant to noise.

Many of the methods to introduce a priori knowledge of shapes use the curvature in the respective points on the contour, and assume that the energy in the shape is minimized when the contour becomes identical to the curvature [29] (Fig. 5) [70]. The problem with such methods is how to guarantee rotation invariance of the contour shapes.

Other relevant methods include a method to conduct high-speed and stable contour detection using a physical model [71], a method to describe the shape in accordance with grammar rules and detect contours that agree with the grammar [72], and a method to detect contours that correspond to the variations of the topology [73].

(3) High speed calculation methods
Snakes are extremely costly in computation, if its minimum value search problem in the multidimensional space is calculated without modification.

Amini et al. [74] proposed a method to use dynamic programming [75] for calculating the discretized energy functions of snakes at high speed. With this method, the points that move sequentially are determined by dynamic programming, based on the variations of the energy functions when the points that constitute contour lines move to the 8-neighbors.

Williams et al. [76] proposed the greedy algorithm which is faster. As with the method of Amini et al., this method determines the movements of movable points that constitute contour lines based on the decrease of the energy function. At present, this method is used widely for high-speed calculations.

(4) Snakes based on the geometric model
Snakes began to attract attention after it was proposed independently by Caselles [77] and Malladi [78] based on the geometric model. Unlike the conventional snakes that modify contours by minimizing the energy function defined by the contour shape and the image, snakes based on the geometric model detect contours by obtaining the moving velocity of the contours based on the space measure that is defined using the image.

This method is characterized by the fact that calculations are not hindered by topological variations, which may occur in the middle of calculations of convergence and split to several objects, since the calculations themselves are not dependent upon the topology. This method has been expanded to three dimensional

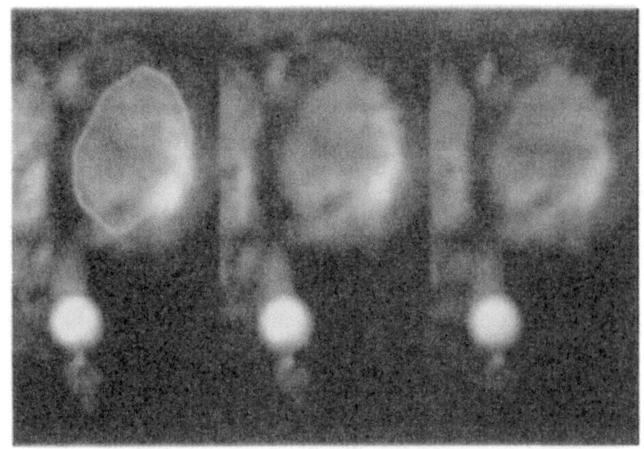

MRI image sequence and a sample contour.

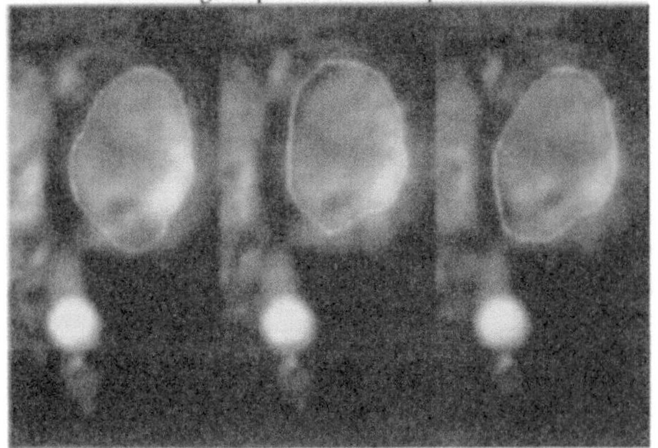

Contour tracking by the conventional snake.

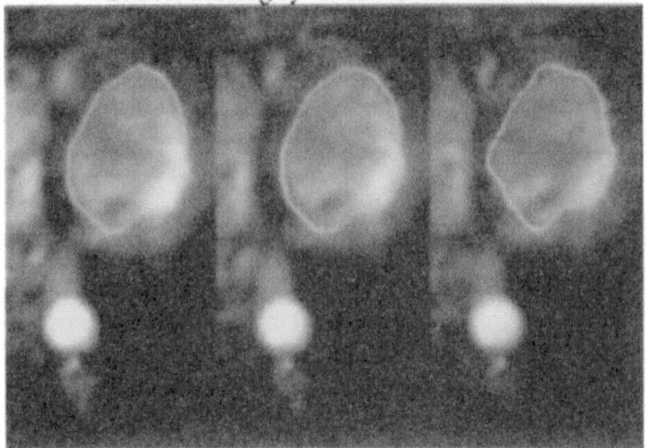

Contour tracking by M-snake.

Fig. 5 Sample contour model (M-snake [29]).

images [79], minimum value calculations [80], and multiple-dimensional images [81].

4. UTILIZATION OF THE GENETIC ALGORITHM

As mentioned previously, optimization by simple reiterative numerical solutions often results in local minima. To avoid this problem, attempts are being made to take the coarse-to-fine strategy based on multi-scales [13] and to apply simulated annealing [15]. However, the search space cannot be expanded infinitely by these methods. To deal with CVIP problems in the real world, it may be necessary to incorporate a method that is capable of searching a wider space.

A candidate of such a searching method is the genetic algorithm (GA) [82], which is attracting attention as a new method for CVIP. The GA is an algorithm for learning, adaptation, and optimum search that is realized by simulating the mechanism of biological evolution. One branch aims to create artificial life by focusing on learning and adaptation, while another believes that GA is simply one of several optimum searching methods.

4.1 What Is the Genetic Algorithm?

With GA, the respective solutions are initially regarded as "individuals" that constitute a "population" that is a set of solutions; each solution is considered as "choromosome." Then, a new solution is created by the "crossover" of two arbitrary solutions. Depending on the situation, the solution may be renewed by "mutation." Then, iterative processing of "selection" based on the "fitness" between the new and old solutions is conducted. Specifically, this method makes the entire population "evolve" to the optimum solution in the final stage by eliminating individuals with low fitness by selection. The crossover, mutation, selection etc., are called genetic operations, which are generally conducted stochastically based on random numbers.

The advantage of GA may be the capability of searching the optimum solution from a wide search space, since the solutions are dealt with as a population, and the search is conducted parallel while maintaining the diversity of many initial solutions. Initial solutions are obtained in the first place, followed by various intermediate solutions. Analogically, it is as if many people start to climb a mountain from various points; children who are born by crossover also start to climb the mountain from unexpected places; and consequently somebody reaches the mountain top, after repeating such a process. This is the property that distinguishes GA from simulated annealing, which is famous as a global optimum search method.

4.2 Examples of Applying GA to CVIP

4.2.1 Automatic Generation of the Image Processing Module Sequence

The first example of applying GA to CVIP may be the automatic generation of an image processing module sequence, which is an optimum search problem that cannot be solved numerically. Use of GA is conceived based on the capability of processing images to a fairly satisfactory level by combining limited numbers and types of image processing modules, assigning constraints of data types when connecting the modules, and replacing and combining part of the sequence that can intuitively be understood as genetic operations.

Shortly after 1975, when Holland at the University of Michigan proposed GA, the Environmental Research Institute of Michigan (ERIM) reported an attempt to use GA as a method to determine the contents of processing in the respective stages of a cytocomputer (a pipeline image processor) that was under development at the institute at the time [83]. This was a search problem of morphological operation sequences for extracting the target shapes. With random search, trials need to be made 1.3×10^{11} times; however, it was reported that the proper program was discovered using GA after approximately 5,000 trials.

References [84] and [85] reported that the image processing modules were expanded to general gray-scale image processing. Other relevant studies include an attempt to narrow the searching area by introducing hierarchy [86].

4.2.2 Search in the Parameter Space

If the target object to be detected from an image can be expressed by parameters, it can be extracted by conducting a search in the parameter space. For instance, if a straight line is expressed as $p = x \cos \theta + y \sin \theta$, the detection of the straight line can be substituted to the search of the parameters (p, θ). Although Hough transform is commonly used, the framework of GA can also be used if these parameters (p, θ) are coded as chromosomes. Reference [87] reported on line detection by GA using (p, θ). The fitness was evaluated based on the number of points on the straight line expressed by the parameters.

In pattern matching of binary shapes [88], GA is applied to a problem to search a shape that is identical to the given model figure of the image. With this method, chromosomes are expressed by the classes of parallel shift, rotation, and scale factors; the model is geometrically transformed by parameters; and the fitness is evaluated by whether or not the same shape is found in the position where the transformation was conducted.

Superquadrics is attracting attention as a model that expresses a three-dimensional shape by a comparatively small number of parameters. To fit this model to 3D depth image, it is sufficient to merely solve a problem that minimizes the fitting errors of the depth and the normal vector between the observed data and the model. However, a strategy for obtaining a global minimum solution needs to be combined with superquadrics, since the commonly used deterministic non-linear least squares method derives only a local minimum solution. Interestingly, some favorable results have been obtained [87] using the same data that was also used when applying simulated annealing [61] (as discussed in Section 3.1).

4.2.3 Shape Model Fitting

With shape model fitting, a model that expresses the shape of a target object is fit to the properties in an image for recognizing the position of the target object, to which GA can be used as a searching strategy. With this method, some sort of information regarding the positions of image properties in real space is expressed as chromosomes and coding is well devised for the respective problems, unlike the method to consider the classes of parameters as chromosomes in Section 4.2.2.

In the extraction of a straight line [89], for example, the respective pixels have chromosomes that consist of the directions and the coordinates of the edges. Pixels that inherit the trait of their parents are generated at the internally dividing points of the parents (two edge points) by the crossover of the neighboring pixels. With the primitive extraction of shapes [90], chromosomes are constructed by the values of the coordinates at three points in the case of circle detection. The idea is that a more global circle can be detected by crossing over two circles, i.e., one made by giving three points in the image and the other by giving three different points. Reference [91] reported the polygonal approximation of a closed curve by GA.

4.2.4 Image Reconstruction

Several attempts have been made to use GA as an optimization method in the application of regularization to CVIP. In stereo matching of binocular stereoscopic vision the disparity distribution where the gray scale level becomes identical in the corresponding points on the left and right is obtained. Here, the two-dimensional active contour model, active net [67], which is capable of conducting local optimization, and GA are combined (Fig. 6) [92]. In the three-dimensional shape reconstruction from gray-scale images [93], GA is used to minimize the error between a shadow image in an estimated three-dimensional shape (under the given lighting and picturing conditions) and a shadow image that is obtained by observation.

In both cases, the estimated two-dimensional data (image, disparity distribution, and depth distribution) are expressed as chromosomes without coding, and the crossover is conducted by two-dimensional region reshuffling. The problem can be solved only if the two-dimensional data generated by genetic operations meet the constraint conditions particular to the problem and can be evaluated quantitatively. Specifically, the reconstruction can be solved as a forward problem not as an inverse problem. This may be the most advantageous point in using GA for image reconstruction.

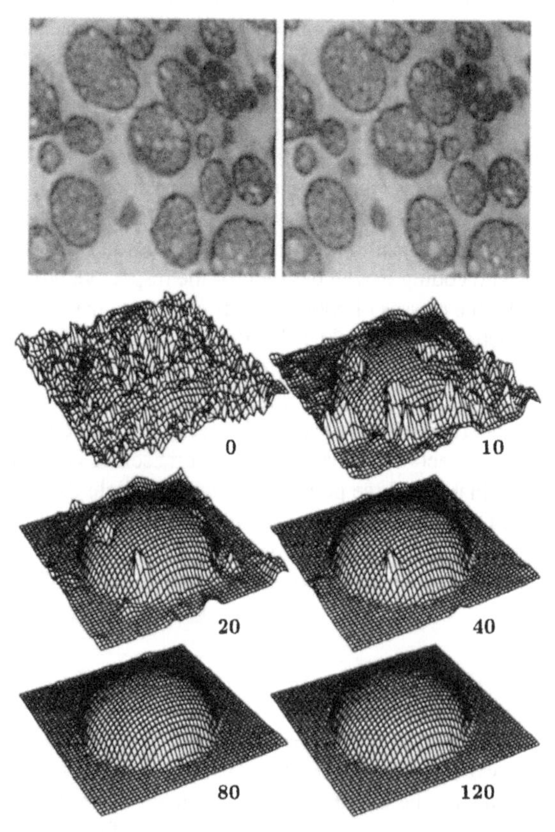

Fig. 6 Stereo matching using GA and active net [92]. Left and right images, iterations 0, 10, 20, 40, 80, 120.

5. CONCLUSION

In this chapter, the authors presented general views of image processing and computer vision based on optimization. It should be noted that diverse approaches, too numerous to be covered here, have been used. Nonetheless, a big question that remains to be answered is: Why can the visual function of "finding the most probable solution from an image" be done by humans quickly, accurately, and naturally, but is so difficult for a machine?

REFERENCES

[1] N. Otsu, "On a Family of Spline Functions for Smooth Curve Fitting," Technical Report of Institute of Electronics and Communication Engineers (IECE), PRL79-47, vol.79, no.165, pp.1-8, 1979 (in Japanese).

[2] G. Wahba and J. Wendelberger, "Some New Mathematical Methods for Variational Objective Analysis Using Splines and Cross Validation," Monthly Weather Review, vol.108, pp.1122-1143, 1980.

[3] Y. Isomichi, "Inverse-Quantization Method for Digital Signals and Images: Area-Approxmation Type," Trans. IECE, vol.J64-A, no. 4, pp.285-292, 1981 (in Japanese).

[4] Y. Isomichi, "Inverse-Quantization Method for Digital Signals and Images: Point-Approximation Type," Trans. IECE, vol.J63-A, no.11, pp.815-821, 1980 (in Japanese).

[5] D.L. Waltz, "Understanding Line Drawings of Scenes with Shadows, " in The Psychology of Computer Vision, P.H. Winston (ed.), McGraw-Hill, New York, pp.19-91, 1975.

[6] A. Rosenfeld, R.A. Hummel, and S.W. Zucker, "Scene Labeling by Relaxation Operations", IEEE Trans. SMC, vol.6, no.6, pp.420-433, 1976.

[7] T. Poggio, V. Torre, and C. Koch, "Computational Vision and Regularization Theory," Nature, vol.317, no.6035, pp.314-319, 1985.

[8] S.E. Farlman, G.E. Hinton, and T.J. Sejnowski, "Massively Parallel Architectures for AI: NETL, Thistle, and Boltzman Machines," Proc. AAAI-83, pp.109-113, 1983.

[9] G.E. Hinton and T.J. Sejnowski, "Optimal Perceptual Inference," Proc. IEEE CVPR'83, pp.448-453, 1983.

[10] S. Geman and D. Geman, "Stochastic Relaxation, Gibbs Distribution, and the Bayesian Restoration of Images," Trans. on PAMI, vol. 6, no. 6, pp.721-741, 1985.

[11] L.C. Briggs, "Machine Contouring Using Minimum Curvature," Geophysics, vol.39, no.1, pp.39-48, 1974.

[12] W.E.L. Grimson, "An Implementation of a Computational Theory of Visual Surface Interpolation," CVGIP, vol.22, pp.39-69, 1983.

[13] A.P. Witkin, D. Terzopoulos, and M. Kass, "Signal Matching through Scale Space," Int. Journal Computer Vision, vol.1, no.2, pp.133-144, 1987.

[14] S. Kirkpatrick, C.D. Gelatt Jr., and M.P. Vecchi, "Optimization by Simulated Annealing," Science, vol.220, no.4598, pp.671-680, 1983.

[15] V. Granville, M. Krivanek, and J.P. Rasson, "Simulated Annealing: A Proof of Convergence," IEEE Trans. PAMI, vol.16, pp.652-656, 1994.

[16] M.S. Landy, "A Brief Survey of Knowledge Aggregation Methods," Proc. 8th ICPR, pp.248-252, 1986.

[17] D. Lee and G.W. Wasilkowski, "Discontinuity Detection and Thresholding: A Stochastic Approach," Proc. IEEE CVPR '91, pp.208-214, 1991.

[18] D. Keren and M. Werman, "Probabilistic Analysis of Regularization," IEEE Trans. PAMI, vol.15, no. 10, pp.982-995, 1993.
[19] D. Keren and M. Werman, "A Bayesian Framework for Regularization," ICPR '94, pp.72-76, 1994.
[20] S.T. Barnard, "Stochastic Stereo Matching Over Scale," Int. Journal Computer Vision, vol.3, no.1, pp.17-32, 1989.
[21] B.K.P. Horn and B.G. Shunk, "Determining Optical Flow," AI, vol.17, pp.185-203, 1981.
[22] K. Ikeuchi and B.K.P. Horn, "Numerical Shape from Shading and Occluding Boundaries," AI, vol. 17, pp.141-184, 1981.
[23] D. Marr and T. Poggio, "A Theory of Human Stereo Vision," Proc. Royal Soc. London Ser. B 204, pp.301-328, 1979.
[24] A. Blake, "Relaxation Labeling: The Principle of Least Disturbance," Pattern Recognition Letters, vol.1, nos.5-6, pp.385-391, 1983.
[25] D. Terzopoulos, "Multilevel Computational Processes for Visual Surface Reconstruction," CVGIP, vol. 24, pp.52-96, 1983.
[26] T. Pajdla and V. Hlavac, "Surface Discontinuities in Range Images," 4th. ICCV, pp.524-528, 1993.
[27] T. Poggio and C. Koch, "Ill-posed Problems in Early Vision: From Computational Theory to Analogue Networks," Proc. Royal Soc. London Ser. B 226, pp.303-323, 1985.
[28] A.N. Tikhonov and V.Y. Arsenin, "Solutions of Ill-Posed Problems," Winston, Washington DC, 1977.
[29] A. Amano and Y. Sakaguchi, M. Minoh, and K.Ikeda, "Snakes Using a Sample Contour Model," Proc. 1st ACCV, pp.538-541, 1993.
[30] R. Hummel and R. Moniot, "Solving Ill-Conditioned Problems by Minimizing Equation Error," 1st ICCV, pp.527-533, 1987.
[31] D. Keren and M. Werman, "Variations on Regularization," ICPR'90, pp.93-98, 1990.
[32] A.M. Thompson, J.C. Brown, J.W. Kay, and D.M. Titterington, "A Study of Methods of Choosing the Smoothing Parameter in Image Restoration by Regularization," IEEE Trans. PAMI, vol.13, no.4, pp.326-339, 1991.
[33] N. Mukawa, "Vector Field Reconstruction Using Variance-based Resolution Control and Its Application to Optic Flow Estimation," (in Japanese) Trans. of IEICE of Japan D-II, vol.J78-D-II, no.7, pp.1028-1038, 1995 (published in English in Systems and Computers in Japan, vol.27, No.13, 1996).
[34] D. Terzopoulos, "Regularization of Inverse Visual Problems Involving Discontinuities," Trans. on PAMI, vol.8, no. 4, pp.413-424, 1986.
[35] D. Lee and T. Pavlidis, "One-Dimensional Regularization with Discontinuities," Trans. PAMI, vol.10, no.6, pp.822-829, 1988.
[36] A. Blake and A. Zisserman, "Some Properties of Weak Continuity Constraints and the GNC Algorithm," Proc. IEEE CVPR'86, pp.656-667, 1986.

[37] S.Z. Li, "Reconstruction without Discontinuities," 3rd. ICCV, pp.709-712, 1990.
[38] R.L. Stevenson, B.E. Schmitz, and E.J. Delp, "Discontinuity Preserving Regularization of Inverse Visual Problems," IEEE Trans. SMC, vol.24, no.3, pp.455-469, 1994.
[39] C. Schnörr, "Unique Reconstruction of Piecewise Smooth Images by Minimizing Strictly Convex Nonquadratic Functionals," Journal Mathematical Imaging and Vision, vol.4, pp.189-198, 1994.
[40] M. Shizawa, "Extension of Standard Regularization Theory into Multivalued Functions: Reconstruction of Smooth Multiple Surfaces," Trans. IEICE, vol.J76-DII, p.1146-1156, 1994 (in Japanese).
[41] D. Terzopoulos, "The Computation of Visible-Surface Representations," IEEE Trans. PAMI, vol.10, no.4, pp.417-438, 1988.
[42] W.E.L. Grimson and T. Pavlidis, "Discontinuity Detection for Visual Surface Reconstruction," CVGIP, vol.30, pp.316-330, 1985.
[43] A. Blake, "The Least-disturbance Principle and Weak Constraints," Pattern Recognition Letters, vol.1, nos.5-6, pp.393-399, 1983.
[44] A. Blake, "Comparison of the Efficiency of Deterministic and Stochastic Algorithms for Visual Reconstruction," IEEE Trans. PAMI, vol.11, no.1, pp.2-12, 1989.
[45] S. Muraki, N. Yokoya, and K. Yamamoto, "3D Surface Reconstruction from Contour Line Image by a Regularization Method," Proc. SPIE Vol.1395, Close-Range Photogrammetry Meets Machine Vision, pp.226-233, Sept. 1990.
[46] N. Yokoya, "Surface Reconstruction Directly from Binocular Stereo Images by Multiscale-Multistage Regularization," ICPR'92, I, pp.642-646, 1992.
[47] R. March, "A Regularization Model for Stereo Vision with Controlled Continuity," Pattern Recognition Letters, vol.10, pp.259-263. 1989,
[48] R. March, "Computation of Stereo Disparity Using Regularization," Pattern Recognition Letters, vol.8, pp.181-187, 1988,
[49] T.E. Boult and L. Chen, "Analysis of Two New Stereo Algorithms," Proc. IEEE CVPR'88, pp.177-182, 1988.
[50] T.E. Boult, "What is Regular in Regularization?" 3rd. ICCV, pp.644-651, 1990.
[51] M.A. Snyder, "The Mathematical Foundations of Smoothness Constraints: A New Class of Coupled Constraints," IUW, pp.154-161, 1990.
[52] M.A. Snyder, "On the Mathematical Foundations of Smoothness Constraints for the Deterimination of Optical Flow and for Surface Reconstruction," IEEE Trans. PAMI, vol.13, no.11, pp.1105-1114, 1991.
[53] M.J. Black and P. Anandan, "A Framework for the Robust Estimation of Optical Flow," 4th. ICCV, pp.231-236, 1993.
[54] S. Raghavan, N. Gupta, and L. Kanal, "Computing Discontinuity-Preserved Image Flow," ICPR'92, A, pp.764-767, 1992.

[55] C. Schnörr, "Computation of Discontinuous Optical Flow by Domain Decomposition and Shape Optimization," Int. Journal Computer Vision, vol.8, no.2, pp.153-165, 1992.
[56] C. Schnörr, "Determining Optical Flow for Irregular Domains by Minimizing Quadratic Functionals of a Certain Class," Int. Journal Computer Vision, vol.6, no.1, pp.25-38, 1991.
[57] D. Geiger and A. Yuille, "A Common Framework for Image Segmentation," Int. Journal Computer Vision, vol.6, no.3, pp.227-243, 1991.
[58] T.E. Boult and M. Lerner, "Energy-Based Segmentation of Very Sparse Range Surfaces," IUW, pp.565-572, 1990.
[59] G. Chen and H. Yang, "Edge Detection by Regularized Cubic B-Spline Fitting," IEEE Trans. SMC, vol.25, no.4, pp.636-643, 1995.
[60] M. Gokmen and C.C. Li, "Edge Detection and Surface Reconstruction Using Refined Regularization," IEEE Trans. PAMI, vol.15, pp.492-499, 1993.
[61] N. Yokoya, M. Kaneta, and K. Yamamoto, "Recovery Of Superquadric Primitives from a Range Image Using Simulated Annealing," ICPR92, (I), pp.168-172, 1992.
[62] S. Muraki, "Volumetric Shape Description of Range Data using Blobby Model," ACM Computer Graphics, vol.25, no.4, pp.227-235, 1991.
[63] M. Kass, A. Witkin, and D. Terzopoulos, "Snakes: Active Contour Models," Int. Journal Computer Vision, vol.1, no.3, pp.321-331, 1988.
[64] A.L. Yuille, P.W. Hallinan, and D.S. Cohen, "Feature Extraction from Faces Using Deformable Templates," Int. Journal Computer Vision, vol.8, no.2, pp.99-111, 1992.
[65] S.R. Gunn and M.S. Nixon, "Snake Head Boundary Extraction Using Global and Local Energy Minimisation," ICPR'96, pp.581-585, 1996.
[66] R. Chung and C. Ho, "Using 2D Active Contour Models for 3D Reconstruction from Serial Sections," ICPR'96, pp.849-853, 1996.
[67] K. Sakaue and K. Yamamoto, "Active Net Model and Its Application to Region Extraction," The Journal of the Institute of Television Engineers of Japan, vol.45, no.10, pp.1155-1163, 1991 (in Japanese).
[68] S. Araki, N. Yokoya, and H. Takemura, "Tracking of Multiple Moving Objects Using Split-and-merge Contour Models Based on Crossing Detection," Proc. Vision Interface '97, pp.65-72, Kelowna, British Columbia, Canada, 1997.
[69] C. Davatzikos and J.L. Prince, "Convexity Analysis of Active Contour Problems," Proc. IEEE CVPR'96, pp.674-679, 1996.
[70] N. Ueda and K. Mase, "Tracking Moving Contours Using Energy-Minimizing Elastic Contour Models," ECCV'92, pp.453-457, 1992.
[71] M. Hashimoto, H. Kinoshita, and Y. Sakai, "An Object Extraction Method Using Sampled Active Contour Model," Trans. of IEICE of Japan D-II, vol.J77-D-II, no.11, pp.2171-2178, 1994 (in Japanese).

[72] B. Olstad and A.H. Torp, "Encoding of a priori Information in Active Contour Models," IEEE Trans. PAMI, Vol.18, No.9, pp.863-872, 1996.
[73] T. Cclnerney and D. Terzopoulos, "Topologically Adaptable Snakes," 5th ICCV, pp.840-845, 1995.
[74] A.A. Amini, S. Tehrani, and T.E. Weymouth, "Using Dynamic Programming for Minimizing the Energy of Active Contours in the Presence of Hard Constraints," 2nd ICCV, pp.95-99, 1988.
[75] A.A. Amini, T. Weymouth, and R. Jain, "Using Dynamic Programming for Solving Variational Problems in Vision," Trans. PAMI, vol.12, no.9, pp.855-867, 1990.
[76] D.J. Williams and M. Shah, "A Fast Algorithm for Active Contours and Curvature Estimation," CVGIP, vol.55, no.1, pp.14-26, 1992.
[77] V. Caselles, R. Kimmel, and G. Sapiro, "Geodesic Active Contours," 5th ICCV, pp.694-699, 1995.
[78] R. Malladi, J.A. Sethian, and B.C. Vemuri, "Shape Modeling with Front Propagation: A Level Set Approach," IEEE Trans. PAMI, vol.17, pp.158-175, 1995.
[79] S. Kichenassamy, A. Kumar, P. Olver, A. Tannenbaum, and A. Yezzi, "Gradient Flows and Geometric Active Contour Models," 5th ICCV, pp.810-815, 1995.
[80] L.D. Cohen and R. Kimmel, "Global Minimum for Active Contour Models: A Minimum Path approach," Proc. IEEE CVPR'96, pp.666-673, 1996.
[81] G. Sapiro, "Vector-valued Active Contours," Proc. IEEE CVPR'96, pp.680-685, 1996.
[82] D.E. Goldberg, "Genetic ALgorithms in Search, Optimization, and Machine Learning," Addison-Wesley, 1989.
[83] A.M. Gillies, "An Image Processing Computer Which Learns by Example," SPIE, vol.155, pp.120-126, 1978.
[84] K. Yamamoto, K. Sakaue, H. Matsubara, and K. Yamagishi, "Miracle-IV: Multiple Image Recognition System Aiming Concept Learning: Intelligent Vision," ICPR'88, pp.818-821, 1988.
[85] H. Matsubara, K. Sakaue, and K. Yamamoto, "Learning a Visual Model and an Image Processing Strategy from a Series of Silhouette Images on MIRACLE-IV," Symbolic Visual Learning, K. Ikeuchi and M. Veloso, eds., Oxford, pp.171-191, 1997.
[86] I. Yoda, "Automatic Generation of Directional Erosion and Dilation Sequence by Genetic Algorithms," Proc. ACCV'95, II, pp.103-107, 1995.
[87] H. Iba, H. deGaris, and T. Sato,"A Bug-based Search Strategy for Problem Solving," ETL-TR92-24, Electrotechnical Laboratory, 1992.
[88] T. Tagao, T. Agui, and H. Nagahashi, "Pattern Matching of Binary Shapes Using a Genetic Method," Trans. of IEICE of Japan D-II, vol.J76-D-II, no.3, pp.557-565, 1993 (in Japanese).

[89] T. Nagao, T. Takeshi, and H. Nagahashi, "Extraction of Straight Lines Using a Genetic Algorithm," Trans. of IEICE of Japan D-II, vol.J75-D-II, no.4, pp.832-834, 1992 (in Japanese).

[90] G. Roth and M.D. Levine, "A Genetic Algorithm for Primitive Extraction", Proc. of the 4th International Conference on Genetic Algorithm, pp.487-494, 1991.

[91] T. Yamagishi and T. Tomikawa, "Polygonal Approximation of Closed Curve by GA," Trans. of IEICE of Japan D-II, vol.J76-D-II, no.4, pp.917-919, 1993 (in Japanese).

[92] K. Sakaue, "Stereo Matching by the Combination of Genetic Algorithm and Active Net (in Japanese)," Tans. IEICE, vol.J77-D-II, no.11, pp.2239-2246, 1994. (published in English in Systems and Computers in Japan, vol.27, no.1, pp.40-48, Scripta Technica, Inc., 1996.)

[93] H. Saito, N. Tsunashima, "Estimation of 3-D Parametric Models from Shading Image Using Genetic Algorithms," ICPR'94, A, pp.668-670, 1994.

[94] J.M. Coughlan and S.J. Ferreira, "Finding Deformable Shapes Using Loopy Belief Propagation," Proceedings of ECCV2002, pp.453-468, 2002.

3

Motion and Shape from Sequences of Images Under Feature Correspondences

Jun Fujiki
National Institute of Advanced Industrial Science and Technology,
Tsukuba, Japan

SUMMARY

The reconstruction of camera motion and object shape from multiple images is a fundamental and important problem in computer vision. This chapter highlights the problem of recovery of camera motion and object shape under some camera projection model from point correspondences especially, using epipolar geometry and the factorization method.

1. INTRODUCTION

The reconstruction of camera motion and object shape from multiple images is a fundamental and important problem in computer vision. This reconstruction is used for various fields such as man-machine interface, virtual reality systems and auto-control robot systems.

This problem generally consists of three steps. The first step is the segmentation of attended objects from images. The second step is the making of correspondences of features such as points, lines, and so on. The last step is to recover camera motion and the shape of the attended object from feature correspondences. This chapter highlights the last step, that is, the features are already corresponded.

To recover the camera motion and the object shape, the characteristics of the camera (camera model) must be considered. Theoretically, a perspective camera is appropriate to represent the camera model because it describes the pin-hole camera precisely. However, the recovering algorithms by the perspective camera have some difficulties because the algorithms are generally composed of non-linear equations, which tend to be unstable in noise, numerical computation, and initial values.

To overcome these difficulties, affine approximations of the perspective camera are considered for the camera model. Of course the accuracy of the reconstruction by the affine approximation camera is inferior to that by the perspective camera because true images are perspective images. However, the affine approximation camera has some advantages.

One of the biggest advantages is that the recovering algorithms are consisted of linear equations. This advantage means the algorithm works fast and is stable in numerical computation. Another advantage is that the reconstruction by the affine approximation camera is used for the initial values of the algorithms for the perspective camera.

Selecting features must also be considered. The features consist of points, lines, quadratic curves, pieces of planes, and so on. This chapter mainly considers the point because the point is the most fundamental feature.

When the camera is not calibrated, that is, when the intrinsic parameters of the camera are unknown, the recovered shape and motion have some ambiguity, that is, the Euclidean structure of the shape is not determined uniquely. This is because the metric of the motion space and that of the shape space cannot be fixed (13,90). Therefore, the intrinsic parameter must be known when considering the Euclidean structure. This chapter mainly gives a general view of the Euclidean reconstruction of camera motion and object shape under point correspondences by a calibrated perspective camera and a calibrated affine approximation camera.

2. CAMERA MODEL

This section outlines the camera models that are the perspective camera and its affine approximation. The camera model is the transformation from a three-dimensional space to a two-dimensional space. It is difficult to say what camera model is best to apply. Kinoshita and Lindenbaum (52) gave one solution of which camera model was appropriate to use between the perspective camera and affine approximation camera via geometric AIC (45).

2.1 Perspective Camera

Let $\mathbf{X}_{fp} = \begin{pmatrix} X_{fp} \\ Y_{fp} \\ Z_{fp} \end{pmatrix}$ and $\mathbf{x}_{fp} = \begin{pmatrix} x_{fp} \\ y_{fp} \end{pmatrix}$ be the coordinate of the p-th feature point in the f-th camera coordinate system and the f-th image coordinate system, respectively, and l be a focal length of the camera. The perspective camera (Fig. 1) is the theoretical model of pin-hole camera and the representation of the camera is

$$\lambda_{fp} \tilde{\mathbf{x}}_{fp} = \begin{pmatrix} l & 0 & 0 & 0 \\ 0 & l & 0 & 0 \\ 0 & 0 & 1 & 0 \end{pmatrix} \tilde{\mathbf{X}}_{fp} \tag{1}$$

where $\tilde{\mathbf{X}}_{fp} = \begin{pmatrix} \mathbf{X}_{fp} \\ 1 \end{pmatrix}$ and $\tilde{\mathbf{x}}_{fp} = \begin{pmatrix} \mathbf{x}_{fp} \\ 1 \end{pmatrix}$ are the three-dimensional homogeneous camera coordinate and two-dimensional homogenous image coordinate of the p-th feature point in the f-th camera, respectively, and λ_{fp} is some scalar called projective depth.

The well-known relationship $x_{fp} = l\frac{X_{fp}}{Z_{fp}}$ and $y_{fp} = l\frac{Y_{fp}}{Z_{fp}}$ are derived by eliminating λ_{fp} from Eq.1 .

Under the perspective projection, two distinct images of eight points are sufficient to determine the camera motion and the object shape up to some projective transformation.

The inverse problem of the perspective projection is not stable for numerical calculation because it represents non-linear equations. To overcome this difficulty, an affine projection model, which is an affine approximation of the perspective projection model, is used.

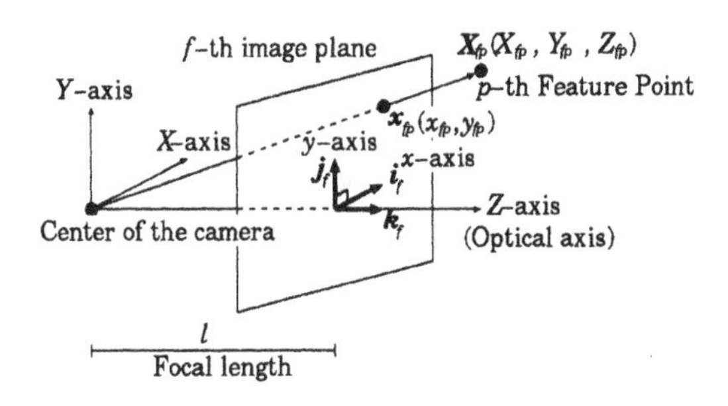

Fig. 1. Perspective camera.

2.2 Affine Approximation Camera

Affine approximation of the perspective camera (Fig. 2) is a kind of affine projection from a three-dimensional space to a two-dimensional space. The property of the affine projection is parallelism (72,89,116), which means parallel lines in three-dimensional space are projected to parallel lines in the image.

The representation of the affine projection is

$$\mathbf{x}_{fp} = A_f \mathbf{X}_{fp} + \mathbf{u}_f \quad \Leftrightarrow \quad \tilde{\mathbf{x}}_{fp} = \begin{pmatrix} A_f & \mathbf{u}_f \\ \mathbf{0}_3^T & 1 \end{pmatrix} \tilde{\mathbf{X}}_{fp} \tag{2}$$

where A_f is a 2×3 matrix and $\mathbf{u}_f$ is a two-dimensional vector, which represents translation of the image.

2.2.1 Paraperspective Camera

The paraperspective camera (74,78,79) is a first-order approximation of perspective camera and derives from Taylor expansion of the perspective camera.

Let $\mathbf{X}_{f*}$ be a coordinate on the f-th camera coordinate system of some specific feature point (*-th feature point) called a referenced feature point and $\mathbf{x}_{f*}$ be an image coordinate of $\mathbf{X}_{f*}$.

The first-order approximation of perspective camera at $\mathbf{X}_{f*}$ and/or $\mathbf{x}_{f*}$ is

$$\begin{aligned} \mathbf{x}_{fp} &= \frac{l}{Z_{f*}^2} \begin{pmatrix} Z_{f*} & 0 & -X_{f*} \\ 0 & Z_{f*} & -Y_{f*} \end{pmatrix} \mathbf{X}_{fp} + \mathbf{x}_{f*} \\ &= \frac{1}{Z_{f*}} \begin{pmatrix} l & 0 & -x_{f*} \\ 0 & l & -y_{f*} \end{pmatrix} \mathbf{X}_{fp} + \mathbf{x}_{f*} \end{aligned} \tag{3}$$

and this is the representation of the paraperspective camera. This approximation is effective when the depth of each feature point is the same as the average depth.

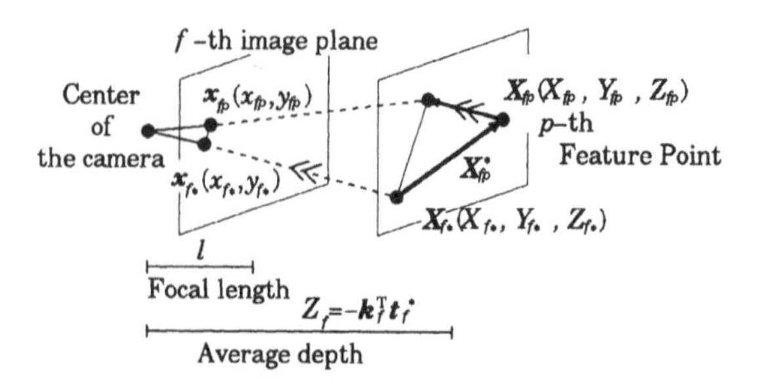

Fig. 2. Paraperspective camera.

Note that the referenced feature point is usually set to the center-of-mass of feature points because it is well known that affine projection maps center-of-mass into center-of-mass.

Under the paraperspective camera, three distinct images of four non-coplanar points are sufficient to determine the camera motion and the object shape uniquely up to a reflection and scale ratio (18). The algorithm to recover the motion and the shape from three or more paraperspective views was presented by Poelman and Kanade (43,79) and is known as the factorization method.

2.2.2 Scaled Orthographic Camera

The scaled orthographic (weak perspective) camera (Fig. 3) is a special case of the paraperspective camera. When the referenced feature point $\mathbf{X}_{f*}$ is close to the optical axis through all images, $\mathbf{x}_{f*}$ is located near the origin of the image coordinate through all images. Then $\mathbf{x}_{f*}$ is regarded as the zero vector and a representation of the paraperspective camera is rewritten as the representation of a scaled orthographic camera:

$$\mathbf{x}_{fp} = \frac{l}{Z_{f*}} \begin{pmatrix} 1 & 0 & 0 \\ 0 & 1 & 0 \end{pmatrix} \mathbf{X}_{fp} . \tag{4}$$

Under the scaled orthographic projection, three distinct images of four non-coplanar points are sufficient to determine the camera motion and the object shape uniquely up to a reflection and scale ratio. The algorithm to recover the motion and the shape from three scaled orthographic views was presented by Ostuni and Dunn (76), and Xu and Sugimoto (117). All algorithms are resolved into the orthographic case essentially by computing the ratio of the scale factor. Ostuni and Dunn (76) compute the scale factor by the roll-pitch-yaw representation of rotation matrix and Xu and Sugimoto (117) compute the scale factor by epipolar geometry. The algorithm to recover from three or more views was presented by Tomasi and Kanade (101,102) and Weinshall and Tomasi (111,112), and is known as the factorization method.

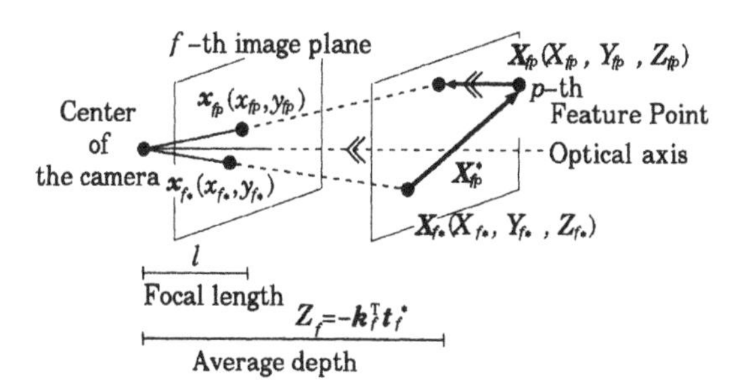

Fig. 3. Scaled orthographic camera.

2.2.3 Orthographic Camera

The orthographic camera (Fig. 4) is the simplest model between the affine approximation cameras, and is a special case of the scaled orthographic camera. When the distance between the camera center and the referenced feature point is almost the same through all images, Z_{f*} is regarded as constant ($= Z$), and the representation of the scaled orthographic camera is rewritten as the representation of the orthographic camera:

$$\mathbf{x}_{fp} = \frac{l}{Z}\begin{pmatrix} 1 & 0 & 0 \\ 0 & 1 & 0 \end{pmatrix}\mathbf{X}_{fp}. \tag{5}$$

Under the orthographic projection, three distinct images of four non-coplanar points are sufficient to determine the camera motion and the object shape uniquely up to a reflection (108).

Hu and Ahuja (33) classify the uncertainty of reconstruction when three distinct images of four non-coplanar points are not obtained. The algorithm to recover the motion and the shape from three orthographic views was presented by Huang and Lee (34) and Kamata (44), and to recover from three or more views are presented by Tomasi and Kanade (101,102) and is known as the factorization method.

The relationship between presented three affine approximation cameras is as follows: The scaled orthographic camera is the orthographic camera with a scale factor, and the paraperspective camera is the scaled orthographic camera with a position effect.

2.2.4 Affine Camera

To apply the above three affine approximation camera, the mapping from pixel coordinate (e.g., CCD position) to image coordinate must be known. This means the camera must be calibrated. When the camera is not calibrated (the intrinsic parameters of the camera are unknown), the affine projection from three-dimensional space to the image coordinate has no restriction.

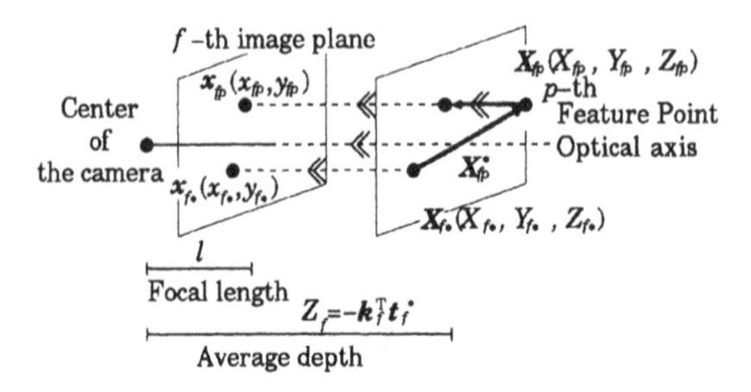

Fig. 4. Orthographic camera.

The affine approximation camera for an uncalibrated camera (89,90) is called an affine camera and its representation is

$$\mathbf{x}_{fp} = A_f \ \mathbf{X}_{fp} + \mathbf{u}_f \tag{6}$$

where there is no assumption for the components of a 2×3 matrix A_f and two-dimensional vector $\mathbf{u}_f$.

Under the affine camera (including the orthographic, the scaled orthographic, and the paraperspective cameras), two distinct images of four non-coplanar points are sufficient to recover the affine structure of the object (18,49).

In comparison with the orthographic, the scaled orthographic and the paraperspective cameras, an affine camera has no metric information to reconstruct the Euclidean structure. Therefore, the affine camera derives only an affine reconstruction of three-dimensional space and the affine camera is not the extension of the orthographic, the scaled orthographic and the paraperspective camera. Then a metric affine projection (MAP) camera, which is the extension of the orthographic, the scaled orthographic, and the paraperspective camera, is presented (19,22,23). The MAP camera was first called generalized affine projection camera (19), however, generalized affine projection is confusing with general affine projection (72), so the name was changed.

2.2.5 Metric Affine Projection Camera

A metric affine projection (MAP) camera (19,22,23) is used to unify the orthographic, scaled orthographic and paraperspective camera. The MAP camera has metric constrains to reconstruct Euclidean structure and its representation is

$$\mathbf{x}^*_{fp} = \frac{1}{\lambda_{f*}} B_f \ \mathbf{X}^*_{fp}, \qquad \mathbf{X}_{f*} = \frac{\lambda_{f*}}{l}\begin{pmatrix} \mathbf{x}_{f*} \\ l \end{pmatrix} \tag{7}$$

where $\mathbf{X}^*_{fp} = \mathbf{X}_{fp} - \mathbf{X}_{f*}$ and $\mathbf{x}^*_{fp} = \mathbf{x}_{fp} - \mathbf{x}_{f*}$ are relative camera coordinates and relative image coordinates from a referenced feature point, λ_{f*} is the depth parameter including global scale parameters, and B_f are the characteristics of a camera consists of known components. The second term of Eq.7 means the referenced feature point is projected by a perspective camera and the term is used to reconstruct camera position.

Let the singular value decomposition (SVD) of B_f be $R_f\Sigma_f D_f$, then the MAP is rewritten as

$$\mathbf{x}^*_{fp} = \frac{1}{\lambda_{f*}} R_f\Sigma_f D_f \ \mathbf{X}^*_{fp} = R_f\Sigma_f \ \frac{1}{\lambda_{f*}}\begin{pmatrix} 1 & 0 & 0 \\ 0 & 1 & 0 \end{pmatrix} \widetilde{D}_f \mathbf{X}^*_{fp} \tag{8}$$

where a 3×3 matrix $\widetilde{Y} = \begin{pmatrix} \mathbf{Y}_1^T \\ \mathbf{Y}_2^T \\ (\mathbf{Y}_1 \times \mathbf{Y}_2)^T \end{pmatrix}$ is derived from a 2×3 matrix $Y = \begin{pmatrix} \mathbf{Y}_1^T \\ \mathbf{Y}_2^T \end{pmatrix}$. Therefore, the MAP is decomposed into two transformations as

$$\mathbf{x}^*_{fp} = R_f \Sigma_f \, \chi^*_{fp}, \quad \chi^*_{fp} = \frac{1}{\lambda_{f*}} \begin{pmatrix} 1 & 0 & 0 \\ 0 & 1 & 0 \end{pmatrix} \widetilde{D}_f \mathbf{X}^*_{fp} . \tag{9}$$

In this decomposition, χ^*_{fp} is a scaled orthographic projection of $\mathbf{X}^*_{fp}$ onto the row space of D_f, called the virtual image plane, and $\mathbf{x}^*_{fp}$ is some known affine transformation of χ^*_{fp}.

Therefore, a MAP camera, as well as a paraperspective camera, is resolved to a scaled orthographic camera (19,22,23) and any algorithm for a scaled orthographic camera can be applied to the MAP camera. Then, three distinct images of four non-coplanar points are sufficient to determine the camera motion and the object shape uniquely up to a reflection and scale ratio for a MAP camera. Moreover, self-calibration of an affine camera is realized by estimating the second affine transformation (83).

2.3 Other Cameras

On the other hand, as the approximation of the perspective camera, an orthoperspective camera (3), a quasi-perspective camera (51), a camera by Sengupta et al. (88), and the second-order approximation of the perspective camera (118) are presented. However, these models are not used frequently because these model are non-linear approximation of the perspective projection.

3. EPIPOLAR GEOMETRY

Epipolar geometry (14,116) is the geometrical relationship between corresponding features in two images. Let $\mathbf{x}_{ip}$ and $\mathbf{x}_{jp}$ be the corresponding point sequences on the i-th image and the j-th image. There exists a geometric constraint $f(\mathbf{x}_{ip}, \mathbf{x}_{jp}) = 0$ between $\mathbf{x}_{ip}$ and $\mathbf{x}_{jp}$ free from p.

The constraint is represented by a 3×3 essential matrix E_{ij}, which is determined only by the relative position of two cameras and is

$$\widetilde{\mathbf{x}}_{ip}^T \, E_{ij} \, \widetilde{\mathbf{x}}_{jp} = 0. \tag{10}$$

This equation is called an epipolar constraint and/or a bilinear constraint. The epipolar constraints derives the relation of the image coordinate of the same feature point between two distinct images. Once the value of the image coordinate $\mathbf{x}_{ip}$ (or $\mathbf{x}_{jp}$) is given, the epipolar constraint (Eq.10) represents the equation of the line called the epipolar line, on the other image plane. Therefore, if E_{ij} is computed, the candidates of the corresponding point of the feature point on one image are lying on the epipolar line of the other image. Then the making correspondence problem is resolved into a one-dimensional search.

The general epipolar constraint, which includes both the perspective model and the affine approximation model, is reported in (116,120,121).

The algorithm of the estimation of the epipolar geometry is very documented in (6,29,75,90,116,119,121). The estimation by voting technique (98,99) is also presented. Luong and Faugeras (61) analyzed the stability of the fundamental matrix for the perspective camera. Especially, Hartley (29) reported that the normalizing of the input image data, that is, the origin shift and the scaling are very effective to realize the numerical stability of fundamental matrix estimation. This normalization is used not only for the epipolar geometry but also for the whole field in computer vision.

3.1 World Coordinate

When projected images of an object without a background are given, it is not determined that either the camera is moving, the object is moving, or both are moving. This is because the projected image is determined only by the relative position of the camera and the object. In other words, the object is supposed to be fixed without loss of generality. Under this supposition, it is convenient to fix the coordinate to the object and that is called the world coordinate.

Let $\mathbf{s}_p$ be the world coordinate of the p-th feature point, $\mathbf{t}_f$ be the f-th camera position, and $C_f = \begin{pmatrix} \mathbf{i}_f^T \\ \mathbf{j}_f^T \\ \mathbf{k}_f^T \end{pmatrix}$ be the f-th camera direction.

Note that $\mathbf{i}_f$ and $\mathbf{j}_f$ correspond to normal vectors along the x and the y axes in the f-th image plane, respectively, and $\mathbf{k}_f$ correspond to a normal vector along the optical axis in the f-th image.

The relationship between camera coordinates and world coordinates is

$$\mathbf{s}_p = \mathbf{t}_f + C_f^T \mathbf{X}_{fp} \quad \Leftrightarrow \quad \mathbf{X}_{fp} = C_f(\mathbf{s}_p - \mathbf{t}_f) \ . \tag{11}$$

3.2 Epipolar Geometry For Perspective Camera

Let $\mathbf{t}_i$ and $\mathbf{t}_j$ be two camera positions respectively, and $\mathbf{s}_p$ be a world coordinate of the p-th feature point. The plane that through these three points $\mathbf{t}_i$, $\mathbf{t}_j$ and $\mathbf{s}_p$ is called an epipolar plane. Another definition of the epipolar plane is that the plane through a feature point and spanned by two projected directions of the feature point for two cameras.

The linear space spanned by $\mathbf{s}_p - \mathbf{t}_i = C_i^T \begin{pmatrix} \mathbf{x}_{ip} \\ l \end{pmatrix}$, $\mathbf{s}_p - \mathbf{t}_j = C_j^T \begin{pmatrix} \mathbf{x}_{jp} \\ l \end{pmatrix}$ and $\mathbf{t}_{ij} = \mathbf{t}_j - \mathbf{t}_i$ is a two-dimensional space. Then the volume of parallel pipe spanned by the three vectors equals zero. By expanding the volume, the following holds

$$\tilde{\mathbf{x}}_{ip}^T E_{ij} \tilde{\mathbf{x}}_{jp} = 0, \quad E_{ij} = \begin{pmatrix} I_2 & \mathbf{0}_2 \\ \mathbf{0}_2^T & l \end{pmatrix} C_i [\mathbf{t}_{ij}]_\times C_j^T \begin{pmatrix} I_2 & \mathbf{0}_2 \\ \mathbf{0}_2^T & l \end{pmatrix} \tag{12}$$

where I_n is an n-dimensional identity matrix, $\mathbf{0}_n$ is an n-dimensional zero vector, and a 3×3 skew-symmetric matrix $[\mathbf{y}]_\times$ is derived from a three-dimensional vector $\mathbf{y}$ that satisfies $[\mathbf{y}]_\times \mathbf{z} = \mathbf{y} \times \mathbf{z}$ ($\times$ denotes exterior product) for any three-dimensional vector $\mathbf{z}$.

Eq.12 is called an epipolar constraint and/or a bilinear constraint, and E_{ij} (which is free from p) is called an essential matrix. To compute the essential matrix E_{ij}, eight corresponding points are needed.

Note that, by rescaling the focal length to 1, the essential matrix is represented in simple form as $E_{ij} = [\mathbf{t}]_\times C$ where $\mathbf{t} = \mathbf{t}_i - C_i C_j^T \mathbf{t}_j$ and $C = C_i C_j^T$.

Once an image coordinate $\tilde{\mathbf{x}}_{ip}$ (or $\tilde{\mathbf{x}}_{jp}$) is given, the epipolar constraint represents the equation of the line on the other image plane. The line is called the epipolar line and the line is through the projected image of the center of another camera. this projected image is called epipole. Shortly speaking, the epipolar line is always through the epipole (14,116)

Another way to derive the epipolar constraint is as follows.

Eq.1 is rewritten as

$$\lambda_{fp} \tilde{\mathbf{x}}_{fp} = \begin{pmatrix} l & 0 & 0 & 0 \\ 0 & l & 0 & 0 \\ 0 & 0 & 1 & 0 \end{pmatrix} \begin{pmatrix} C_f & -\mathbf{t}_f \\ \mathbf{0}_3^T & 1 \end{pmatrix} \tilde{\mathbf{s}}_p = \begin{pmatrix} \mathbf{p}_{f1}^T \\ \mathbf{p}_{f2}^T \\ \mathbf{p}_{f3}^T \end{pmatrix} \tilde{\mathbf{s}}_p = P_f \tilde{\mathbf{s}}_p \tag{13}$$

where $\tilde{\mathbf{s}}_p = \begin{pmatrix} \mathbf{s}_p \\ 1 \end{pmatrix}$. Therefore $\begin{pmatrix} x_{fp}\mathbf{p}_{f3}^T - \mathbf{p}_{f1}^T \\ y_{fp}\mathbf{p}_{f3}^T - \mathbf{p}_{f2}^T \end{pmatrix} \tilde{\mathbf{s}}_p = \mathbf{0}_2$ holds for all p. Then the essential matrix E_{ij} is derived from

$$\det \begin{pmatrix} x_{ip}\mathbf{p}_{i3}^T - \mathbf{p}_{i1}^T \\ y_{ip}\mathbf{p}_{i3}^T - \mathbf{p}_{i1}^T \\ x_{jp}\mathbf{p}_{j3}^T - \mathbf{p}_{j1}^T \\ y_{jp}\mathbf{p}_{j3}^T - \mathbf{p}_{j1}^T \end{pmatrix} = 0 \quad \Leftrightarrow \quad \tilde{\mathbf{x}}_{ip}^T \, E_{ij} \, \tilde{\mathbf{x}}_{jp} = 0 . \tag{14}$$

3.3 Affine Epipolar Geometry

One of the characteristics of an affine approximation camera is parallelism. That is, all points are projected along the same direction to an image plane. Therefore, the set of the epipolar plane for affine approximation camera is a set of parallel planes. Then the set of the epipolar lines of another image on one image is the set of parallel lines. In this case, the epipole does not exist and/or the epipole is a point at infinity.

In the affine approximation model $\mathbf{x}_{fp} = A_f C_f \, \mathbf{X}_{fp} + \mathbf{u}_f$, the representation of the essential matrix E_{ij} is $E_{ij} = \begin{pmatrix} O_{2,2} & (A_i^+)^T C_i \mathbf{c}_{ij} \\ \mathbf{c}_{ji} C_j^T A_j^+ & e \end{pmatrix}$ where $O_{2,2}$ is a 2×2 zero matrix, Y^+ is a Moore-Penrose inverse matrix of Y, $\mathbf{c}_{ij} = -\mathbf{c}_{ji}$ is a vector perpendicular to both projection directions of the i-th and the j-th images, and e is some scalar. The concrete representation for paraperspective, scaled orthographic, and orthographic cameras are shown in Xu and Zhang (116).

Note that the upper left four components of E_{ij} are all zero, then the epipolar constraints for the affine approximation model is rewritten explicitly as

$$\left((A_i^+)^T C_i \mathbf{c}_{ij}\right)^T \mathbf{x}_{ip} + \left((A_j^+)^T C_j \mathbf{c}_{ji}\right)^T \mathbf{x}_{jp} + e = 0 \tag{15}$$

[the same representation is derived in Huang and Lee (34) and Kurata et al. (54,55,57) in another way]. Then it needs four corresponding non-coplanar points to determine the affine epipolar constraint.

When the coefficients of epipolar constraint are computed from image points as

$$\mathbf{f}_i^T \mathbf{x}_{ip} + \mathbf{f}_j^T \mathbf{x}_{jp} + e = 0 \quad \Leftrightarrow \quad \mathbf{f}_i^T \mathbf{x}_{ip}^* + \mathbf{f}_j^T \mathbf{x}_{jp}^* = 0 , \tag{16}$$

there holds

$$\left\| A_i^T \mathbf{f}_i \right\| = \left\| A_j^T \mathbf{f}_j \right\| . \tag{17}$$

Especially for MAP,

$$\lambda_{i*} : \lambda_{j*} = \left\| B_i^T \mathbf{f}_i \right\| : \left\| B_j^T \mathbf{f}_j \right\| \tag{18}$$

holds (24,76,116). Then the ratio of the depth parameters can be computed and the MAP (including paraperspective and scaled orthographic projection) is resolved to orthographic projection.

The other representation of an epipolar constraint in Eq.16 is $\begin{pmatrix} \mathbf{f}_i \\ \mathbf{f}_j \end{pmatrix}^T \begin{pmatrix} \mathbf{x}^*_{ip} \\ \mathbf{x}^*_{jp} \end{pmatrix} = 0$, then the epipolar equation represents that the inner product between the vector arraying the coefficients of the epipolar equation, and the p-th joint image which is made by arraying the trajectory of the p-th feature point and is documented in Triggs (103). Therefore, the vector arraying the coefficients of the epipolar equation is the left kernel of the set of joint images, that is, a non-zero vectors belonging to the left kernel of $\begin{pmatrix} W_i^* \\ W_j^* \end{pmatrix}$ where $W_f^* = (\mathbf{x}^*_{1p}, \cdots, \mathbf{x}^*_{Fp})$.

3.4 Rectification And Euler Angle Representation

Because a MAP camera is resolved to an orthographic camera, it is sufficient to consider the orthographic camera without loss of generality.

In the orthographic camera, epipolar lines of each of two images are perpendicular to the intersection of two images. Therefore, epipolar lines are regarded as a common horizontal line of two images. Then rotating the two images to set common horizontal lines horizontally makes it easier to make correspondences of feature points because corresponded points are lying horizontal. This operation is called rectification (5,54,55,57,87). Rectification is closely related to the Euler angle representation of rotation matrix (Fig. 5) which changes one orthographic camera to another orthographic camera (24,54,55,57).

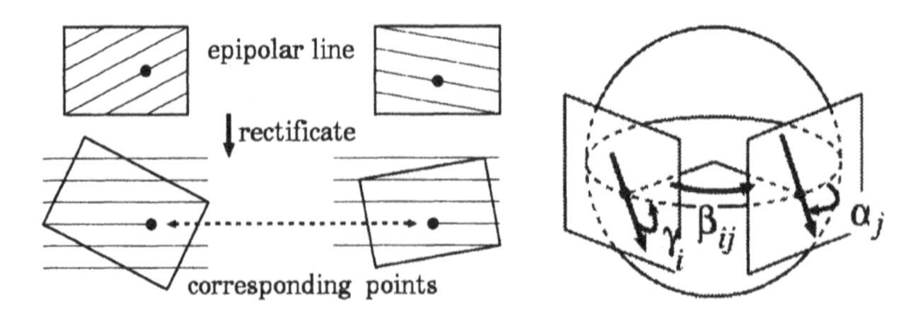

Fig. 5. Rectification and Euler angle representation

Let $C^{ij} = R_z(\alpha_i)R_y(\beta_{ij})R_z(\gamma_j)$ be an Euler angle representation of rotation matrix which changes the i-th orthographic camera to the j-th orthographic camera. The epipolar constraint only determines the two angles α_i and γ_j, and cannot determine β_{ij} which represents the angle between the two images.

This result means when there are two images, only an affine structure of three-dimensional space is reconstructed (49,90), and the three-dimensional space is Euclidean reconstructed up to some affine transformation of which parameterized one-parameter β_{ij} (18,117) [equivalent result for Koenderink and van Doon representation (49) of rotation matrix was obtained by Quan and Ohta (86)].

This result is easy to understand by mapping each orthographic image to a point on a unit sphere and a unit tangent vector along horizontal line at the point (24,93). In this representation, an epipolar plane though the center of the sphere and two points on the sphere which corresponded to each image, and an epipolar line is the grate circle through the points (the intersection between the epipolar plane and the sphere). In the epipolar geometry of orthographic cameras, the meaning of the Euler angle representation of rotation matrix is how to move the vector corresponding to the horizontal line of the i-th image to the vector corresponding to the horizontal line of the j-th image.

As already explained, two affine projection images are not sufficient to recover Euclidean structure. However, arbitrarily oriented views are synthesized from the linear combination of two affine projection images (71,109). This result was interpreted from a point of view by the factorization method for two images (54,55,57).

3.5 Extension Of Epipolar Geometry

3.5.1 For Perspective Camera

Recently, the geometrical constraints for feature correspondences under multiple views were investigated. The geometry among three or four images is called trilinear constraint (15,16,30,91,92) or quadlinear constraint (15,103).

Trilinear constraints are derived similar to Eq.14 from the determinant of a 4×4 matrix of each row is chosen as different four rows of a 6×4 matrix $\begin{pmatrix} x_{ip}\mathbf{p}_{i3}^T - \mathbf{p}_{i1}^T \\ y_{ip}\mathbf{p}_{i3}^T - \mathbf{p}_{i1}^T \\ x_{jp}\mathbf{p}_{j3}^T - \mathbf{p}_{j1}^T \\ y_{jp}\mathbf{p}_{j3}^T - \mathbf{p}_{j1}^T \\ x_{kp}\mathbf{p}_{k3}^T - \mathbf{p}_{k1}^T \\ y_{kp}\mathbf{p}_{j3}^T - \mathbf{p}_{k1}^T \end{pmatrix}$ over three images (at least one of each suffix i, j, k is chosen), and quadlinear constraints are derived from the determinant of a 4×4 matrix of each row is chosen as four different rows of a 8×4 matrix that organizing the same way as trilinear constraints.

However, the geometry among five or more images does not exist (15) because four rows of a 10×4 matrix is not over five images.

3.5.2 For Affine Camera

Trilinear and quadlinear constraints for an affine camera (86,100) were also derived. Similar to an epipolar constraint, the coefficients of trilinear constraints are derived from a non-zero vectors belonging to the left kernel of $\begin{pmatrix} W_i^* \\ W_j^* \\ W_k^* \end{pmatrix}$, of which at least two components correspond to different images equal to zero.

Quadlinear constraints are derived in the same way as quadlinear constraints and the geometry among five or more images does not exist for the same reason as the perspective camera.

3.5.3 Other Features

There is no geometrical constraint of corresponding lines between two images. The geometrical constraint of corresponding lines among three images was investigated by Hartley (30).

Karl and Heyden (50) derived the geometric constraints for corresponding lines and corresponding conics for an affine camera. Quan (82,84) derived the geometric constraints for corresponding conics for a perspective camera.

4. THREE DIMENSIONAL RECONSTRUCTION FOR AFFINE APPROXIMATION CAMERA

For a calibrated affine approximation camera, three distinct images of four non-coplanar points are necessary and sufficient to determine Euclidean reconstruction of the camera motion and the object shape, uniquely up to a reflection (18,22,23,108).

The pair of one three-dimensional reconstruction and another reflected three-dimensional reconstruction is called Necker reversal. Both reconstructions theoretically exist and true reconstruction cannot be chosen from two reconstructions only by corresponding points.

When this necessary and sufficient condition is not satisfied, Euclidean reconstruction is not obtained. In the case of two images of four non-coplanar points is obtained, only affine structure is reconstructed as already explained, and for the other cases, that is, all points are coplanar. Hu and Ahuja (33) classified the uncertainty of reconstruction.

For three or more images, many Euclidean reconstruction algorithms are presented for an orthographic camera (34,44,101,102,115), a scaled orthographic camera (76,86,93,101,102,111,112,117), a paraperspective camera (2,43,79) and a MAP camera (19,20,22,23).

These algorithms are mainly classified into two types. One is geometrical algorithm and the other is algebraic algorithm.

Roughly speaking, the main characteristic of a geometrical algorithm is to recover camera motion first by resolving itself to the algorithm for an orthographic camera, and recover object shape after that. Epipolar geometry-based algorithms are representative of the geometrical algorithm.

On the other hand, the main characteristic of an algebraic algorithm is recovering camera motion and object shape simultaneously by reconstruct affine structure first and determine the affine transformation which is a freedom of two images reconstruction and define the metric of three-dimensional space after that. The factorization method is a representative of the geometrical algorithm.

4.1 Geometrical Algorithm

4.1.1 Epipolar Based Algorithm

All algorithms based on epipolar geometry are resolved into the orthographic case by computing the ratio of the scale factor. Ostuni and Dunn (76) compute the scale factor by the roll-pitch-yaw representation of rotation matrix. Xu and Sugimoto (117) and Shimshoni et al. (93) compute the scale factor by epipolar geometry. Then it is sufficient to consider an orthographic camera without loss of generality.

Shimshoni et al. (93) resolved the determination of three rotation matrices to the determination of three lengths of a spherical triangle, of which three angles are known.

Let δ_f be a clockwise angle from an epipolar line of the f-1st image to an epipolar line of the f+1st image on the f-th image (mod 3), and δ_f is the angle of a spherical triangle at a point corresponding to the f-th image (Fig. 6).

Three angles of a spherical triangle determine the three lengths of a spherical triangle by spherical trigonometry, then the spherical triangle is determined. Each vertex of the spherical triangle is corresponded to the position of the image, then three rotation matrices are determined.

Xu and Sugimoto (117) determined the three rotation matrices by Euler angle representation of a rotation matrix.

Let $C^{ij} = R_z(\alpha_i)R_y(\beta_{ij})R_z(\gamma_j)$ be an Euler angle representation of a rotation matrix that changes an i-th orthographic camera to a j-th orthographic camera, with an epipolar equation determining the two angles α_i and γ_j.

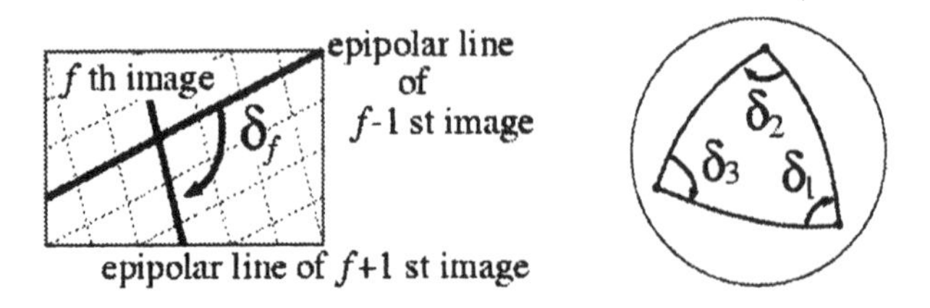

Fig. 6. Epipolar lines and the spherical triangle.

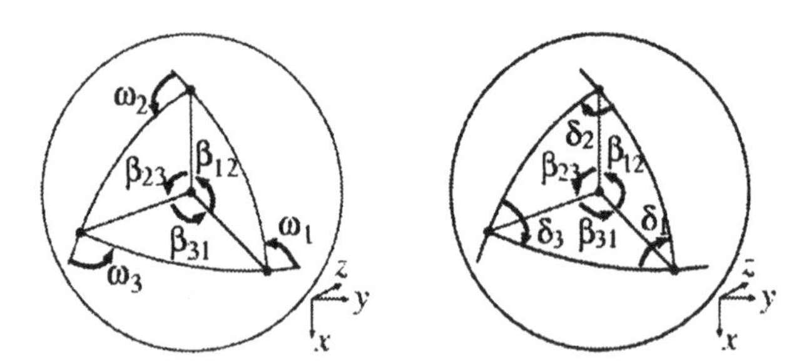

Fig. 7. Xu and Sugimoto (left) and Shimshoni et al. (right).

There holds $C^{23}C^{12} = (C^{31})^T$ and comparing the (2,2) and (2,3) components of both sides of the equation,

$$\cos\beta_{f+1,f-1} = \frac{-\cos\omega_f + \cos\omega_{f+1}\cos\omega_{f-1}}{\sin\omega_{f+1}\sin\omega_{f-1}}, \tag{19}$$

$$\frac{\sin\beta_{f+1,f-1}}{\sin\omega_f} = \frac{\sin\beta_{f-1,f}}{\sin\omega_{f+1}} = \frac{\sin\beta_{f,f+1}}{\sin\omega_{f-1}} \tag{20}$$

where $\omega_f = \alpha_f + \gamma_f$.

Comparing to Shimshoni et al. (93) and Xu and Sugimoto (117), there holds $\omega_f + \delta_f = \pi$ and Eqs.19-20 are equivalent to the law of cosine and law of sine of spherical trigonometry (Fig. 7). Necker reversal of these two algorithms is corresponded to the sign of $\sin\beta_{ij}$ (β_{ij} and $-\beta_{ij}$).

Through spherical trigonometry, the algorithm of Huang and Lee (34) also provides the determination of three lengths of a spherical triangle from three angles of a spherical triangle (24).

In addition, when three points on a sphere corresponding to three images lies on some great circle of the sphere, the three points do not form a spherical triangle. In this situation, Euclidean structure is not reconstructed by the methods based on epipolar geometry. However, the factorization method can recover Euclidean structure even in this situation.

4.1.2 Other Geometrical Algorithms

Let C^{1f} be a rotation matrix which changes the first orthographic camera to the f-th orthographic camera. C^{1f} is parameterized by only one parameter β_{1f} ($f \neq 1$) from the epipolar geometry [note that $\beta_{11} = 0$ holds because C^{11} is an identity matrix].

When setting a world coordinate to the camera coordinate of the first image, let ψ_p^* be a relative joint image of the p-th feature point which is arraying the relative image coordinates of the p-th feature point $\mathbf{y}_{fp}^* = \lambda_{f*} \chi_{fp}^*$ resolved to an orthographic image through three images.

The relation between ψ_p^* and the relative world coordinate of the p-th feature point $\mathbf{s}_p^*$ is $\psi_p^* = \begin{pmatrix} \mathbf{y}_{1p}^* \\ \mathbf{y}_{2p}^* \\ \mathbf{y}_{3p}^* \end{pmatrix} = \begin{pmatrix} (I_2 \ \ \mathbf{0}_2) \\ (I_2 \ \ \mathbf{0}_2) C^{12}(\beta_{12}) \\ (I_2 \ \ \mathbf{0}_2) C^{13}(\beta_{13}) \end{pmatrix} \mathbf{s}_p^* = M\mathbf{s}_p^*$

Let W^* and S^* be defined as $W^* = (\psi_1^*, \ldots, \psi_P^*)$ and $S^* = (\mathbf{s}_1^*, \ldots, \mathbf{s}_P^*)$, there holds

$$W^* = MS^* \quad \Rightarrow \quad \text{span } W^* = \text{span } M \tag{21}$$

where "span Y" is the linear space spanned by columns of a matrix Y.

Shapiro et al. (89) derived simultaneous equations of β_{12} and β_{13} by replacing the condition of Eq.21 to the same condition as

$$W^* = (\text{proj } M) W^*$$

where "proj Y" is the projection matrix to the linear space spanned by columns of a matrix Y. The simultaneous equations of β_{12} and β_{13} are also derived by replacing the condition of Eq.21 to the same condition as

$$M = (\text{proj } W^*) M .$$

Quan and Ohta (86) derived simultaneous equations of β_{12} and β_{13} by replacing the condition of Eq.21 to $\mathbf{a}^T M = \mathbf{0}_6^T$ where vector $\mathbf{a}$ belongs to the left kernel of W^* computed by trilinear constraints. Because at least two components of $\mathbf{a}$ must equal 0, there are four simultaneous equations are derived.

These two methods are equivalent because kernel is a residue of projection.

4.2 Factorization Method

The factorization method is a robust and efficient method for accurately recovering the motion and the shape from a sequence of images. The characteristics of the method is stable in numerical computation because the algorithm consists of linear computation and the algorithm treats all images uniformly. The former characteristic ensure the convergence of the algorithm, that is, there is no need to pay attention to initialization. The latter characteristic means the reconstruction is not dependent on some specific image that may have large errors.

The factorization method was first presented by Tomasi and Kanade (101,102) for the orthographic camera, and the method was extended to the methods for the scaled orthographic camera (43,111,112), the paraperspective camera (43,78,79), the MAP camera (19,22,23) and the perspective camera (7,8,11,12,26,95,104-107).

On the other hand, the factorization method was extended for multi-body (9,10,27), articulated object (73), line segment (4,70,85) and deformable object based on fixed cameras (96,97). The recursive factorization methods (20,21,69) were also investigated.

4.3 Factorization Method For MAP Camera

The essence of the factorization method is the relation among the image coordinates, the projection matrix, and the world coordinates of the feature points and is described in a quite simple form as

(measurement matrix) = (motion matrix) × (shape matrix)

where measurement matrix is organized by image coordinates of the feature points, motion matrix is organized by camera direction, and shape matrix is organized by the world coordinates of the feature points.

4.3.1 Algorithm Of Factorization Method

The representation of a MAP camera for relative coordinates in the world coordinates is

$$\mathbf{x}^*_{fp} = A_f C_f \mathbf{s}^*_p = \frac{1}{\lambda_{f*}} B_f C_f \mathbf{s}^*_p = \frac{1}{\lambda_{f*}} R_f \Sigma_f \begin{pmatrix} 1 & 0 & 0 \\ 0 & 1 & 0 \end{pmatrix} \widetilde{D}_f C_f \mathbf{s}^*_p \tag{22}$$

where $\mathbf{s}^*_p = \mathbf{s}_p - \mathbf{s}_*$ is a relative coordinate of the p-th feature point in the world coordinate.

Let measurement matrix W^*, shape matrix S^* and motion matrix M be defined as

$$W^* = \begin{pmatrix} W^*_1 \\ \vdots \\ W^*_F \end{pmatrix}, \quad W^*_f = (\mathbf{x}^*_{f1}, \ldots, \mathbf{x}^*_{fP}), \tag{23}$$

$$M = \begin{pmatrix} M_1 \\ \vdots \\ M_F \end{pmatrix}, \quad M_f = A_f C_f, \quad S^* = (\mathbf{s}_1^*, \ldots, \mathbf{s}_P^*). \tag{24}$$

Eqs.22-24 derive the following equation as

$$W^* = MS^*. \tag{25}$$

Note that there is no shape information in the motion matrix M, and there is no motion information in the shape matrix S^*. That is, the measurement matrix W^* is separated of motion information and shape information by decomposing the left side of Eq.25 to the product of two matrices M and S^*. Therefore, the reconstruction of the camera motion and the object shape is equivalent to the decomposition of W^* into the multiplication of two matrices M and S^*.

The factorization method realizes the decomposition by a two-step procedure. The first step is the reconstruction of an affine structure and the second step is an upgrading the affine structure to an Euclidean structure.

The first step, that is, the recovering of affine structure, is realized by any decomposition of W^* into

$$W^* = \hat{M}\hat{S}^* \tag{26}$$

where $\hat{M}$ and $\hat{S}^*$ are the same sized matrix as M and S^*, respectively.

This is because the row space of both S^* and $\hat{S}^*$ are equivalent to the row space of W^*. Therefore there exists a non-singular 3×3 matrix A which satisfies

$$M = \hat{M}A, \; S^* = A^{-1}\hat{S}^*. \tag{27}$$

Then $\hat{S}^*$ represents the affine structure of the shape. The decomposition of the left side of Eq.26 is realized when there are two distinct images are obtained. This is a well-known fact that the affine structure of the shape is determined by two distinct images.

The measurement matrix is organized by perspective projected image coordinates, then the measurement matrix has an affine approximation error. The measurement matrix also has a tracking error and quantization error of pixel units. Therefore, the rank of W^* usually exceeds four, on the contrary to the theoretical rank of W^* being three. Then an appropriate measurement matrix (of rank three) must be estimated from the observed measurement matrix.

Singular value decomposition (SVD) is commonly used for the first step of the factorization method because SVD gives the least squares estimation of W^* by taking three large singular values of SVD and the other singular values are regarded as zero. However, the cost of SVD of large matrices is very high. Then Mahamud and Hebert (62) proposed iterative randomized cascade basis reduction (IRCBR) to reduce the cost of SVD of large matrices.

Note that SVD is not an essential process in the factorization method (18,46) theoretically. Kuo et al. (53), Kurata et al. (56), and Hwang et al. (37) presented the factorization method without using SVD and/or least squares estimation.

After the first step, the affine structure must be upgraded to an Euclidean structure. This upgrade is realized by the determination of the matrix A defined in Eq.27.

The constraints to determine A are derived from the orthonormality of the basis of the camera coordinate system, that is C_f are orthogonal matrices.

Let $\hat{M} = \begin{pmatrix} \hat{M}_1 \\ \vdots \\ \hat{M}_F \end{pmatrix}$, the orthonormality is represented as

$$M_f M_f^T = A_f C_f C_f^T A_f^T = A_f A_f^T . \tag{28}$$

Then the constraints for $Q = AA^T$ are represented as

$$\hat{M}_f Q \hat{M}_f^T = A_f A_f^T = \frac{1}{\lambda_{f*}^2} B_f B_f^T . \tag{29}$$

These constraints are called metric constraints because the constraints determine the metric of three-dimensional space.

The metric constraints (Eq.29) are linear homogeneous equations for the component of Q because each λ_{f*} is an unknown parameter (only the ratio among λ_{f*} can be determined by epipolar geometry). Then only the ratio among the component of Q is determined. By determining the global scale parameter, the value of each component of Q is uniquely determined. Usually, the value of the component of Q is determined to satisfy the depth parameter of the first image equals one, that is, $\lambda_{1*} = 1$.

Another determination method for the matrix Q is that resolving a MAP camera to an orthographic camera. From Eq.9 and Eq.11,

$$\boldsymbol{\chi}_{fp}^* = \Sigma_f^{-1} R_f^T \mathbf{x}_{fp}^*, \quad \boldsymbol{\chi}_{fp}^* = \frac{1}{\lambda_{f*}} \begin{pmatrix} 1 & 0 & 0 \\ 0 & 1 & 0 \end{pmatrix} \tilde{D}_f C_f \mathbf{s}_p^* . \tag{30}$$

In Eq.30, the ratio of λ_{f*} is determined from epipolar geometry. When a global scale parameter is fixed as setting $\lambda_{1*} = 1$, the values λ_{f*} are uniquely determined. Then let $\mathbf{y}_{fp}^* = \lambda_{f*} \boldsymbol{\chi}_{fp}^*$, $\mathbf{y}_{fp}^*$ is regarded as the orthographic image on a virtual image plane.

By using $\mathbf{y}_{fp}^*$ for the measurement matrix, the metric constraints are resolved to

$$\hat{M}_f Q \hat{M}_f^T = I_2 \quad \Leftrightarrow \quad (\hat{M}_f \otimes \hat{M}_f) \operatorname{cs} Q = \operatorname{cs} I_2 . \tag{31}$$

Then Q is determined by

$$\operatorname{cs} Q = (\hat{\Xi}^T \hat{\Xi})^{-1} \hat{\Xi}^T \boldsymbol{\iota} \tag{32}$$

where $\Xi = \begin{pmatrix} \hat{M}_1 \otimes \hat{M}_1 \\ \vdots \\ \hat{M}_F \otimes \hat{M}_F \end{pmatrix}$ and $\boldsymbol{\iota} = \begin{pmatrix} \operatorname{cs} I_2 \\ \vdots \\ \operatorname{cs} I_2 \end{pmatrix}$.

Let LL^T be the Cholesky decomposition of the positive symmetric matrix Q, that is, $LL^T = Q$, the general solution of the matrix A is $A = LU$ where U is any orthogonal matrix. Then the general solution of the motion matrix M and the shape matrix S^* are parameterized by U as

$$M = \hat{M} L U , \quad S^* = U^T L^{-1} \hat{S}^* \tag{33}$$

and the camera direction is recovered by $C_f = (\tilde{A}_f)^{-1} \tilde{M}_f$.

The freedom of M and S^* represented by U is equivalent to the freedom of taking the world orthonormal coordinate of three-dimensional space. Thus, all the general solutions are equivalent up to a reflection, which corresponds to $\det U = \pm 1$.

Note that the matrix Q calculated by this procedure sometimes is not obtained as symmetric and positive definite in reality because there exists observation error and affine approximation error. In this case, some numerical process is needed in the computation of Q as (83).

4.3.2 Geometry Of Factorization Methods

To consider the geometry of factorization method, it is sufficient to consider for only the orthographic camera because any MAP camera is resolved to an orthographic camera.

Each metric constraints (Eq.31) gives the sub-metric on two-dimensional row space of $\hat{M}_f$ and the reconstruction is the recovering of the three-dimensional metric from multiple two-dimensional sub-metrics. In this interpretation, the ratio of depth parameters of two images is derived from the metric consistency of the intersection of two sub-metrics corresponded to two images.

To understand the consistency intuitively, considering the quadratic form

$$\mathbf{r}^T Q \mathbf{r} = 1$$

is convenient. In this situation, the quadratic form represents an ellipsoid because Q should be positive definite, and each metric corresponds to an ellipse, which is the section of the ellipsoid made by the plane passing through the center of the ellipsoid.

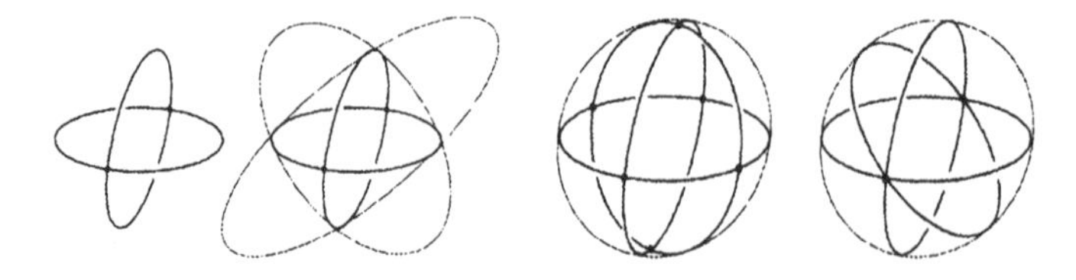

Fig. 8. How many ellipses are needed to determine a unique ellipsoid ?

Then the reconstruction problem is resolved to the recovering of the ellipsoid from the multiple sections of the ellipsoid. From this point of view, it is easy to understand the well-known fact that two images are not sufficient and three distinct images are sufficient to reconstruct the shape and the motion because two images are not sufficient to determine an ellipsoid uniquely and three distinct images are sufficient to determine an ellipsoid uniquely (Fig. 8).

4.4 Weighted Factorization Method

One of the characteristics of the factorization method is treating all images uniformly. This characteristic is easy to understand from how to organize a measurement matrix. However, this uniformity is disadvantageous when very noisy images are obtained. To overcome the disadvantage, weighted factorization methods are presented.

Aguiar and Moura (2) used the weights that are proportional to the reciprocal of the standard deviation of noise generated to image coordinates of the feature point. That is, the weights normalize the noise. Irani and Anandan (41) generalized the method when the noise of the image coordinates have correlation.

Aanaes, et al. (1) used weights iteratively by using a robust function to reduce noise from the measurement matrix.

4.5 Multi-Body Factorization Method

When considering the structure from motion problem for multi-body objects, there are two approaches taken. One approach is to segment each object first, and another approach is to segment and reconstruct simultaneously.

Weber and Malik (110) segmented objects by optical flow under a scaled orthographic camera.

Wolf and Shashua (114) used the direct product of epipolar constraints called the segmentation matrix to segment objects.

Gear (25) proposed a multi-body grouping method by using the row echelon form, which is closely related to Gauss-Jordan elimination, and realizes the segmentation of linear space into independent linear subspaces. Gear also considered to the numerical stability of the proposed method.

Costeira and Kanade (9,10) separated the multi-body by defining the shape interaction matrix. The property of the matrix is that the (i, j)-element should be zero when the i-th feature point and the j-th feature point belong to different objects. Kanatani (47) pointed out the proof of the property of the interaction matrix have some problems and gave the complete proof. The computational cost to segment the multi-body on the method is too high, then, Ichimura and Tomita (38) proposed the segmentation method to reduce the costs.

Kanatani (48) presented a segmentation algorithm by using an interaction matrix (9,10) and the geometric AIC (45).

Ichimura (39,40) combined a shape interaction matrix and discriminant analysis to improve the accuracy and robustness of segmentation.

Fitzgibbon and Zisserman (17) proposed a method which reconstructs independently moving objects and calibrates camera parameters simultaneously under uncalibrated perspective camera. The key idea of the method is the intrinsic camera parameters must be the same for each independently moving object.

Han and Kanade (27) presented the factorization method for moving objects with constant velocity. The key idea of their method is the factorization of velocity vectors.

Nagasaki el al. (73) estimated joint constraints among multiple bodies from the motion parameter of each object by evaluating the rank of the measurement matrix.

Held (31) used QR factorization of shape space to merge feature points on the same object. This method is called incremental factorization.

4.6 Recursive Factorization Method

When considering the recursive factorization method, the suppression of computational time is needed because the computational time grows proportionally to the number of images. Moreover, consideration of the fixation of world coordinates is an important point because the reconstruction of the motion and the shape is generally the reconstruction of relative coordinates of them.

Morita and Kanade (69) suppressed the computational time by replacing the computation of SVD into the approximation of an eigen value problem and realizing the fixation of world coordinates by fixing the orthonormal basis of the row space of the shape space. The shape space estimation of the method developed by Li and Brooks (60).

Fujiki and Kurata (20-23) suppressed the computational time by extracting motion information, which is equivalent to the metric information of three-dimensional space by principal component analysis (PCA) and realizing the fixation of world coordinates by computing the orthogonal matrix connects two shape matrices. The extraction of motion information by PCA treats all images as uniformly as possible, which is one of the inherent advantages of the factorization method. Then the method gives stable reconstruction even if some images contain errors. Although reconstruction may be less than stable by the batch process factorization method (79,102), the reduction of processing time more than compensates for this instability.

The method extended by Kurata et al. (58,59), and this method selects good reconstructed points by Lmedes criterion to suppress the reconstruction error due to occlusion.

4.7 Other Factorization Methods For Affine Camera

Morris and Kanade (70) presented the unified factorization method for points, line segments and planes by defining the uncertainty of points, line segments and planes, which are represented as Gaussian probability density functions.

Quan and Kanade (85) presented the factorization method for line segments by regarding the reconstruction of the direction of line segments as the structure from motion problem in two-dimensional space. The minimal work case of the algorithm was investigated by Åstöm et al. (4).

Tan et al. (96,97) presented the factorization method for a deformable object based on fixed cameras. The key idea of this method is the relative position of cameras are independent from time when the cameras are fixed.

Aguiar and Moura (2) presented a rank-1 factorization method for an orthographic camera, which is easy to extend to a MAP camera. In an orthographic camera, the first two coordinates of feature points are known from the first image when setting a world coordinate to the camera coordinate of the first image. When the measurement matrix is projected to the row space of the first two rows of the shape matrix, the rank of residue should equal one. That is, the factorization method is resolved to the decomposition of a rank one matrix.

5. THREE DIMENSIONAL RECONSTRUCTION FOR PERSPECTIVE CAMERA

5.1 Factorization Method For Perspective Camera

From Eq.13, the measurement matrix for a perspective camera and its decomposition is described as

$$\begin{pmatrix} \lambda_{11}\mathbf{x}^*_{11} & \cdots & \lambda_{1P}\mathbf{x}^*_{1P} \\ \vdots & \ddots & \vdots \\ \lambda_{F1}\mathbf{x}^*_{F1} & \cdots & \lambda_{FP}\mathbf{x}^*_{FP} \end{pmatrix} = \begin{pmatrix} P_1 \\ \vdots \\ P_F \end{pmatrix} (\tilde{\mathbf{s}}_1, \ldots, \tilde{\mathbf{s}}_P) \tag{34}$$

However, the decomposition of the left side to the right side is difficult because the projective depths λ_{fp} are unknown parameters. Therefore, to decompose the measurement matrix, the projective depths must be estimated. After the projective depths are estimated, the left side of Eq.34 is determined and the decomposition, that is, the reconstruction is realized. However, the decomposition has an ambiguity represented by a projective matrix (13,28). To reconstruct the ambiguity represented by a projective matrix (13,28). To reconstruct the Euclidean structure, the projective matrix must be computed.

The factorization method for perspective cameras has two steps. The first step is the recovery of the projective depth and the arbitrary decomposition of the measurement matrix represents projective reconstruction. The second step is the determination of the projective matrix to upgrade the projective reconstruction to Euclidean reconstruction. This situation is similar to the factorization method for a MAP camera.

Triggs (104), Sturm and Triggs (95), Deguchi and Triggs (11,12) presented a factorization method for uncalibrated perspective views via epipolar geometry without iterative computation. These methods give the projective reconstruction of three-dimensional space. To acquire Euclidean reconstruction, five points of Euclidean coordinates must be known. The key point of the method is to estimate the projective depths by computing a fundamental matrix of several pairs of two images in the framework of epipolar geometry.

However, the method requires high accuracy of the fundamental matrices although the estimation of the fundamental matrices from point correspondences between two images is quite sensitive even if the input data is normalized as presented by Hartley(29). Moreover, the accuracy of the estimated projective depths depend on how to choose the pairs of two images, that is, this method does not treat all images uniformly which is one of the inherent advantages of the factorization method. Thus when one unreliable image is chosen for computing the fundamental matrix, the reconstruction becomes unreliable.

Ueshiba and Tomita (105,106) presented another factorization method for uncalibrated perspective views, which treats all images uniformly and simultaneously. The method estimates the projective depths iteratively starting from the estimation of the paraperspective projection by the conjugate gradient method, which is a non-linear optimum method under the criterion that the rank of the measurement matrix for the perspective camera is four. In addition, the papers also presented the recovery of the Euclidean structure for a calibrated camera by metric constraints for the perspective camera.

Ukita and Shakunaga (107) advanced Ueshiba and Tomita's method to the method for the perspective camera with a variable focal length. The advancement of the idea is to neglect the metric constraint, which include the focal length as a parameter in the recovery of the Euclidean structure. Therefore, the value of the focal length is not needed to compute the Euclidean reconstruction. However, four images are needed to reconstruct the Euclidean structure. Furthermore, this paper also provides a frame selection method to remove the frame that failed to reconstruct by epipolar geometry, and a segmentation method, which includes error detection and error correction in the feature tracking sequence. The unstability of the method revised by Hachiya and Shakunaga (26).

Mahamud and Hebert (63) introduced the constraints of the projection matrix and shape matrix independently, which are called dual subspace constraints, and estimated projection matrix and shape matrix iteratively.

Christy and Horaud (7,8,36) presented the iterative method for a calibrated camera to estimate the projective depths by estimating the image coordinates of a scaled orthographic or paraperspective camera from the image coordinates of a perspective camera. The method gives projective depths and Euclidean reconstruction simultaneously. This method has rewritten in the framework of the factorization method (21).

The outlines of the method for a paraperspective camera is as follows.

Let $\mathbf{x}_{fp}^{\text{per}}$ be the image coordinates of a perspective projection and $\mathbf{x}_{fp}^{\text{para}}$ be the image coordinates of a paraperspective projection of the p-th feature point in the f-th image, and let $\mathbf{x}_{fp}^{*\text{per}}$ and $\mathbf{x}_{fp}^{*\text{para}}$ be the relative image coordinates from $\mathbf{x}_{fp}^{*\text{per}} = \mathbf{x}_{fp}^{*\text{para}}$.

The relationship between the image coordinates of the paraperspective projection and those of the perspective projection are derived from Fig.9 as

$$\mathbf{x}_{fp}^{*\text{para}} = \mu_{fp}^{*}\mathbf{x}_{fp}^{*\text{per}}, \quad \mu_{fp}^{*} = \frac{\lambda_{fp}}{\lambda_{f*}} = 1 + \frac{k_f^T s_p^*}{\lambda_{f*}}. \tag{35}$$

Now, the measurement matrix containing projective depths is defined as

$$W^* = \begin{pmatrix} \mu_{11}\mathbf{x}_{11}^* & \cdots & \mu_{1P}\mathbf{x}_{1P}^* \\ \vdots & \ddots & \vdots \\ \mu_{F1}\mathbf{x}_{F1}^* & \cdots & \mu_{FP}\mathbf{x}_{FP}^* \end{pmatrix} . \tag{36}$$

where $\mathbf{x}_{fp}^*$ are observed quantities, that is, the perspective image coordinates.

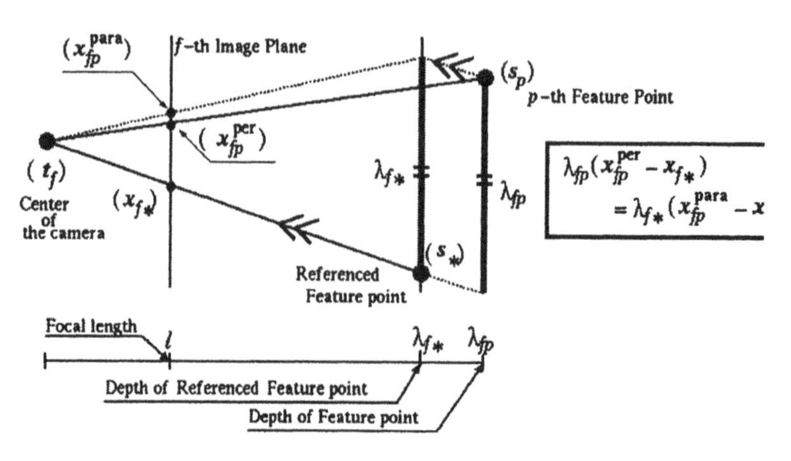

Fig. 9. Perspective image and paraperspective image.

Note that the measurement matrix does not use homogeneous coordinates contrary to other perspective factorization methods.

The algorithm for the Cristy and Horaud method is as shown below.

STEP1: Let $\mu^{*}_{fp} = 1$ and perform the Euclidean reconstruction with the paraperspective factorization method. Name one reconstruction of the motion and the shape to $M^{(+)}$, $S^{*(+)}$ and its reflected reconstruction of the motion and the shape to $M^{(-)}$, $S^{*(-)}$ respectively. Update $\mu^{*(\pm)}_{fp}$ by Eq.36 for each solution (hereafter, the same symbols are used in the same manner).

STEP2: Update the measurement matrix $W^{*(\pm)}$.

STEP3: Perform Euclidean reconstruction with paraperspective factorization.

STEP4: Update $M^{(\pm)}$, $S^{*(\pm)}$ by choosing the consistent solution from a pair of Euclidean reconstructions.

STEP5: Update $\mu^{*(\pm)}_{fp}$ by Eq.35 for each solution.

STEP6: If $\mu^{*(\pm)}_{fp}$ does not converge, return to STEP2.

STEP7: Select the solution which most closely matches the observed image coordinates out of $M^{(+)}$, $S^{*(+)}$ and $M^{(-)}$, $S^{*(-)}$.

The algorithm was extended recursively by Fujiki and Kurata (21). The extended method is a combination of the original method and the recursive factorization for a MAP camera (20,22,23).

5.2 The Other Algorithms

Hartley et al. (28) presented the projective reconstruction from two images of uncalibrated perspective cameras by decomposition of an essential matrix.

Quan (80,81) presented the projective reconstruction from three images of an uncalibrated perspective camera by invariants computed from six points in three images.

Hartley (30) gave the projective reconstruction for a perspective projection by using point and line correspondences through a trifocal tensor.

Heyden and Åstöm (32) proposed the reconstruct method for perspective projection varying focal lengths.

McLauchlan et al. (66-68) presented a recursive method for estimating motion and shape through both affine projection and perspective projection using a variable state-dimension filter. However, reconstruction is very sensitive to the latest images.

6. MISSING DATA

There are mainly two approaches to estimate missing data caused by tracking errors or occlusion. One approach is estimating the missing data by the intersection of epipolar lines, which is well-known. Another approach is estimating the component of measurement matrix by rank condition.

Tomasi and Kanade (102) presented an iterative algorithm to estimate missing data. First, selecting the largest rectangular subset of the matrix with no missing data. Second, selecting an additional column or row that has missing data, and the missing data is estimated by the linear combination of the columns or the rows of the rectangular subset. After this procedure, the size of the largest rectangular subset of the matrix with no missing data goes larger, and this procedure finishes when all missing data are estimated. Finally, the estimation is refined using the steepest descent minimization. However, this method does not use reliable data uniformly. When the depth of the feature points of a selected subset matrix is biased, the estimations of the feature points with different depths might be unreliable. Furthermore, this method costs too much to find the largest subset matrix.

Maruyama and Kurumi (65) treated missing data as parameters and estimated the parameter iteratively to minimize the Frobenius distance from three-dimensional shape space.

Poelman and Kanade (43,77) proposed the weighted factorization method to overcome this difficulty. This method set the missing data to zero and reduced the occlusion effect by adding the weight that comes from uncertainty to each column. This method is convenient because the method has no iteration. However, the weight is added only for points and this method cannot reflect the uncertainty of each frame.

Huang and Nishida (35) estimated the missing data by selecting a rectangular subset of the matrix which consists of the feature points that are thought to located nearby. The method improved the disadvantage of Tomasi and Kanade (102). However this method still costs too much to choose the subset matrix.

Shum et al. (94) estimated the measurement matrix by minimizing the cost function for SVD, which consists of only observed data. The method first decomposes the measurement matrix of which the missing component is set to zero by SVD. After that, revise the representation of SVD iteratively by minimizing the sum of squares distance of observed data through the gradient decent method. The minimization is formulated as a kind of weighted least squares method and the minimization treats all observed data uniformly. Poelman (78) also presented a similar method as the paper.

Jacobs (42) estimated the three-dimensional column space of the measurement matrix in order to fill missing data. When a column of the measurement matrix has k missing data, the column represents the k-dimensional affine space by filling arbitrary real numbers as missing data. Because the three-dimensional column space of the measurement matrix is a subspace of the sum of three arbitrary affine spaces corresponded to each column, the three-dimensional column space is the intersection of the sum of affine spaces.

7. UNIFYING DIFFERENT CAMERAS

Zhang et al. (122) proposed the epipolar geometry between a perspective camera and a scaled orthographic camera.

Marugame et al. (64) presented the Euclidean reconstruction from one scaled orthographic image and multiple perspective images. The key idea of the method is that a scaled orthographic camera preserves the ratio of the length of parallel line segments.

REFERENCES

(1) H. Aanaes, R. Fisker, K. K. Åstöm, J. M. Carstensen, ``Robust Factorization," IEEE Trans. PAMI, vol.24, no.9, pp.1215-1225, SEP. 2002.
(2) P. M. Q. Aguiar and J. M. F. Moura, ``Factorization as a rank 1 problem," Proc. CVPR'99, pp.178-184, 1999.
(3) J.Y. Aloimonos,``Prespective approximations," Image and Vision Computing, vol.8, no.3, pp.179-192, Aug.1990.
(4) K. Åstöm, A. Heyden, F. Kahl and M. Oskarsson, ``Structure and motion from lines under affine projections," Proc. ICCV'99, pp.285-292, 1999.
(5) N. Ayache and F. Lustman, ``Trinocular stereo vision for robotics," IEEE Trans. PAMI, vol.13, no.1, pp.73-85, Jan. 1991.
(6) G. Chesi, A. Garulli, A. Vicino and R. Cipolla, ``Estimating the fundamental matrix via constrained least-squares: a convex approach," IEEE Trans. PAMI, vol.24, no.3, pp.397-401, March 2002.
(7) S. Christy and R. Horaud, ``Euclidean reconstruction: from paraperspective to perspective," Proc. 4th ECCV, vol.2, pp.129-140, 1996.
(8) S. Christy and R. Horaud, ``Euclidean shape and motion from multiple perspective views by affine iterations," IEEE Trans. PAMI, vol.18, no.11, pp.1098-1104, Nov. 1996.
(9) J. Costeira and T. Kanade, ``A multi-body factorization method for motion analysis," Proc. 5th ICCV, pp.1071-1076, May 1995.
(10) J.P. Costeira and T. Kanade, ``A multibody factorization method for independently moving objects," IJCV, vol.29, no.3, pp.159-179, 1998.
(11) K. Deguchi, ``Factorization Method for Structure from Multiple Perspective Images," CVIM 106-6, pp.35-42, 1997.7.24, (in Japanese).
(12) K. Deguchi and B. Triggs, ``Factorization method for structure from perspective images," Technical Report of IEICE, PRMU 96-139, pp.81-88, Jan. 1997, (In Japanese).
(13) O.D. Faugeras, ``What can be seen in three dimensions with an uncalibrated stereo rig ?," Proc. 2nd ECCV, Santa Margherita Ligure, pp.563-578, May 1992.
(14) O.D. Faugeras, ``Three-dimensional computer vision: a geometric viewpoint," MIT press, Cambridge, MA, 1993.
(15) O. Faugeras and B. Mourrain, ``On the geometry and algebra of the point and line correspondences between N images," Proc. 5th ICCV, pp.951-956, 1995.

(16) O. Faugeras and L. Robert, ``What Can Two Images Tell Us About a Third One ?," IJCV, vol.18, pp.5-19, 1996.
(17) A.W. Fitzgibbon and A. Zisserman, ``Multibody structure and motion: 3-d reconstruction of independently moving objects," Proc. ECCV'00, 2000.
(18) J. Fujiki and T. Kurata, ``The metric of affine shape and motion : the intuitive interpretation in terms of the factorization method," Proc. SPIE97, San Diego, Vision Geometry VI, vol.3168-22, pp.206-217, Aug. 1997.
(19) J. Fujiki and T. Kurata, ``An mathematical analysis of the factorization method for generalized affine projection model," Technical Report of IEICE, PRMU 97-142, pp.101-108, Nov. 1997, (in Japanese).
(20) J. Fujiki, T. Kurata and M. Tanaka, ``Iterative factorization method for object recognition," Proc. SPIE98, San Diego, Vision Geometry VI, vol.3454-18, pp.192-201, July 1998.
(21) J. Fujiki and T. Kurata, ``Recursive factorization method for the paraperspective model based on the perspective projection," Proc. ICPR'00, pp.406-410, Sep. 2000.
(22) J. Fujiki, ``3d reconstruction from sequences of 2d images under point correspondences - an mathematical analysis of the factorization method -," Proc. Institute of statistical mathematics, vol.49, no.1, pp.77-108, 2001, (in Japanese).
(23) J. Fujiki and T. Kurata, ``A Recursive factorization method for the metric affine projection model," Trans. IEICE Japan, vol.J84-D-II, no.8, pp.1663-1673, Aug. 2001, (In Japanese).
(24) J. Fujiki, ``Epipolar geometry for metric affine projection model," Technical Report of IEICE, PRMU 2002-138, pp.55-60, Dec. 2002, (in Japanese).
(25) C.W. Gear, ``Multibody grouping from motion images," IJCV, vol.29, no.2, pp.133-150, 1998.
(26) T. Hachiya and T. Shakunaga, ``A factorization method for perspective structure and motion from multiple images with variable focal length," Technical Report of IEICE, PRMU2000-68, pp.49-56, Sep. 2000, (in Japanese).
(27) M. Han and T. Kanade, ``Reconstruction of a scene with multiple linearly moving objects," Proc. CVPR'00, 2000.
(28) R. Hartley, R. Gupta and T. Chang, ``Stereo from uncalibrated cameras," Proc. CVPR'92, Urvana Champaign, pp.761--764, June 1992.
(29) R.I. Hartley, ``In defence of the eight-point algorithm," IEEE Trans. PAMI, vol.19, no.6, pp.580-592, June 1997.
(30) R.I. Hartley, ``Lines and points in three views and the trifocal tensor," IJCV, vol.22, no.2, pp.125-140, 1997.
(31) A. Held, ``Piecewise shape reconstruction by incremental factorization," Proc. BMVC'96, 1996.
(32) A. Heyden and K. Åstöm, ``Euclidean reconstruction from image sequences with varying and unknown focal length and principal point," Proc. CVPR'98, pp.438-443, 1997.
(33) X. Hu and N.Ahuja, ``Mirror uncertainty and uniqueness condition for determining shape and motion from orthographic projection," IJCV, vol.13, no.3, pp.295-309, 1994.
(34) T.S. Huang and C.H. Lee, ``Motion and structure from orthographic projections," IEEE Trans. PAMI, vol.11, no.5, pp.536-540, May 1989.
(35) Y.J. Huang and H. Nishida, ``Reconstruction of measurement matrices for recovering shape and motion from long image sequences," Proc. IAPR Workshop on MVA98, Makuhari, Chiba, pp.463-466, Nov. 1998.

(36) R. Horaud, F. Dornaika, B. Lamiroy and S. Christy, ``Object pose: the link between Weak Perspective, Paraperspective, and full perspective," IJCV, vol.22, no.2, pp.173-189, 1997.
(37) K. Hwang, N. Yokoya, H. Takemura and K. Yamazawa, ``A factorization method using 3-d linear combination for shape and motion recovery," Proc. ICPR'98, vol.2, pp.959-963, Aug. 1998.
(38) N. Ichimura and F. Tomita, ``Motion segmentation based of feature selection from shape matrix," Trans. IEICE Japan, vol.J81-D-II, no.12, pp.2757-2766, Dec. 1998, (in Japanese).
(39) N. Ichimura, ``Motion segmentation based on factorization method and discriminant criterion," Proc. ICCV'99, pp.600-605, 1999.
(40) N. Ichimura, ``A robust and efficient motion segmentation based on orthogonal projection matrix of shape space," Proc. CVPR'00, pp.446-452, 2000.
(41) M. Irani and P. Anandan, ``Factorization with uncertainty," Proc. ECCV'00, pp.539-553, 2000.
(42) D. Jacobs, ``Linear fitting with missing data: application to structure-from-motion and to characterizing intensity images," Proc. CVPR'97, pp.206-212, 1997.
(43) T. Kanade, C.J. Poelman and T. Morita, ``A factorization method for shape and motion recovery," Trans .IEICE Japan, vol. D-II, vol.J76-D-II, no.8, pp.1497-1505, Aug. 1993, (in Japanese).
(44) H. Kamada, M. Siohara and Y. Hao, ``Simultaneous recovering motion and shape of 3d moving," Trans. IEICE Japan, vol.J76-D-II, no.8, pp.1554-1561, Aug. 1993, (in Japanese).
(45) K. Kanatani, ``Statistical optimization for geometric computation : theory and practice," Elsevier Science, 1996.
(46) K. Kanatani, ``Factorization without factorization: from orthographic to perspective," Technical Report of IEICE, PRMU98-26, June 1998, (in Japanese).
(47) K. Kanatani, ``Factorization without factorization: multibody segmentation," Technical Report of IEICE, PRMU98-117, pp.37-43, Nov. 1998, (in Japanese).
(48) K. Kanatani, ``Motion segmentation by subspace separation and model selection," Proc. ICCV'01, pp.301-306, 2001.
(49) J.J. Koenderink and A.J.van Doorn: ``Affine Structure from Motion", J. Opt. Soc. Am. A, vol.8, no.2, pp.377-395, 1991.
(50) F. Kahl and A. Heyden, ``Affine structure and motion from points, lines and conics," IJCV, vol.33, no.3, pp.163-180, 1999.
(51) K. Kinoshita and J. Sato, ``Quasi-perspective camera model," MIRU98, pp.I-249-I-254, July 1998.
(51) K. Kinoshita and M. Lindenbaum, ``Camera model selection based on Geometric AIC," CVPR00, pp.514-519, 2000.
(53) H. Kuo, K. Yamazawa, H. Iwasa, H. Takemura and N. Yokoya, ``A new factorization method based on 3d linear combination for shape and motion recovery," Proc. 55th General Conf. of IPSJ, pp.2-313-2-314, 1997, (In Japanese).
(54) T. Kurata, J. Fujiki and K. Sakaue, ``Affine epipolar geometry via factorization method and its application," Proc. SPIE, San Diego, vol.3457-13, July 1998.
(55) T. Kurata, J. Fujiki and K. Sakaue, ``Affine epipolar geometry via factorization method," Proc. ICPR'98, Brisbane, pp.862-866, Aug. 1998.
(56) T. Kurata, J. Fujiki and K. Sakaue, ``A factorization method for generalized affine projection model using robust estimation," Technical Report of IEICE, PRMU, Jan. 1999, (in Japanese).

(57) T. Kurata, J. Fujiki and K. Sakaue, ``Affine epipolar geometry via factorization method," Trans. IPSJ Journal, vol.40, no.8, pp.3188-3197, Aug. 1999, (in Japanese).
(58) T. Kurata, J. Fujiki, M. Kourogi and K. Sakaue, ``A fast and robust approach to recovering structure and motion from live video frames," Proc. CVPR'00, pp.528-535, June 2000.
(59) T. Kurata, J. Fujiki, M. Kourogi and K. Sakaue, ``A fast and robust approach to recovering structure and motion from live image sequences," Trans. IEICE Japan, vol.J84-D-II, no.12, pp.2515-2524, Dec. 2001, (in Japanese).
(60) Y. Li and M.J. Brooks, ``An efficient recursive factorization method for determining structure from motion," Proc. CVPR'99, 1999.
(61) Q.-T. Luong and O.D. Faugeras, ``The fundamental matrix: theory, algorithms, and stability analysis," IJCV, vol.17, pp.43-75, 1996.
(62) S. Mahamud and M. Hebert, ``Efficient recovery of low-dimensional structure from high-dimensional data," Proc. ICCV99, pp.592-599, 1999.
(63) S. Mahamud and M. Hebert, ``Iterative projective reconstruction from multiple views," Proc. CVPR00, pp.430-437, 2000.
(64) A. Marugame, J. Katto and M. Ohta, ``Structure recovery with multiple cameras from scaled orthographic and perspective views," IEEE Trans. PAMI, vol.21, no.7, pp.628-633, JULY 1999.
(65) M. Maruyama and S. Kurumi, ``Bidirectional optimization for reconstructing 3d shape from an image sequence with missing data," Proc. ICIP99, pp.120-124, 1999.
(66) P.F. McLauchlan, I.D. Reid and D.W. Murray, ``Recursive affine structure and motion from image sequences," Proc. ECCV'94, pp.217-224, 1994.
(67) P.F. McLauchlan and D.W. Murray, ``A unifying framework for structure and motion recovery from image sequences," Proc. 5th ICCV, pp.314-320, 1995.
(68) P.F. McLauchlan, ``A batch/recursive algorithm for 3d scene reconstruction," Proc. CVPR'00, pp.738-743, 2000.
(69) T. Morita and T. Kanade, ``A sequential factorization method for recovering shape and motion from image streams," IEEE Trans. PAMI, vol.19, no.8, pp.858-867, Aug. 1997.
(70) D.D. Morris and T. Kanade, ``A unified factorization algorithm for points, line segments and plane with uncertainty models," Proc. ICCV98, 1998.
(71) Y. Mukaigawa, Y. Nakamura and Y. Ohta, ``synthesis of facial views with arbitrary poses and expressions using multiple facial images," Trans. IEICE Japan, vol.J80-D-II, no.6, pp.1555-1562, June 1997, (in Japanese).
(72) J.L. Mundy and A. Zisserman, eds, ``Geometric invariance in computer vision," MIT Press, 1992.
(73) T. Nagasaki, T. Kawashima and Y. Aoki, ``Structure estimation of an articulated object from motion image analysis based on factorization method," Trans. IEICE Japan, vol.J81-D-II, no.3, pp.483-492, March 1998, (in Japanese).
(74) Y. Ohta, K. Maenobu and T. Sakai, ``Obtaining surface orientation from texels under perspective projection," Proc. 7th International Joint Conf. AI, pp.746-751, 1981.
(75) S.I. Olsen, ``Epipolar line estimation," Proc. ECCV92', pp.307-311, 1992.
(76) J. Ostuni and S. Dunn, ``Motion from three weak perspective images using image rotation," IEEE Trans. PAMI, vol.18, no.1, pp.64-69, Jan. 1996.
(77) C.J. Poelman and T. Kanade, ``A paraperspective factorization method for shape and motion recovery," CMU-CS-93-219, Dec. 1993.
(78) C.J. Poelman, ``The paraperspective and projective factorization method for recovering shape and motion," CMU-CS-95-173, July 1995.

(79) C.J. Poelman and T. Kanade, ``A paraperspectve factorization method for shape and motion recovery," IEEE Trans. PAMI, vol.19, no.3, pp.206-218, March 1997.
(80) L. Quan, ``Invariants of 6 points from 3 uncalibrated images," Proc. 3rd ECCV, vol. 2, pp.459-470, 1994.
(81) L. Quan, ``Invariants of six points and projective reconstruction from three uncalibrated images," IEEE Trans. PAMI, vol.17, no.1, pp.34-46, Jan. 1995.
(82) L. Quan, ``Invariant of a pair of non-coplanar conics in space: definition, geometric interpretation and computation," Proc. 5th ICCV, pp.926-931, 1995.
(83) L. Quan, ``Self-calibration of an affine camera from multiple views," IJCV, vol.19, no.1, pp.93-105, 1996.
(84) L. Quan, ``Conic reconstruction and correspondence from two views," IEEE Trans. PAMI, vol.18, no.2, pp.151-160, Feb. 1996.
(85) L. Quan and T. Kanade, ``Affine structure from line correspondences with uncalibrated affine cameras," IEEE Trans. PAMI, vol.19, no.8, pp.834-845, Aug. 1997.
(86) L. Quan and Y. Ohta, ``A new linear method for Euclidean motion/structure from three calibrated affine views," Proc. CVPR'98, pp.172-177, (1998).
(87) S.M. Seitz and C.R. Dyer, ``Complete scene structure from four point correspondences," Proc. 5th ICCV, pp.330-337, 1995.
(88) K. Sengupta, I. Poupyrev, T. Sakaguchi and J. Ohya, ``A new camera projection model and its application in reprojection," MIRU98, pp.I-243-I-248, July 1998.
(89) L.S. Shapiro, A. Zisserman and M. Brady, ``3d motion recovery via affine epipolar geometry," IJCV, vol.16, pp.147-182, 1995.
(90) L.S. Shapiro, ``Affine analysis of image sequences," Camblidge Univ. Press, 1995.
(91) A. Shashua, ``Algebraic functions for recognition," IEEE Trans. PAMI, vol.17, no.8, pp.779-789, Aug. 1995.
(92) A. Shashua and M. Werman, ``Trilinearity of three perspective views and its associated tensor," Proc. 5th ICCV, pp.920-925, 1995.
(93) I. Shimshoni, R. Basri and Rivlin, ``A Geometric interpretation of weak-perspective motion," IEEE Trans. PAMI, vol.21, no.3, pp.252-257, March 1999.
(94) H.Y. Shum, K. Ikeuchi and R. Reddy, ``Principal component analysis with missing data and its application to polyhedral object modeling," IEEE Trans. PAMI, vol.17, no.9, pp.854-867, Sept. 1995.
(95) P. Sturm and B. Triggs, ``A factorization based algorithm for multi-image projective structure and motion," Proc. 4th ECCV, vol.2, pp.709-720, 1996.
(96) J.K. Tan, S. Ishikawa, H. Ikeuchi, I. Nohara and S. Hata, ``Recovering human motion by factorization," Proc. 3rd Symposium on Sensing via Image Inform., Yokohama, C-14, pp. 111-114, June 1997.
(97) J.K. Tan, S. Kawabata and S. Ishikawa, ``An efficient technique for motion recovery based on multiple views," Proc. IAPR Workshop on MVA98, Makuhari, Chiba, pp.270-pp.273, Nov. 1998.
(98) C.-K. Tang, G. Medioni and M.-S. Lee, ``Epipolar geometry estimation by tensor voting 8d," Proc. ICCV'99, pp.502-509, (1999).
(99) C.-K. Tang, G. Medioni and M.-S. Lee, ``*N*-dimensional tensor voting and application to epipolar geometry estimation," IEEE Trans. PAMI, Vol.23, No.8, pp.829-844, Aug. 2001.
(100) T.Thórhallsson and D.W.Murray, ``The tensors of three affine views," Proc. CVPR'99, pp.450-456, (1999).

(101) C. Tomasi and T. Kanade, ``The factorization method for the recovery of shape and motion from image streams," Proc. Image Understanding Workshop, pp.459-472, San Diego, Jan. 1992.

(102) C. Tomasi and T. Kanade, ``Shape and motion from image streams under orthography: a factorization method," IJCV, vol.9, no.2, pp.137-154, 1992.

(103) B. Triggs, ``Matching constraints and the joint image," Proc ICCV'95, pp.338-343, 1995.

(104) B. Triggs, ``Factorization method for projective structure and motion," Proc CVPR'96, San Fransisco, pp.845-851, 1996.

(105) T. Ueshiba and F. Tomita, ``A factorization method for multiple perspective views using affine projection as the initial camera model," CVIM 107-1, pp.1-8, Sept. 1997, (in Japanese).

(106) T. Ueshiba and F. Tomita, ``A factorization method for projective and Euclidean reconstruction from multiple perspective views via iterative depth estimation," Proc. 5th ECCV, vol.1, pp.296-310, Freiburg, Germany, June 1998.

(107) N. Ukita and T. Shakunaga, ``3d reconstruction from wide-range image sequences using perspective factorization method," Technical Report of IEICE, PRMU 97-276, pp.81-88, March 1998, (in Japanese).

(108) S. Ullman, ``The interpretation of visual motion," MIT Press, Cambridge, Mass., 1979.

(109) S. Ullman and R. Basri, ``Recognition by linear combinations of models," IEEE Trans. PAMI, vol.13, no.10, pp.992-1005, Oct. 1991.

(110) J. Weber and J. Malik, ``Rigid body segmentation and shape description from dense optical flow under weak perspective," IEEE Trans. PAMI, vol.19, no.2, pp.139-143, Feb. 1997.

(111) D. Weinshall and C. Tomasi, ``Linear and incremental acquisition of invariant shape models from image sequences," Proc. 4th ICCV, pp.675-682, 1993.

(112) D. Weinshall and C. Tomasi, ``Linear and incremental acquisition of invariant shape models from image sequences," IEEE Trans. PAMI, Vol.17, No.5, pp.512-517, May 1995.

(113) M. Wilczkowiak, E. Boyer and P. Sturm, ``Camera calibration and 3d reconstruction from single images using parallelepipeds," Proc. ICCV'01, 2001.

(114) L. Wolf and A. Shashua, ``Two-body segmentation from two perspective views," Proc. CVPR'01, vol.1, pp.263-270, 2001.

(115) Y. Xirouhakis and A. Delopoulos, ``Least square estimation of 3d shape and motion of rigid objects from their orthographic projections," IEEE Trans. PAMI, Vol.22, No.4, pp.393-399, Apr 2000.

(116) G. Xu and Z. Zhang, ``Epipolar geometry in stereo, motion and object recognition - a unified approach," Kluwer Academic Publishers, 1996.

(117) G. Xu and N. Sugimoto, ``A linear algorithm for motion from three weak perspective images using Euler angles," IEEE Trans. PAMI, Vol.21, No.1, pp.54-57, Jan, 1999.

(118) H. Yu, Q. Chen, G. Xu and M. Yachida, ``3d shape and motion by SVD under higher-order approximation of perspective projection," Proc. ICPR'96, pp.456-460, 1996.

(119) Z. Zhang, R. Deriche, O. Faugeras and Q.T. Luong, ``A robust technique for matching two uncalibrated images through the recovery of the unknown epipolar geometry," Artificial Intelligence 78, pp.87-119, 1995.

(120) Z. Zhang and G. Xu, ``A general expression of the fundamental matrix for both perspective and affine cameras," Proc. IJCAI'98, pp.1502-1507, 1998.

(121) Z. Zhang, ``Determining the epipolar geometry and its uncertainly: a review," IJCV, vol.27, no.2, pp.161-195, 1998.
(122) Z. Zhang, P. Anandan and H.-Y. Shum, ``What can be determined from a full and a weak perspective image?," Proc. ICCV'99, pp.680-687, (1999).

4

Compression and Representation of 3-D Images

Takeshi Naemura
Stanford University, Stanford, CA

Masahide Kaneko
The University of Electro-Communications, Tokyo, Japan

Hiroshi Harashima
The University of Tokyo, Tokyo, Japan

SUMMARY

This chapter surveys various studies on compression and representation of 3-D images. There are several types of 3-D images such as stereo pairs, multi-view images, volumetric images, holograms, geometric models, and light fields. Most of the work in this field has concentrated on reducing the redundancy that is unique to individual types of 3-D images. For example, several techniques concerned with the concept of disparity compensation have been developed for the compression of stereo pairs and multi-view images. In addition to disparity compensation, the concept of an epipolar plane image (EPI) has been playing an important role in the compression of multi-view images. These techniques, however, depend heavily on individual camera configurations. Accordingly, to deal with several multi-view configurations and other types of 3-D images comprehensively, a new platform for 3-D image representation was introduced. This platform aimed to outgrow the framework of 3-D "image" communication and to open up a new field that would be called 3-D "spatial" communication. Here, the ray-based method in particular has a wide range of applications, including the integration of virtual and physical worlds as well as the efficient transmission of the physical world.

1. INTRODUCTION

Three-dimensional (3-D) imaging is a promising technology expected to go beyond the conventional framework of visual communication. Several kinds of 3-D displays have been developed to enhance the reality of visual experience, and they have been progressing rapidly. The realization of 3-D image communication, however, requires the development of various support technologies such as 3-D image capture, 3-D image representation, 3-D image transmission, and 3-D image handling. The recent state of 3-D TV research, especially in terms of human factors and systems was reported in (1).

This chapter surveys the results of major studies in the field of 3-D image communication focusing on the following technical interests:

- compression of 3-D images with the aim of transmitting a huge amount of data efficiently, and
- display-independent representation of 3-D images to make communication more flexible.

The display-independent feature shows promise as a means of outgrowing the framework of 3-D "image" communication and opening up a new field of technology that is called 3-D "spatial" communication.

Sections 2 to 4 of this chapter survey various coding methods specialized for individual types of 3-D images, namely, stereo pairs, multi-view images, panoramic images, volumetric images, and holograms. Section 5 then reviews the 3-D image coding paradigm of display-independent representation.

2. STEREO IMAGE COMMUNICATION

Stereo image communication uses two images of the same scene captured from different perspectives. Since stereoscopic systems can easily provide depth sensation, they have been widely used for various kinds of applications including education, medicine, entertainment, and virtual reality.

2.1 Disparity Estimation

One of the most important ideas in the study of 3-D image processing is that of "disparity." Disparity estimation is concerned with the analysis of geometric differences between two or more views of the same scene. It is well-known, however, that disparity estimation is an ill-posed problem. Thus, a smoothness constraint is generally introduced in the disparity field such as in relaxation labeling by Barnard and Thompson (2) and dynamic programming by Ohta and Kanade (3). There are two types of disparity estimation algorithms: feature-based and intensity-based. The feature-based algorithm attempts to match significant abstract features, such as edges, to estimate the disparity field. If a feature-based disparity estimation is used for coding, it is necessary to transmit both the coordinates of the features and the corresponding disparities. Since this is somewhat inefficient for coding applications, intensity-based methods are generally employed.

As for stereoscopic video, the relationship between motion and disparity fields can be used to estimate both vector fields more effectively. Recognizing this fact, the stereo-motion fusion was investigated by Waxman et al. (4)(5). The simplest way of insuring coherent disparity and motion analysis is to predict one vector field by a linear combination of others. For sequential stereo pairs, the following equation holds for the same 3-D point projected on four 2-D images (see Figure 1):

$$\boldsymbol{d}_{t+1} - \boldsymbol{d}_t = \boldsymbol{m}_l - \boldsymbol{m}_r,$$

where $\boldsymbol{d}_{t+1}$ and $\boldsymbol{d}_t$ stand for disparities at time $t+1$ and t, respectively, and $\boldsymbol{m}_l$ and $\boldsymbol{m}_r$ are left and right motion vectors as described by Tamtaoui and Labit (6)(7). In other words, disparity information greatly simplifies motion estimation as reported by Kim and Aggarwal (8). On the other hand, Gambotto showed that stereo matching between current frames can be predicted from previous matching and the

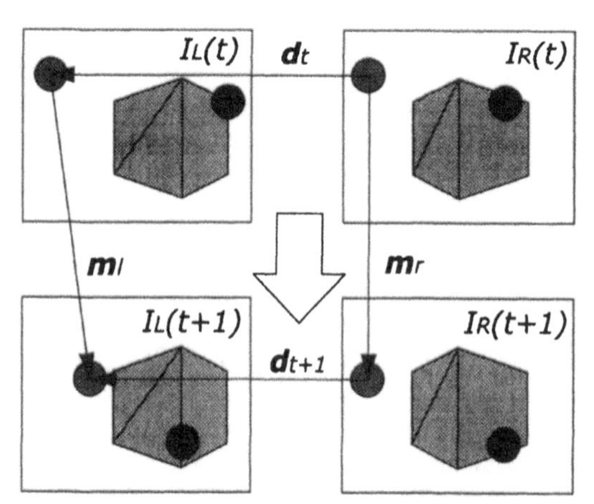

Figure 1 Coherent motion and disparity analysis.

result of motion estimation (9). Many approaches have been proposed for motion-stereo fusion research, e.g. multi-resolution dynamic programming by Liu and Skerjanc (10), simultaneous motion-disparity estimation and segmentation by Altunbasak et al. (11), and a maximum a posteriori estimator by Malassiotis and Strintzis (12).

The details of several techniques investigated for this purpose are reviewed in (13) and (14). Waldowski used disparity information to enhance the quality of region of interest (15).

2.2 Basics of Stereo Image Coding

A stereo pair exhibits spatial correlation between the left and right images. Lukacs pointed out that this correlation can be exploited by predictive coding schemes to transmit or store the stereo pair using slightly more data than one image of the pair (16). Actually, most work on stereo image compression has focused on disparity compensated prediction (DC) that is similar to motion compensated prediction (MC) widely used in the field of 2-D video coding. The DC algorithm can be described as follows:

1. The left image is coded independently of the right image and transmitted to the receiver.
2. The right image is divided into a number of blocks. For each block, a similar block in the encoded left image is assigned (in other words, the disparity of each block is estimated). Disparity data is then transmitted to the receiver.
3. The blocks in the right image are predicted on the basis of their similar blocks as shown in Figure 2. Finally, the prediction error (the residual image) is transmitted to the receiver.

Since the scheme of DC is similar to that of MC, many products for MC used in 2-D video coders can be applied to DC. It is important, however, to be aware of the subtle differences between DC and MC. These differences are listed below.

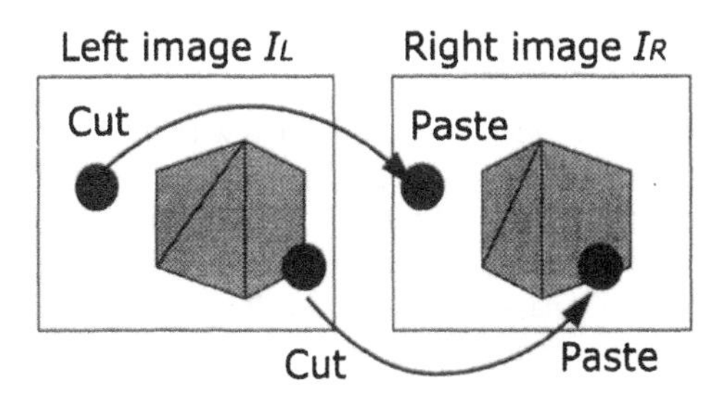

Figure 2 Disparity compensation.

- The disparity d consists of two components: a horizontal component dx and a vertical component dy. When the optical axes of stereo cameras are aligned in parallel, dy is always zero and dx is always positive, whereas in a 2-D video sequence, the image of an object can undergo vertical and horizontal displacements in both directions, as well as rotation.
- In a stereo pair, every object in the left image is displaced in the right image, while only a few objects change their locations from frame to frame in a typical video sequence.
- In a video sequence, the intensities of corresponding points in sequential frames can be assumed to be equal. This assumption, however, is often false in a stereo pair, because of the individual differences between the left and right cameras.
- In order to effectively compress the amount of data without subjectively intolerable distortions, it is possible to exploit the way in which the brain processes a stereo pair.

2.3 Improvements in Stereo Image Communication

In the 1980s,

- the basic idea of DC was proposed by Lukacs (16),
- the adaptive combination of DC and MC was evaluated by Schupp (17),
- the DC approach and the 3-D discrete cosine transform (DCT) approach were compared by Dinstein et al. (18)(19), and
- the statistical characteristics of stereo pairs and residual images were investigated by Yamaguchi et al. (20).

Since then, several improvements have been carried out in this field, as summarized, for instance, in (21). They are briefly reviewed below.

2.3.1 MPEG Compatible Coding

The temporal scalability feature of MPEG-2 can be employed when the left sequence is coded on the lower layer and the right one is transmitted as the enhancement layer. The left bit-stream offers the basic non-stereoscopic signal and results in full stereoscopic video when combined with the right one. For example, all enhancement frames can be configured to be B-pictures (bidirectionally predictive-coded pictures) as shown in **Figure 3**. Thus, the disparity vectors can be transmitted as prediction vectors, which are embedded in the basic MC framework of MPEG-2. Tseng and Anastassiou (22) and Puri et al. (23) reported the effectiveness of the MPEG-based approaches.

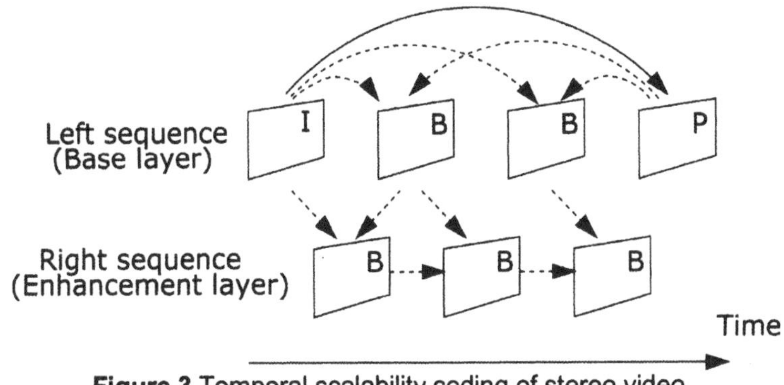

Figure 3 Temporal scalability coding of stereo video.

The RACE DISTIMA project featured the development of MPEG-2 compatible hardware demonstrators for data compression and transmission of stereoscopic video (24).

2.3.2 Human Visual System

Psychophysical experiments showed that a stereo pair with one high resolution image and another lower resolution image is sufficient to provide depth sensation. From this point of view, mixed-resolution coding methods for a stereo pair were proposed by Perkins (25) and Sethuraman et al. (26). Mixed-resolution coding is easy to implement, and it can reduce the bit rate by 46% with respect to a system that employs no coding (25). On the other hand, Sawhney et al. proposed an image-based rendering method for synthesizing a high resolution left eye image from a set of high resolution right eye images and a low resolution left eye image (27). This should be helpful for mixed-resolution coding methods.

The property of spatio-temporal contrast sensitivity in the human visual model by Pei and Lai (28), and visual pattern image coding (VPIC), which codes images using a predefined set of visual patterns significant for the human visual system by Craievich and Bovik (29), can also be exploited. Stelmach et al. reported, however, that the subjective quality of a stereo image sequence fell approximately midway between the quality of the left and right images (30), and for spatial low-pass filtering, the binocular percept is dominated by the high-quality image (31), while temporal low-pass filtering produces unacceptable results (32). It is still important to investigate the characteristics of the human visual system to achieve efficient compression of 3-D image data.

2.3.3 Suppressing Residuals

Disparity-compensated stereo residuals contain more structure than motion-compensated pairs in a video sequence. Incompleteness in disparity estimation appears as horizontal alignment errors that cause the vertical features in the residual. From this point of view, Moellenhoff and Maier studied specialized coding methods for the residual image produced by disparity compensation (33)(34). They evaluated their performance by a stereo-unique metric (the deviation of the disparity map after compression from the original disparity map) as well as mean square error (MSE). Since this new metric illustrates the ability of compressed stereo imagery to maintain depth information, it seems to reflect human perception.

Outgrowing the conventional framework of DC, more effective prediction methods were proposed to suppress the residuals.

In order to suppress the differences in the gain of the left and right cameras, disparity-compensated transform-domain predictive coding (DCTDP) was proposed by Perkins (25). Let R be the transform (e.g. DCT coefficients) of the right image subblock and L be the transform of the matching subblock from the encoded left image. R is predicted by a linear predictor of the form

$$\hat{R} = A \otimes L + B$$

where the array multiplication operation $\otimes$ is performed by multiplying the (i, j)th element of the first matrix by the (i, j)th element of the second matrix. The matrices A and B are chosen based on the statistics obtained from a training set so that each coefficient is predicted using a minimum variance predictor. DCTDP coding typically provides left picture gains on the order 1.0 to 1.5 dB over independent coding of the left picture using the same number of bits (25).

On the other hand, Aydinoglu and Hayes proposed the subspace projection technique (SPT) to combine DC and transform coding of the residual into a single framework by using an adaptive incomplete transform (35)(36). The idea is to code left images independently and to code right images using a locally adaptive transform T, which is applied to each $m \times m$ block $\boldsymbol{br}$ of the right image. The transform T is chosen to be a projection operator that projects each $m \times m$ block $\boldsymbol{br}$ onto a subspace $\boldsymbol{S}$ of $R^{m \times m}$. This subspace is spanned by a set of N vectors. The first k vectors in the spanning set are block-dependent vectors and can be obtained from left images using disparity compensation. The remaining $(N\text{-}k)$ vectors are fixed, $\boldsymbol{F} = \{f\,i\}_1^{N-k}$, and are chosen so that they approximate the low frequency components of the block by forming a polynomial approximation of vector $\boldsymbol{br}$. For low bit rates, i.e., around 0.05 bits/pixel, the SPT is about 2 dB better than the disparity compensation.

In this category, Seo and Sadjadi proposed sequential orthogonal subspace updating (SOSU) method to project an image block onto a subset of best-basis vectors (37)(38). In their experiment, the basis vectors were selected one by one from the neighboring 64 blocks, as well as 62 typical edge blocks, forming an image-dependent set of basis vectors. Simulation results demonstrated the effectiveness of the SOSU scheme (35.57 dB, 0.7 bits/pixel) in comparison with the DC-DCT schemes (32.05 dB, 0.73 bits/pixel) and their former proposal (39).

2.3.4 Other Advances

There are several other contributions to stereo image compression:

- multiresolutional region-based DC by Sethuraman et al. (40),
- 3-D MC method by Grammalidis et al. (41),
- NTSC-compatible representation by Labonte et al. (42)(43),
- rate-distortion optimization for motion and disparity field estimation by Tzovaras and Strintzis (44)(45),
- dependent bit allocation framework by Woo and Ortega (46)(47),
- Markov random field (MRF) model for disparity compensation (48) and for segmentation of disparity field (49) by Woo and Ortega, and
- modified overlapped block disparity compensation by Woo and Ortega (50)(51).

On the other hand, the object-based and model-based methods (52) constitute recent trends in stereo image coding. In general, perceptually better image quality can be achieved by segmentation-based prediction than by the popular block-based prediction. The object-based coding methods play an important role in encoding the segmented region corresponding to an object in 3-D space. In particular, the model-based methods, which describe a scene in a structural way, are well-suited for 3-D image coding. It is very significant that these methods are useful for interpolating intermediate views between the left and right views (in other words, synthesizing multi-view images). From this point of view, it can be said that studies on stereo image coding are coming to involve multi-view image coding. Further explanations of this new field of technology are presented in the following section.

3. MULTI-VIEW IMAGE COMMUNICATION

A set of multi-view images is comprised of several images of the same scene acquired from different perspectives. While the depth perception from a stereo pair is derived from binocular disparity, that from multi-view images is obtained from both binocular disparity and parallax caused by the observer's motion. Thus, multi-view images can enhance the 3-D visual effect by providing a look-around capability.

As the number of viewpoints (cameras) increases, however, the amount of data expands and the input system becomes more complex. The view interpolation technique, which synthesizes images observed from intermediate viewing angles between cameras, will help reduce the complexity of the input system and enhance the 3-D visual effects (see Figure 4).

In Figure 4, an intermediate point P_s is interpolated according to the equation $P_rP_s = s\ P_rP_l$, where P_rP_l is the disparity and s [0, …, 1]. This technique is also useful for the efficient representation of multi-view images, because a large

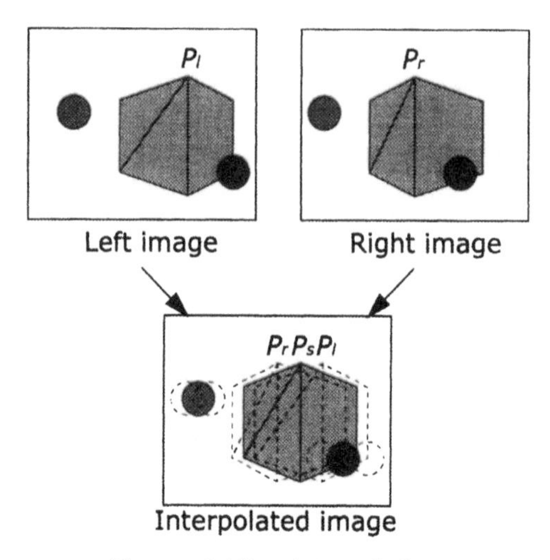

Figure 4 View interpolation.

number of views can be compressed into a smaller number of key views by which all other original views can be interpolated.

Consequently, there are two issues that should be addressed in a coding system of multi-view images:

- how to compress the huge amount of data required for the autostereoscopic visual effect, and
- how to interpolate intermediate views.

3.1 Building upon Stereo Image Communication

It is straightforward to apply several techniques developed for stereo image coding or 2-D video coding to multi-view image coding. Several attempts have been reported, such as the application of H.261 coders by Naemura et al. (53), SPT (subspace projection technique) by Aydinoglu and Hayes (54)(55), and MPEG-2 for a 2-D array of images by Kiu et al. (56).

As for view interpolation, several methods have been proposed, some of which are listed below:

- pioneering research on an optical parallel processing system for a stereo pair by Oshima and Okoshi (57),
- a method for horizontally aligned cameras by Sasaki and Furukawa (58),
- a method for symmetric trinocular cameras on a 2-D plane by Skerjanc and Liu (59),

- an interactive method using range data (depth information) by Chen and Williams (60),
- an image alignment approach to deal with nonparallel cameras by Wang and Wang (61), and
- an application of Bayesian winner-take-all reconstruction for higher quality interpolation by Mansouri and Konrad (62).

In any case, a non-smooth disparity field will degrade the perceptual quality of interpolated views. The block-based method may suffer from blocking artifacts. Moreover, for stereo video sequences, Hendriks and Marosi showed that the temporal consistency of disparity fields is effective for suppressing the annoying artifacts of interpolated images (63). Thus, the estimation of a dense disparity field, in which each pixel has its own disparity information, is essential for view interpolation. In such a context, the following methods were investigated:

- dynamic programming methods by Tseng and Anastassiou (64) and Grammalidis and Strintzis (65)(66), and
- modified block matching algorithm by Izquierdo (67)(68).

The methods of compressing a dense disparity field were also proposed. They can be evaluated by the quality of interpolated images at the receiver. Some of them are listed below.

- MPEG-based methods
  - The MPEG-2 standard with 3-D computer graphics techniques by Tseng and Anastassiou (69).
  - Incomplete 3-D layered representation using the video object plane (VOP) of the MPEG-4 standard by Ohm and Muller (70), Ekmekci (71), and Kauff et al. (72).
  - Extension of MPEG-4 sprite coding by Grammalidis et al. (73).
- 3-D model-based methods
  - Simple convex-surface models by Fujii and Harashima (74), Papadimitriou and Dennis (75), and Ohm and Izquierdo (76).
  - A 3-D layered model by Naemura et al. (77)(78).
  - Joint 3-D motion/disparity compensation using a wire-frame model by Tzovaras et al. (79).
  - Globally rigid and locally deformable 3-D wire-frame models by Malassiotis and Strintzis (80)(81) and Kompatsiaris et al. (82).
- Others
  - A nonlinear interpolator, which generates a full-sized map that partially preserves original disparity edge information from block-based disparity data by Pagriari et al. (83).

- The chain map method, which combines several disparity maps into one with very low inherent redundancy by Redert and Hendriks (84).
- Delaunay triangulation of a disparity field by Fan and Ngan (85).
- Interframe coding of disparity fields by Tzovaras et al. (86).

While this field of technology is relatively new, a hardware system for real time stereoscopic video conferencing with viewpoint adaptation has already been realized by Ohm et al. (87).

3.2 Epipolar-Plane Images and Their Application

Multi-view images have a unique structure in reflecting the scene being viewed. In this section, we consider a horizontal lineup of equidistant cameras with parallel optical axes as shown in Figure 5.

The concept of an epipolar plane image (EPI) was proposed by Bolles and Baker (88)(89) for building a 3-D representation of a static scene from a sequence of images. Figure 6 illustrates a simple example of synthesizing EPIs from multi-view images captured by the input system shown in Figure 5. Sorting each view I_v by its viewpoint location v, the solid block representation of multi-view images is constructed as shown on the left side of Figure 6. A horizontal slice of this solid block data is an EPI. It is important that an EPI is essentially composed of straight lines whose slope indicates disparity and that each line contains all information about corresponding points recorded in the multi-view images as shown on the right side of Figure 6.

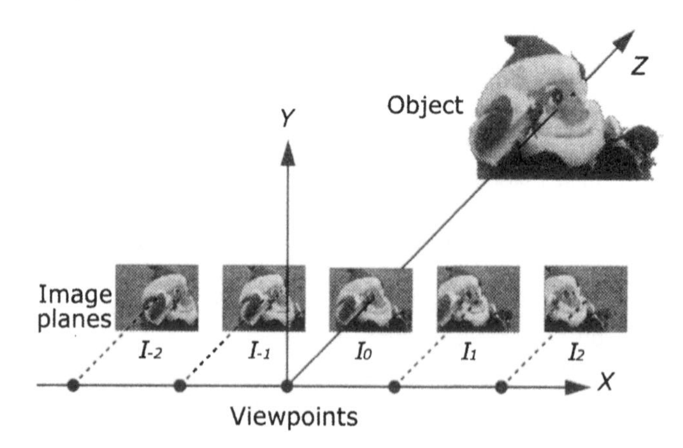

Figure 5 An example of a multi-view camera configuration (images are courtesy of University of Tsukuba).

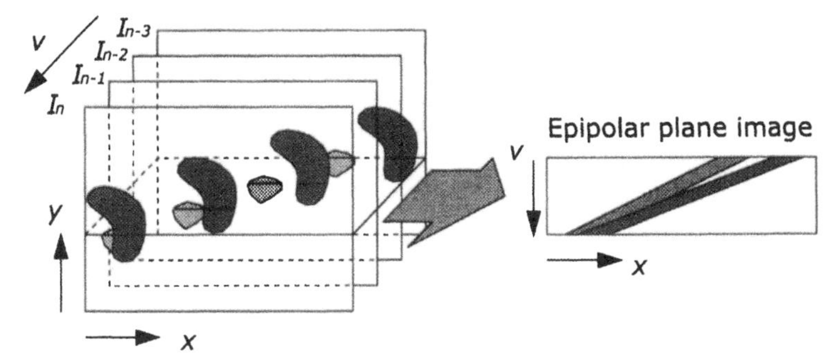

Figure 6 Synthesis of epipolar plane images.

Figure 7 Synthesized 3-D layered model (78).

To achieve both data compression and view interpolation of multi-view images, a structure recovery method based on EPI analysis was investigated by Fujii et al. (90). In particular, a triangular patch model was examined and used due to its compatibility with the DC method based on an affine transformation by Fujii and Harashima (74). In order to cope with occlusions, a method of segmenting the solid block of data reported in (91) was applied to synthesize the triangular-patch models of multi-view images by Naemura et al. (77)(78).

Figure **7** shows the 3-D layered model synthesized from multi-view images (78). The shape and texture of the model is designed to minimize the differences between input images and their corresponding images reconstructed from the model.

At the same time, to avoid the difficulties in structure recovery, a kind of waveform-based method was investigated for both data compression and view interpolation by Naemura and Harashima (92). Since the resolution of the v-axis in the solid block representation of multi-view images is equal to the number of viewpoints, the number can be virtually increased by enhancing the resolution of the v-axis. View interpolation can therefore be achieved by resolution enhancement in epipolar-plane images without recovering the structure. In such a context, fractal-based compression, which is scalable to any resolution, was applied to multi-view compression (92).

4. OTHER 3-D IMAGES

4.1 Panoramic Images

A panoramic image is an orientation-independent representation of a 3-D scene. It contains all the information needed to view the scene with a look-around capability in 360 degrees. This type of image can be created by stitching together overlapping photographs as described by Chen (93). A series of pictures, however, cannot be stitched together without being warped. This warping process is independent of scene components but dependent on camera configuration. Such integration can be seen as a data compression method for several photographs of the same viewpoint.

Stereoscopic panorama is also a new field. Naemura et al. proposed the basic concept (94), Shum and He realized a rendering system that allows users to move their viewpoints (95), and Peleg et al. presented further detailed discussion in (96).

4.2 Volumetric Images

Compression of volumetric image data is imperative for clinical picture archiving and communication systems and for tele-medicine and telepresence networks. A fractal-based approach by Cochran et al. (97) and a wavelet-based approach by Luo and Chen (98) were reported in this field.

4.3 Holographic Images

For compression of holographic data, it was noticed that, in the frequency domain, hologram data can be seen as a pattern related to a perspective view of scene objects. We can consider the case in which a hologram is segmented into elements that are then transformed into the frequency domain. Since the hologram data after segmentation and transformation have characteristics similar to a set of 2-D images, it is possible to apply 2-D image compression techniques to hologram data compression as proposed by Yoshikawa and Tamai (99).

5. THE DISPLAY-INDEPENDENT APPROACH

Most of the studies surveyed in Sections 2 to 4 are specialized for limited combinations of input and output technologies, namely, stereo pairs, multi-view images, panoramic images, volumetric images and holographic images. As for multi-view images, several approaches have been proposed for configuring cameras (the number of cameras, their positions, their directions, their focal lengths, and so on). The next generation communication system, however, should not be restricted to specific 3-D image formats but be flexible for various kinds of 3-D images. Not only image data captured by a camera but also spatial data (such as stereo pairs, multi-view images, integral photographs, holograms, volumetric data, and geometric models of objects) should be used seamlessly in the next-generation visual communication system. For this purpose, it is important to establish a method of representing several kinds of spatial data in the same way that is neutral for any input and output system.

Considering such requirements, the concept of "integrated 3-D visual communication" illustrated in Figure 8 was proposed by Harashima (100). Its key feature is the display-independent intermediate representation of visual data; all the differences between input and output systems will be absorbed by intermediate representation because it is neutral for any format of visual data. Thus, we can select our favorite way of observing transmitted 3-D images regardless of the input method. This function, as a new coding paradigm, will promote the progress of 3-D image communication. Recent progress was summarized in (101).

5.1 Practical Methods of Neutral Representation

To date, two practical methods for the neutral representation of 3-D visual data have been investigated: the ray-based approach described below and the structure-based approach by Fujii and Harashima (74) and Naemura et al. (77)(78).

It is obvious that the structure-based approach is a practical way of representing scene objects. For instance, VRML (virtual reality modeling language) is a standardized method of representing geometric models in the virtual world. This approach is well-suited for handling and displaying objects in the virtual world. On the other hand, since visual sensation is excited by rays of light incident on the retina, each of these rays can be seen as a primitive element of visual cues and any kind of visual data can be decomposed into a set of light ray data. Conversely, arbitrary display-dependent visual data can be synthesized from the set of light ray data. The ray-based approach is useful for displaying objects in the physical world as well as in the virtual world.

To apply the structure-based method to the physical world, we must construct geometric models of every object in the world. It is not easy, however, to measure the accurate shapes, surface textures and reflection models of physical objects. To solve this problem, a great deal of improvement in the field of structure recovery from images will be required. What is essential for visual

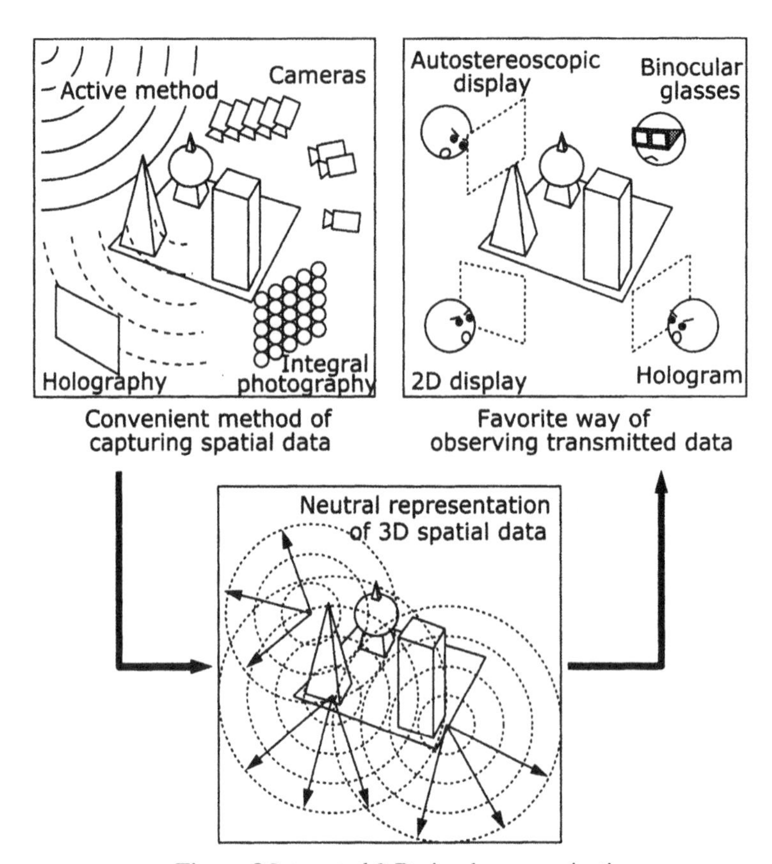

Figure 8 Integrated 3-D visual communication.

communication, however, is not the structural properties of scene objects but the light rays emitted from them. The merit of the ray-based approach is that we can avoid the problems of structure recovery. From this point of view, the following sections will clarify the basic concept and applications of the ray-based approach.

5.2 Ray-Based Representation of Visual Data

The ray-based approach was introduced to the field of 3-D image communication including holographic displays in (102) after the proposal of structure-based representation of light rays (103) by Fujii and Harashima. It has a close connection with the concept of image-based rendering, which has attracted a great deal of attention in the field of computer graphics through pioneering work by Katayama et al. (104), McMillan and Bishop (105), Levoy and Hanrahan (106), Gortler et al. (107) and Yanagisawa et al. (108). Both of these approaches aim to avoid the difficulties in constructing geometric models of objects in the physical world. Since an image can be regarded as a set of light rays, the ray-based approach involves the image-based one.

A set of light rays passing through any point in 3-D space is called a pencil, which is distinguished by its position (X,Y,Z). Each of the light rays which constitute a pencil can be identified by its direction (θ,φ) represented as longitude and latitude angle of propagation. Therefore, by providing a five-dimensional data space denoted as $XYZ\theta\varphi$, we can store the light ray data (power or wavelength) denoted as $f(X,Y,Z,\theta,\varphi)$ at any point in any direction, separately. This data space is called the "plenoptic function" (109) or "ray-space" (102).

A panoramic image, which records the incident light arriving from all directions at point (Xa,Ya,Za), can be represented as follows:

$$f'(\theta,\varphi) = f(X,Y,Z,\theta,\varphi)\,|_{(X,Y,Z)=(Xa,Ya,Za)}$$

This means that a panoramic image corresponds to a 2-D sub-space of the 5-D data space. From this point of view, $f(X,Y,Z,\theta,\varphi)$ can be regarded as a set of panoramic images at any location in 3-D space. In the same way, it can be said that a hologram plane $Z = g(X,Y)$ contains light ray data within a 4-D sub-space as follows.

$$f'(X,Y,\theta,\varphi) = f(X,Y,Z,\theta,\varphi)\,|_{Z=g(X,Y)}$$

In many cases, a 4-D data space is enough to represent 3-D images, though there are certainly some applications requiring a 5-D data space. Further investigations were given by Fujii et al. (110) and Naemura et al. (111)(112). Chai et al. presented a theoretical framework for sampling 4-D data space (113).

5.3 Framework of Ray-Based System

In order to apply the ray-based method to the intermediate representation illustrated in Figure 8, we must consider the following three principal phases: (a) input phase, (b) transmission phase, and (c) output phase.

In the input phase (a), input visual data is converted into light rays and mapped onto a 4-D data space. We can also synthesize light rays from geometric models in the virtual world by the ray-tracing technique. For 2-D image input, camera position (Xa,Ya,Za) and the relationship between image coordinates (x,y) and directions (θ,φ) of rays are required to be given or estimated. In many cases, interpolation of light rays will be required in this phase. Several input methods have been implemented, for instance,

- the light field gantry by Levoy and Hanrahan (106),
- versatile 16-camera array systems by Naemura et al. (114)(115),
- pixel-independent random access image sensors by Ooi et al. (116),
- FPGA-based camera array components by Wilburn et al. (117), and
- an application of integral photography by Naemura et al. (118).

In the transmission phase (b), since the 4-D data space still contains a huge amount of data, an efficient compression method is strongly required. Several methods have been examined as listed below. Since they try to compress different data sets under different conditions from each other, it is hard to compare them thoroughly. I believe, however, it is still useful to see the relationship between image quality (dB) and compressed amount of data (bits/pixel) as a reference.

- vector quantization by Levoy and Hanrahan (106),
    - Buddha light field: 36 dB, 0.069 bits/pixel (106)
    - Lion light field: 27 dB, 0.027 bits/pixel (106)
- fractal-based interpolation and compression by Naemura and Harashima (92),
    - Dog image set: 37.5dB, 0.11 bits/pixel (92)
- 4-D DCT-based compression by Takano et al. (119)(120),
    - Santa image set: 36.6 dB, 0.07 bits/pixel (120)
- disparity-based compression by Fujii et al. (121),
- multiple reference frame prediction and just-in-time rendering by Zhang and Li (122),
- hierarchical approach to disparity-based compression by Magnor and Girod (123)(124)(125) and Tong and Gray (126),
    - Dragon light field: 36 dB, 0.068 bits/pixel (125)
    - Buddha light field: 40 dB, 0.027 bits/pixel (125)
- 4-D wavelet-based progressive compression by Lalonde and Fournier (127) and Magnor et al. (128),
- model-aided and model-based coding and their comparison, combination and analysis by Magnor et al. (129)(130)(131)(132)(133),
- non-orthogonal basis function for compression and interpolation of ray space by Fujii et al. (134),
- concentric mosaic as a 3-D representation of a 4-D light field by Shum and He (95),
- block-based coding of concentric mosaic by Zhang and Li (135), and
- 3-D wavelet compression of concentric mosaics by Luo et al. (136), Wu et al. (137) and Li et al. (138).
    - Lobby concentric mosaic: 36.3 dB, 0.2 bits/pixel (138)
    - Kids concentric mosaic: 33.8 dB, 0.4 bits/pixel (138)

In the output phase (c), the light ray data appropriate for the display system at the receiver is extracted or interpolated from the transmitted data space. Some types of optical effects, such as fogging and shadowing, can also be added.

Naemura et al. implemented a system that can interactively render arbitrary views of real scenes in motion from sixteen video sequences (114)(115). This means that all of the above three phases are processed in real time.

5.4 Integration of Virtual Scenes and Real Objects

Finally, we outline mixed reality applications in which physical objects represented by light ray data are placed in the virtual world represented by geometric models. Figure 9 shows an example of a cyber space called CyberMirage by Katayama et al. (139) and Uchiyama et al. (140). Components inside a building are presented by polygonal models and dresses are presented by light ray data. These objects are implemented so that the observer can manipulate viewpoints in real time on SGI Indigo2 IMPACT or equivalent machines.

Representation of a virtual space using polygonal model data is based on a subset of VRML 1.0. Descriptors of VRML are expanded so that they can merge light ray data with VRML data. Light ray data are objects which can only be observed since they do not have explicit shape models. A scheme is built, however, to move or rotate objects represented by light ray data in the CyberMirage system. CyberMirage with the features mentioned above is useful for representing products realistically in the display of a virtual mall.

Figure 9 CyberMirage.

6. CONCLUSIONS

This chapter presented a broad survey of the results of various studies on 3-D image coding, which is one of the most critical fields for constructing a 3-D image communication system. Particular attention was paid to efficient compression and the display-independent representation of 3-D image data.

Most of the work on 3-D image coding has concentrated on compression methods. In particular, the concept of disparity compensation was developed for the compression of stereo images. To consider many other types of 3-D images, however, a more general and neutral platform for the representation of 3-D images is required. Thus, the concept of neutral representation as a viable alternative to 3-D image coding was introduced with the aim of outgrowing the framework of 3-D "image" communication and to realize "spatial" communication. The ray-based approach to this concept has a wide range of applications, including the efficient transmission of real space and the integration of virtual and real spaces.

The 3-D image coding technology has a close connection with many other fields such as computer graphics (CG), virtual reality (VR), and computer vision (CV). More comprehensive studies related to these technologies will contribute to further development and progress of spatial communication.

REFERENCES

(1) T Motoki, H Isono, I Yuyama. Present status of three-dimensional television research. Proc IEEE 83(7):1009-1021, 1995.

(2) S Barnard, W Thompson. Disparity analysis of images. IEEE Trans Pattern Anal & Machine Intell PAMI-2(4):333-340, 1980.

(3) Y Ohta, T Kanade. Stereo by intra- and inter- scanline search using dynamic programming. IEEE Trans Pattern Anal & Machine Intell PAMI-7(2):139-154, 1985.

(4) A Waxman, S Sinha. Dynamic stereo: Passive ranging to moving objects from relative image flows. IEEE Trans Pattern Anal & Machine Intell PAMI-8(4):406-412, 1986.

(5) A Waxman, J Duncan. Binocular image flows: Steps toward stereo-motion fusion. IEEE Trans Pattern Anal & Machine Intell PAMI-8(6):715-729, 1986.

(6) A Tamtaoui, C Labit. Coherent disparity and motion compensation is 3DTV image sequence coding schemes. Proceedings of IEEE ICASSP'91, 1991, pp 2845-2848.

(7) A Tamtaoui, C Labit. Constrained disparity and motion estimator for 3DTV image sequence coding. Signal Process : Image Commun 4(1):45-54, 1991.

(8) Y Kim, J Aggarwal. Determining object motion in a sequence of stereo images. IEEE J Robotics and Automation RA-3(6):599-614, 1987.

(9) J Gambotto. Determining stereo correspondences and egomotion from a sequence of stereo images. Proceedings of IAPR 10th ICPR, 1990, pp 259-262.

(10) J Liu, R Skerjanc. Stereo and motion correspondence in a sequence of stereo images. Signal Process : Image Commun 5(4):305-318, 1993.

(11) Y Altunbasak, A Tekalp, G Bozdagi. Simultaneous motion-disparity estimation and segmentation from stereo. Proceedings of IEEE ICIP'94, 1994, vol III, pp 73-77.

(12) S Malassiotis, M Strintzis. Joint motion/disparity MAP estimation for stereo image sequences. IEE Proc. Vis Image Signal Process 143(2):101-108, 1996.

(13) U Dhond, J Aggarwal. Structure from stereo - a review. IEEE Trans Syst, Man, & Cybernetics 19(6):1489-1510, 1989.
(14) T Huang, A Netravali. Motion and structure from feature correspondences: A review. Proc. IEEE 82(2):252-268, 1994.
(15) M Waldowski. A new segmentation algorithm for videophone applications based on stereo image pairs. IEEE Trans Commun 39:1856-1868, 1991.
(16) M Lukacs. Predictive coding of multi-viewpoint image sets. Proceedings of IEEE ICASSP'86, 1986, pp 521-524.
(17) W Schupp. A study on the efficient coding of three-dimensional moving pictures. Master's thesis, The Univ of Tokyo, Tokyo, Japan, 1988.
(18) I Dinstein, G Guy, J Rabany, J Tzelgov, A Henik. On stereo image coding. Proceedings of IAPR 9th ICPR, 1988, pp 357-359.
(19) I Dinstein, G Guy, J Rabany, J Tzelgov, A Henik. On the compression of stereo images: Preliminary results. Signal Process 17(4):373-382, 1989.
(20) H Yamaguchi, Y Tatehira, K Akiyama, Y Kobayashi. Stereoscopic images disparity for predictive coding. Proceedings of IEEE ICASSP'89, 1989, pp 1976-1979.
(21) B Choquet, F Chassaing, J Fournier, D Pele, A Poussier, H Sanson. 3D TV studies at CCETT. Proceedings TAO 1st Int Symp 3D Image Commun Technol, 1993, S-1-2.
(22) B Tseng, D Anastassiou. Compatible video coding of stereoscopic sequences using MPEG-2's scalability and interlaced structure. Proceedings of Int Workshop on HDTV'94, 1994, 6-B-5.
(23) A Puri, R Kollarits, B Haskell. Basics of stereoscopic video, new compression results with MPEG-2 and a proposal for MPEG-4. Signal Process: Image Commun 10:201-234, 1997.
(24) M Ziegler, L Falkenhagen, R Horst, D Kalivas. Evolution of stereoscopic and three-dimensional video. Signal Process: Image Commun. 14:173-194, 1998.
(25) M Perkins. Data compression of stereopairs. IEEE Trans Commun 40(4):684-696, 1992.
(26) S Sethuraman, M Siegel, A Jordan. A multiresolution framework for stereoscopic image sequence compression. Proceedings of IEEE ICIP'94, 1994, vol II, pp 361-365.
(27) H Sawhney, Y Guo, K Hanna, R Kumar, S Adkins, S Zhou. Hybrid stereo camera: An IBR approach for synthesis of very high resolution stereoscopic image sequences. Proceedings of ACM SIGGRAPH2001, 2001, pp 451-460.
(28) S Pei, C Lai. An efficient coding algorithm for 3D video with spatio-temporal HVS model and binary correlator disparity estimator. Proceedings of Int Workshop on HDTV'95, 1995, pp. 2B9-2B-16.
(29) D Craievich, A Bovik. Stereo image compression using VPIC, Proceedings of IEEE ICIP'96, 1996, vol II, pp 879-882.
(30) L Stelmach, W Tam. Stereoscopic image coding: Effect of disparate image-quality in left- and right-eye view. Signal Process: Image Commun 14:111-117, 1998.
(31) L Stelmach, W Tam, D Meegan, A Vincent, P Corriveau. Human perception of mismatched stereoscopic 3D inputs. Proceedings of IEEE ICIP2000, 2000, vol I, pp 5-8.
(32) L Stelmach, W Tam, D Meegan, A Vincent. Stereo image quality: Effects of mixed spatio-temporal resolution. IEEE Trans Circuit & Syst Video Technol 10(2):188-193, 2000.
(33) M Moellenhoff, M Maier. DCT transform coding of stereo images for multimedia applications. IEEE Trans Industr Electron 45:38-43, 1998.
(34) M Moellenhoff, M Maier. Transform coding of stereo image residuals. IEEE Trans Image Process 7:804-812, 1998.

(35) H Aydinoglu, M Hayes. Stereo image coding. Proceedings of IEEE ISCAS, 1995, vol I, pp 247-250.
(36) H Aydinoglu, M Hayes. Stereo image coding: A projection approach. IEEE Trans Image Process 7:506-516, 1998.
(37) S Seo, M Sadjadi. 2-D filter-based disparity compensation using sequential orthogonal subspace updating, Proceedings of IEEE ICIP'98, 1998, vol II, pp 603- 607.
(38) S Seo, M Sadjadi. A 2-D filtering scheme for stereo image compression using sequential orthogonal subspace updating. IEEE Trans Circuits Syst Video Technol 11:52-66, 2001.
(39) S Seo, M Sadjadi, B Tian. A least-square-based 2-D filtering scheme for stereo image compression. IEEE Trans Image Process 9:1967-1972, 2000.
(40) S Sethuraman, M Siegel, A Jordan. A multiresolutional region based segmentation scheme for stereoscopic image compression. Proceedings of SPIE Digital Video Compression -- Algorithms and Technologies, 1994, vol 2419, pp 265-274.
(41) N Grammalidis, S Malassiotis, D Tzovaras, M Strintzis. Stereo image sequence coding based on three-dimensional motion estimation and compensation. Signal Process: Image Commun 7(2):129-145, 1995.
(42) F Labonte, C Dinh, P Cohen. A NTSC-compatible compact representation for stereoscopic sequences. Proceedings of IEEE ICIP'96, 1996, vol II, pp 883-886.
(43) F Labonte, C Tam, L Dinh, J Faubert, P Cohen. Spatiotemporal spectral coding of stereo image sequences. IEEE Trans Circuit & Syst Video Technol 9(1):144-155, 1999.
(44) D Tzovaras, M Strintzis. Motion and disparity estimation using rate distortion theory for very low bit rate and multiview image sequence coding. Proceedings of SPIE VCIP'97, 1997, vol 3024, pp 352-359.
(45) D Tzovaras, G Strintzis. Motion and disparity field estimation using rate-distortion optimization. IEEE Trans Circuit & Syst Video Technol 8(2):171-180, 1998.
(46) W Woo, A Ortega. Dependent quantization for stereo image coding. Proceedings of SPIE VCIP'98, 1998, vol 3309, pp 902-913.
(47) W Woo, A Ortega. Optimal blockwise dependent quantization for stereo image coding. IEEE Trans Circuits & Syst Video Technol 9(6):861-867, 1999.
(48) W Woo, A Ortega. Stereo image compression with disparity compensation using the MRF model. Proceedings of SPIE VCIP'96, 1996, vol 2727, pp 28-41.
(49) W Woo, A Ortega. Stereo image compression based on disparity field segmentation. Proceedings of SPIE VCIP'97, 1997, vol 3024, pp 391-402.
(50) W Woo, A Ortega. Modified overlapped block disparity compensation for stereo image coding. Proceedings of SPIE VCIP'99, 1999, vol 3653, pp 570-581.
(51) W Woo, A Ortega. Overlapped block disparity compensation with adaptive windows for stereo image coding. IEEE Trans Circuits & Syst Video Technol 10(2):194-200, 2000.
(52) K Aizawa, T Huang. Model-based image coding: Advanced video coding techniques for very low bit-rate applications. Proc IEEE 83(2):259-271, 1995.
(53) T Naemura, K Yoshioka, T Fujii, Y Nakaya, H Harashima. Disparity compensated predictive coding of multi-view 3-D images. Proceedings of 1993 ITE Annual Convention, 1993, 24-4 (in Japanese).
(54) H Aydinoglu, M Hayes. Compression of multi-view images. Proceedings of IEEE ICIP'94, 1994, vol II, pp 385-388.
(55) H Aydinoglu, M Hayes. Multi-view image coding using local orthogonal bases. Proceedings of SPIE VCIP'97, 1997, vol 3024, pp 340-351.

(56) M Kiu, X Du, R Moorhead, D Banks, R Machiraju. Two dimensional sequence compression using mpeg. Proceedings of SPIE VCIP'98, 1998, vol 3309, pp 914-921.
(57) K Oshima, T Okoshi. Synthesis of an autostereoscopic 3-D image from binocular stereoscopic images. Appl Opt 18(4):469-476, 1979.
(58) Y Sasaki, T Furukawa. A synthesis algorithm of arbitrary intermediate-viewing-angle images from several pairs of stereograms. Trans IEICE J63-D(9):813-814, 1980 (in Japanese).
(59) R Skerjanc, J Liu. A three camera approach for calculating disparity and synthesizing intermediate pictures. Signal Process: Image Commun 4(1):55-64, 1991.
(60) S Chen, L Williams. View interpolation for image synthesis. Proceedings of ACM SIGGRAPH'93, 1993, pp 279-288.
(61) R Wang, Y Wang. Multiview video sequence analysis, compression, and virtual viewpoint synthesis. IEEE Trans Circuits & Syst Video Technol 10(3):397-410, 2000.
(62) A Mansouri, J Konrad. Bayesian winner-take-all reconstruction of intermediate views from stereoscopic images. IEEE Trans Image Process 9(10):1710-1722, 2000.
(63) E Hendriks, G Marosi. Recursive disparity estimation algorithm for real time stereoscopic video application. Proceedings of IEEE ICIP'96, 1996, vol II, pp 891-894.
(64) B Tseng, D Anastassiou. A theoretical study on an accurate reconstruction of multiview image based on the viterbi algorithm. Proceedings of IEEE ICIP'95, 1995, vol II, pp 378-381.
(65) N Grammalidis, M Strintzis. Disparity and occlusion estimation for multiview image sequences using dynamic programming. Proceedings of IEEE ICIP'96, 1996, vol II, pp 337-340.
(66) N Grammalidis, M Strintzis. Disparity and occlusion estimation in multiocular systems and their coding for the communication of multiview image sequences. IEEE Trans Circuits & Syst Video Technol 8(3):328-344, 1998.
(67) M Izquierdo. Stereo matching for enhanced telepresence in three-dimensional videocommunications. IEEE Trans Circuits & Syst Video Technol 7(4):629-643, 1997.
(68) M Izquierdo. Stereo image analysis for multi-viewpoint telepresence applications. Signal Process: Image Commun 11:231-254, 1998.
(69) B Tseng, D Anastassiou. Multiviewpoint video coding with MPEG-2 compatibility. IEEE Trans Circuits & Syst Video Technol 6(4):414-419, 1996.
(70) J Ohm, K Muller. Incomplete 3D representation of video objects for multiview applications. Proceedings of PCS'97, 1997, pp 427-432.
(71) S Ekmekci. Encoding and reconstruction of incomplete 3-D video objects. IEEE Trans Circuits & Syst Video Technol 10(7):1198-1207, 2000.
(72) P Kauff, C Fehn, E Cooke, O Schreer. Advanced incomplete 3D representation of video objects using trilinear warping for novel view synthesis. Proceedings of Picture Coding Symposium 2001, 2001, pp 429- 432.
(73) N Grammalidis, D Beletsiotis, M Strintzis. Sprite generation coding in multiview image sequences. IEEE Trans Circuits & Syst Video Technol 10(2):302-311, 2000.
(74) T Fujii, H Harashima. Data compression and interpolation of multi-view image set. IEICE Trans Inf & Syst E77-D(9):987-995, 1994.
(75) D Papadimitriou, T Dennis. Stereo in model-based image coding. Proceedings of PCS'94, 1994, pp296-299.
(76) J Ohm, M Izquierdo. An object-based system for stereoscopic viewpoint synthesis. IEEE Trans Circuits & Syst Video Technol 7(5):801-811, 1997.
(77) T Naemura, M Kaneko, H Harashima. 3-D object based coding of multi-view images. Proceedings of PCS'96, 1996, pp 459-464.

(78) T Naemura, T Yanagisawa, M Kaneko, H Harashima. 3-D layered representation of multiview images based on 3-D segmentation. J Inst Telev Eng 50(9):1335-1344, 1996 (in Japanese).
(79) D Tzovaras, N Grammalidis, M Strintzis. Object-based coding of stereo image sequences using joint 3-D motion/disparity compression. IEEE Trans Circuits & Syst Video Technol 7(2):312-327, 1997.
(80) S Malassiotis, M Strintzis. Coding of video-conference stereo image sequences using 3D models. Signal Process: Image Commun 9(2):125-135, 1997.
(81) S Malassiotis, M Strintzis. Object-based coding of stereo image sequences using three-dimensional models. IEEE Trans Circuits & Syst Video Technol 7(6):892-905, 1997.
(82) I Kompatsiaris, D Tzovaras, M Strintzis. Flexible 3D motion estimation and tracking for multiview image sequence coding. Signal Process: Image Commun 14:95-110, 1998.
(83) C Pagriari, M Perez, T Dennis. Reconstruction of intermediate views from stereoscopic images using a rational filter. Proceedings of IEEE ICIP'96, 1996, vol II, pp 627-631.
(84) A Redert, E Hendriks. Disparity map coding for 3D teleconferencing applications. Proceedings of SPIE VCIP'97, 1997, vol 3024, pp 369-379.
(85) H Fan, K Ngan. Disparity map coding based on adaptive triangular surface modelling. Signal Process: Image Commun 14:119-130, 1998.
(86) D Tzovaras, N Grammalidis, M Strintzis. Disparity field and depth map coding for multiview 3D image generation. Signal Process: Image Commun 11:205-230, 1998.
(87) J Ohm, K Gruneberg, E Hendriks, M Izquierdo, D Kalivas, M Karl, D Papadimatos, A Redert. A realtime hardware system for stereoscopic videoconferencing with viewpoint adaptation. Signal Process: Image Commun 14:147-171, 1998.
(88) R Bolles, H Baker, D Marimont. Epipolar-plane image analysis: An approach to determining structure from motion. Int J Computer Vision 1:7-55, 1987.
(89) H Baker, R Bolles. Generalizing epipolar-plane image analysis on the spatiotemporal surface. Int J Computer Vision 3:33-49, 1989.
(90) T Fujii, J Hamasaki, M Pusch. Data compression for an autostereoscopic 3D image. Proceedings of Int Symp on Three Dimensional Image Technol & Arts, 1992, pp 171-178.
(91) T Naemura, M Kaneko, H Harashima. 3-D segmentation of multi-view images based on disparity estimation. Proceedings of SPIE VCIP'96, 1996, vol 2727, pp 1173-1184.
(92) T Naemura, H Harashima. Fractal coding of a multi-view 3-D image. Proceedings of IEEE ICIP'94, 1994, vol III, pp 107-111.
(93) S Chen. QuickTimeVR -an image-based approach to virtual environment navigation-. Proceedings of ACM SIGGRAPH'95, 1995, pp 29-38, 1995.
(94) T Naemura, M Kaneko, H Harashima. Multi-user immersive stereo. Proceedings of IEEE ICIP'98, 1998, vol I, pp 903-907.
(95) H Shum, L He. Rendering with concentric mosaics. Proceedings of ACM SIGGRAPH'99, 1999, pp 299-306.
(96) S Peleg, B Ezra, Y Pritch. Omnistereo: Panoramic stereo imaging. IEEE Trans Pattern Anal & Machine Intell 23(3):279-290, 2001.
(97) W Cochran, J Hart, P Flynn. Fractal volume compression. IEEE Trans Visualization and Computer Graphics 2(4):313-321, 1996.
(98) J Luo, C Chen. Coherently three-dimensional wavelet-based approach to volumetric image compression. J Electronic Imaging 7(3):474-485, 1998.

(99) H Yoshikawa, J Tamai. Holographic image compression by motion picture coding. Proceedings of SPIE Practical Holography X, 1996, vol 2652, pp 2-9.
(100) H Harashima: Three-dimensional image coding, future prospect. Proceedings of PCSJ'92, 1992, pp 9-12 (in Japanese).
(101) T Naemura, H Harashima. Ray-based approach to Integrated 3D Visual Communication. Proceedings of SPIE Three-Dimensional Video and Display: Devices and Systems (Critical Reviews of Optical Science and Technology), 2000, vol CR76, pp 282-305.
(102) T Fujii. A basic study on the Integrated 3-D Visual Communication. PhD dissertation, The Univ. of Tokyo, Tokyo, 1994 (in Japanese).
(103) T Fujii, H Harashima. Coding of an autostereoscopic 3-D image sequence, Proceedings of SPIE VCIP'94, 1994, vol 2308, pp 930-941.
(104) A Katayama, K Tanaka, T Oshino, H Tamura. A viewpoint independent stereoscopic display using interpolating of multi-viewpoint images. Proceedings of SPIE Stereoscopic displays and virtual reality systems II, 1995, vol 2409, pp 11-20.
(105) L McMillan, G Bishop. Plenoptic modeling: An image-based rendering system. Proceedings of ACM SIGGRAPH'95, 1995, pp 39-46.
(106) M Levoy, P Hanrahan. Light field rendering. Proceedings of ACM SIGGRAPH'96, 1996, pp 31-42.
(107) S Gortler, R Grzeszczuk, R Szeliski, M Cohen. The Lumigraph. Proceedings of ACM SIGGRAPH'96, 1996, pp 43-54.
(108) T Yanagisawa, T Naemura, M Kaneko, H Harashima. Handling of 3-D objects using ray space. J Inst Telev Eng 50(9):1345-1351, 1996 (in Japanese).
(109) E Adelson, J Bergen. The plenoptic function and the elements of early vision. In: M Landy and J Movshon ed. Computer Models of Visual Processing. MIT Press, 1991, ch 1.
(110) T Fujii, M Kaneko, H Harashima. Representation of 3D spacial information by rays and its application. J Inst Telev Eng 50(9):1312-1318, 1996 (in Japanese).
(111) T Naemura, M Kaneko, H Harashima. Efficient representation of 3-D spatial data based on orthogonal projection of ray data samples. J Inst Image Inf & Telev Eng 51(12):2082-2089, 1997 (in Japanese).
(112) T Naemura, M Kaneko, H Harashima. Orthographic approach to representing 3-D images and interpolating light rays for 3-D image communication and virtual environment. Signal Process: Image Commun 14:21-37, 1998.
(113) J Chai, X Tong, S Chan, H Shum. Plenoptic sampling. Proceedings of ACM SIGGRAPH2000, 2000, pp 307-318.
(114) T Naemura, H Harashima. Real-time Video-based rendering for augmented spatial communication. Proceedings of SPIE VCIP'99, 1999, vol 3653, pp 620-631.
(115) T Naemura, J Tago, H Harashima. Real-time Vide-based modeling and rendering of 3D scenes. IEEE Computer Graphics & Applications 22(2):66-73, 2002.
(116) R Ooi, T Hamamoto, T Naemura, K Aizawa. Pixel independent random access image sensor for real time image-based rendering system. Proceedings of IEEE ICIP2001, 2001, vol II, pp 193-196.
(117) S Wilburn, M Smulski, K Lee, M Horowitz. Light field video camera. Proceedings of SPIE Media Processors2002, 2002, vol 4674, pp 29-36.
(118) T Naemura, T Yoshida, H Harashima. 3-D computer graphics based on integral photography. Opt Express 8:255-262, 2001 (http://www.opticsexpress.org/oearchive/source/30085.htm).

(119) T Takano, T Naemura, M Kaneko, H Harashima. 3-D space coding based on light ray data - local expansion of compressed light ray data. J Inst Image Inf & Telev Eng 52(9):1321-1327, 1998 (in Japanese).
(120) T Takano, T Naemura, H Harashima. 3D space coding using Virtual Object Surface. Systems and Computers in Japan 32(12):47-59, 2001.
(121) T Fujii, T Kimoto, M Tanimoto. Data compression of 3-D spatial information based on ray-space coding. J Inst Image Inf & Telev Eng 52(3):356-363, 1998 (in Japanese).
(122) C Zhang, J Li. Compression of Lumigraph with multiple reference frame (MRF) prediction and just-in-time rendering. Proceedings of IEEE DCC2000, 2000, pp 254-263.
(123) M Magnor, B Girod. Adaptive Block-based light field coding. Proceedings of IWSNHC3DI'99, 1999, pp 140-143.
(124) M Magnor, B Girod. Hierarchical coding of light fields with disparity maps. Proceedings of IEEE ICIP'99, 1999, vol III, pp 334-338.
(125) M Magnor, B Girod. Data compression for light-field rendering. IEEE Trans Circuit & Syst Video Technol 10(3):338-343, 2000.
(126) X Tong, R Gray. Coding of multi-view images for immersive viewing. Proceedings of IEEE ICASSP2000, 2000, pp 1879-1882.
(127) P Lalonde, A Fournier. Interactive rendering of wavelet projected light fields. Proceedings of Graphics Interface'99, 1999, pp 107-114.
(128) M Magnor, A Endmann, B Girod. Progressive compression and rendering of light fields. Proceedings of Vision, Modeling and Visualization, 2000, pp 199-203.
(129) M Magnor, B Girod. Model-based coding of multi-viewpoint imagery. Proceedings of SPIE VCIP2000, 2000, pp 14-22.
(130) M Magnor, P Eisert, B Girod. Model-aided coding of multi-viewpoint image data. Proceedings of IEEE ICIP2000, 2000, vol II, pp 919-922.
(131) B Girod, M Magnor. Two approaches to incorporate approximate geometry into multi-view image coding. Proceedings of IEEE ICIP2000, 2000, vol II, pp 5-8.
(132) M Magnor, P Eisert, B Girod. Multi-view image coding with depth maps and 3-D geometry for prediction. Proceedings of SPIE VCIP2001, 2001, pp 263-271.
(133) M Magnor, B Girod. Sensitivity of image-based and texture-based multi-view coding to model accuracy. Proceedings of IEEE ICIP2001, 2001, vol III, pp 98-101.
(134) T Fujii, T Kimoto, M Tanimoto. Compression and interpolation of ray-space based on nonorthogonal expansion of local EPI pattern. Proceedings of Picture Coding Symposium, 2001, pp 311-314.
(135) C Zhang, J Li. Compression and rendering of concentric mosaics with reference block coded (BRC). Proceedings of SPIE VCIP2000, 2000, vol 4067, pp 43-54.
(136) L Luo, Y Wu, J Li, Y Zhang. Compression of concentric mosaic scenery with alignment and 3D wavelet transform. Proceedings of SPIE Image and Video Commun and Process, 2000, vol 3974, pp 89-100.
(137) Y Wu, L Luo, J Li, Y Zhang. Rendering of 3D wavelet compressed concentric mosaic scenery with progressive inverse wavelet synthesis (PIWS). Proceedings of SPIE VCIP2000, 2000, vol 4067, pp 31-42.
(138) J Li, H Y Shum, Y Q Zhang. On the compression of image based rendering scene: A comparison among block, reference and wavelet coders. International Journ Image and Graphics 1(1):45-61, 2001.
(139) A Katayama, S Uchiyama, H Tamura, T Naemura, M Kaneko, H Harashima. A cyber-space creation by mixed ray space data with geometric models. Trans IEICE J80-D-II(11):3048-3057, 1997 (in Japanese).

(140) S Uchiyama, A Katayama, A Kumagai, H Tamura, T Naemura, M Kaneko, H Harashima. Collaborative Cybermirage: A shared cyberspace with mixed reality. Proceedings of VSMM'97, 1997, pp9-18.

5

Omnidirectional Sensing and Its Applications

Yasushi Yagi
Osaka University, Osaka, Japan

SUMMARY

This chapter presents a critical survey of the literature on omnidirectional sensing. The area of vision applications such as autonomous robot navigation, telepresence, and virtual reality is being expanded by the use of cameras with wide angles of view. In particular, a real-time omnidirectional camera with a single center of projection is suitable for analyzing and monitoring. From its omnidirectional input image, any desired image can be generated, and projected on a designated image plane, such as a pure perspective image or a panoramic image. In this chapter, designs and principles of existing omnidirectional cameras, which can acquire an omnidirectional (360 degrees) field of view, are reviewed and their applications in fields of autonomous robot navigation, telepresence, remote surveillance and virtual reality are discussed.

1. INTRODUCTION

For people, peripheral vision informs us that something is going on. However, our peripheral acuity is not good enough to tell us precisely what it is. A central zone in the retina, called the fovea centralis, handles high-quality vision and is used for understanding the details of an object. The view field of binocular vision is approximately 6 degrees. To recognize the whole shape of an object, peripheral vision with a wide view field is used for combining local fovea images observed by eye movements. This shows that sensing with a wide field of view is an important factor for human perception.

For navigation, a robot's vision must generate a spatial map of its environment for path planning, obstacle (possibly moving) avoidance, and for finding a candidate of interest. For this purpose, a detailed analysis is not necessary, but a high speed rough understanding of the environment around the robot is required. If considered from the standpoint of machine perception, applications such as autonomous navigation and telepresence need a field of view as wide as possible.

Thus, a real-time omnidirectional camera, which can acquire an omnidirectional (360 degrees) field of view at video rate, would have applications in a variety of fields such as autonomous navigation, telepresence, virtual reality and remote monitoring. There have been several attempts to acquire omnidirectional images using a rotating camera, a fish-eye lens, a conic mirror and a spherical mirror. Over the past 15 years, researchers in computer vision, applied optics, and robotics have investigated and reported in a number of papers regarding omnidirectional cameras and their applications. In particular, research applying omnidirectional vision to telepresence, virtual reality, remote monitoring, and surveillance has rapidly increased since the mid 1990's.

In this chapter, designs and principles of existing omnidirectional cameras, which can acquire an omnidirectional (360 degrees) field of view, are reviewed and their applications in fields of autonomous robot navigation, telepresence, remote surveillance, and virtual reality are discussed.

2. OMNIDIRECTIONAL SENSING

Approaches for obtaining omnidirectional images can be roughly classified into three types according to their structure: use of multiple images, use of special lenses and use of convex mirrors. Another criterion of omnidirectional sensor classification is whether it has a single center of projection or not. A single center of projection is an important characteristic of an omnidirectional sensor as from an omnidirectional input image it makes it possible for any desired image to be generated and projected onto any designated image plane, such as a distortion-free perspective image or a panoramic image. This easy generation of a distortion-free perspective or panoramic images allows robot vision to use existing image processing methods for autonomous navigation and manipulation. It also allows a human user to see familiar perspective images or panorama images instead of an unfamiliar omnidirectional input image.

2.1. Use of Multiple Images

2.1.1. Rotating Camera System

A straightforward approach for obtaining omnidirectional images is to rotate the camera around a vertical axis as shown in Figure 1 (a)(1)(2)(3)(4). The camera rotates with a constant angular velocity. Vertical scan lines are taken from different images and stitched together to form an omnidirectional image The horizontal resolution of an obtained omnidirectional image is dependent not on the camera resolution, but on the angular resolution of the rotation. Therefore, one advantage of this method is that we can acquire a high-resolution omnidirectional image if the camera is controlled precisely. A single center of projection can be made by an exact alignment of the rotating camera axis and the center of the projection. However, this method has the disadvantage that it requires a rather long time to obtain an omnidirectional image, which restricts its use to static scenes and rules out dynamic or real-time applications.

Barth developed a comparatively fast panoramic imaging system by rotating a vertical line scan camera around a vertical axis (4). Krishnan and Ahuja developed another idea to improve the performance of the acquisition time of a panoramic image (5). They rotated the camera around the vertical axis with a regular angle and obtained a panoramic image by connecting a sequence of images at each pan position. A useful advantage of a rotating camera is that omnidirectional range data can be acquired while the camera is rotating if the focal point of the camera is at a certain distance from the rotational axis as shown in Figure 1(b) (6)(7). Two panoramic images are obtained by arranging vertical lines through each slit at the sides of the image sensor. Coarse range data can be acquired by matching similar lines from both panoramic images.

Ishiguro built a multiple vision agent robot system that consisted of four cameras rotating independently of each other. These cameras can be used for either omnidirectional imaging or active vision sensing (10). In any case, although precise azimuth information is available in the omnidirectional view obtained by a rotating camera, the time-consuming imaging prevents its application to real-time situations. Another disadvantage of rotating cameras is that they require the use of moving parts, and they are not suitable for robotic applications.

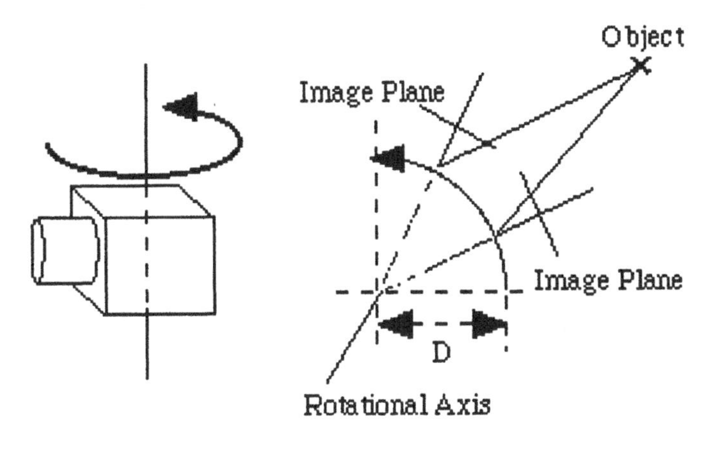

Figure 1 Rotating camera.

2.1.2. Multiple Camera System

One natural method not utilizing rotation is the use of multiple cameras. A panoramic imaging system by using four conventional TV cameras and four triangular mirrors was built by a researcher at Bell Lab., AT&T(11). Researchers at the Nara Institute of Science and Technology developed a high-resolution omnidirectional stereo imaging system that can take images at video-rate (12). The sensor system takes an omnidirectional view with a component constructed of the six cameras and a hexagonal mirror, which is designed so that the vertical lens center of the six cameras are located at a fixed point in the mirror images. Therefore, an obtained omnidirectional image satisfies the relation of a single center of projection. Stereoscopic view can be built by aligning two symmetrical sets of the components on the same vertical line. Such a configuration reduces the 2-D stereo matching problem to a 1D one. The system can acquire high resolution and single viewpoint image at video-rate, however, as pointed out, the alignment and calibration among each of the six cameras is difficult and has not yet been achieved. It is also difficult to build a compact system because acquisition utilizing twelve cameras needs an equal number of AD converters or recorders.

2.1.3. Use of Special Lenses

Fish-eye Lenses

A fish-eye camera, which employs a fish-eye lens instead of a conventional camera lens, can acquire a wide-angle view, as a hemispherical view, in real-time. Morita proposed a motion stereo method for measuring the three-dimensional locations of lines by mapping an input image on spherical coordinates (13). Researchers at the University of Cincinnati applied a fish-eye camera to mobile robot position control by using given targets (14)(15)(16)(17) and following lines (18)(19). The camera, however, can obtain good resolution only for the ceiling area, and has poor resolution on the side and ground views. This means that the image obtained through a fish-eye lens has rather good resolution in the center region but has poor resolution in the peripheral region. Image analysis of the ground (floor) and objects on it is difficult because they appear along the circular boundary where image resolution is poor. Furthermore, it is difficult to generate a complete perspective image, because fish-eye lenses do not have the property of a single center of projection.

Panoramic Annular Lens

Fish-eye lenses are good commercial optics to apply to wide view visualization, but are unsuitable for computer vision because they do not satisfy the need for a single center of projection. Greguss proposed an optical lens system that can acquire a distortion-free panoramic image (20)(21). The optics of the proposed system, named panoramic annular lens (PAL), consists of a single glass block with two reflective and two refractive planes as shown in Figure 2. Thus, PAL optics do not need to be aligned and can easily be miniaturized, whereas other omnidirectional imaging optics must align several optical elements. Figure 3 (a) and (b) show a prototype of PAL optics and an example of an input image, respectively. Further, PAL can be used not only for all round data acquisition and recording but also for panoramic visualization. Panoramic images recorded by PAL optics or computed according to the ray tracing laws of the PAL optic are then reprojected. However, one disadvantage of PAL optics is that it is difficult to increase the field of view of depression.

2.1.4. Use of Convex Mirrors

The basic idea in the use of a convex mirror is that the camera is pointed vertically toward the convex mirror with the optical axis of the camera lens aligned with the mirror axis. By setting the axis of the camera to the vertical, we can acquire a 2π

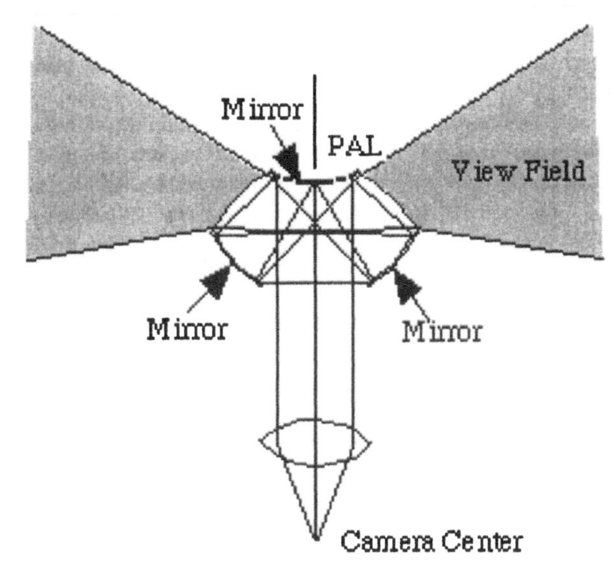

Figure 2 Optical relation of panoramic annular lens.

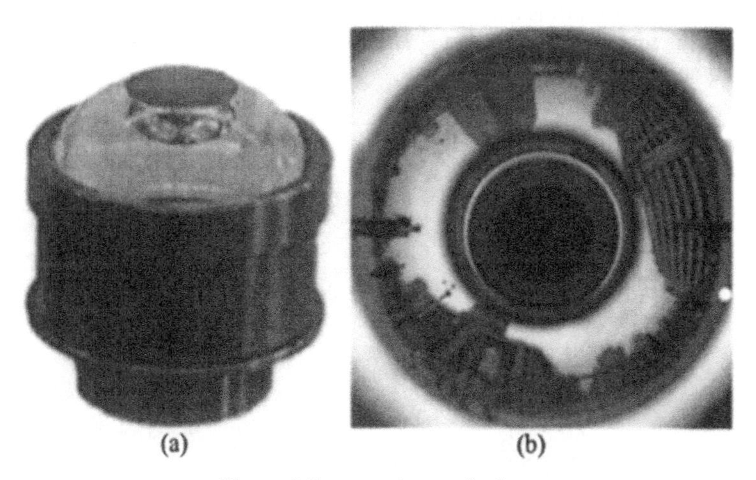

Figure 3 Panoramic annular lens.

view around the camera in real time. However, the anisotropic property of the convex mirror, known as spherical aberration, astigmatism, etc (22), results in a lot of blurring in the input image of the TV camera if the optical system is not well designed. For instance, the curvature of the conical mirror across a horizontal section is the same as that of the spherical mirror. If the camera lens has a large relative aperture, rays coming through the horizontal section from the object point do not focus at a common point, an effect known as spherical aberration. Curvature of the conical mirror on the vertical section is the same as that of a planar mirror while that on the horizontal section is the same as that of a spherical mirror; hence, the rays reflected through vertical or horizontal planes focus at different points. Thus, to reduce blurring and to obtain a clear picture, an optical system that can cover focus points from both optics is needed. In particular, careful design is necessary when using a single convex mirror.

Conical Mirror

An image obtained by a conical mirror has good resolution in the peripheral. Yagi developed an omnidirectional image sensor, COPIS, together with an image processing method, to guide, the navigation of a mobile robot (Figure 4 (a)) (23)(24). As shown in Figure 4 (b), it provides a panoramic view around the sensor. Mapping of the scene onto the image by COPIS involves a conic projection causing vertical edges in the environment to appear as lines radiating from the image center. The vertical lines provide a useful cue in a man-made environment, such as rooms and corridors, containing many objects with vertical edges; (for example, doors, desks and shelves). If a line is vertical, it appears radially in the image plane, as shown in Figure 5. Thus, COPIS can easily find and track the vertical edges by searching consecutive images for radial edges. However, a serious drawback of COPIS is that it does not have a focal point (a center of a viewpoint), and it is therefore impossible to generate a distortion-free perspective image from an omnidirectional input image.

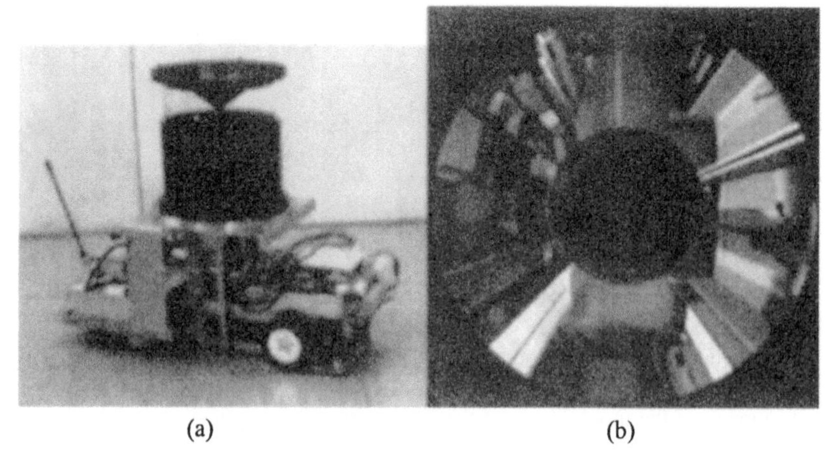

(a) (b)

Figure 4 COPIS.

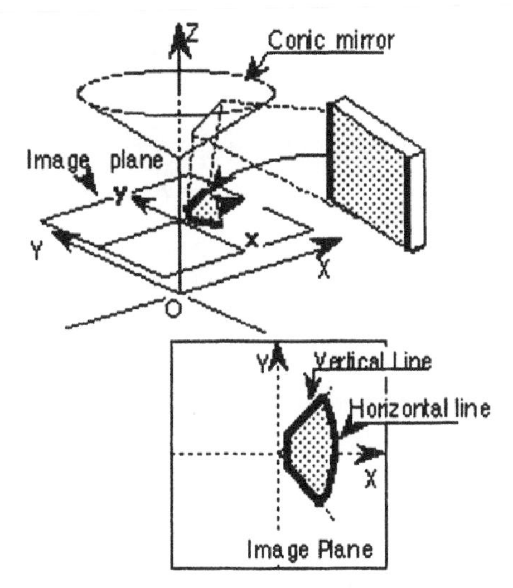

Figure 5 Conic projection.

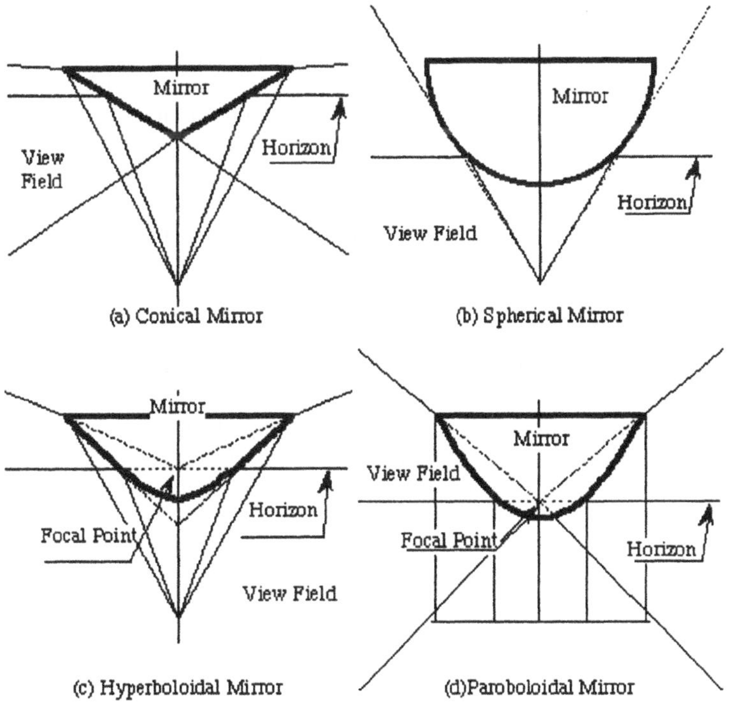

Figure 6 Field of views of omnidirectional image sensors.

Figure 7 HyperOmni Vision.

Spherical Mirror

Similar to the fish-eye lens, an image obtained from a spherical mirror has rather good resolution in the center region but poor resolution in the peripheral region. As shown in Figure 6, a spherical mirror yields an image of the environment around it and its field of view is the widest among sensors with convex mirrors (25)(26). Such a wide view is useful for locating a robot along its route. Hong proposed a method for navigating a robot using the appearance changes of features on the horizon (25)(27). However, the geometrical relation of the single center of projection was not satisfied.

Hyperboloidal Mirror

A hyperboloidal mirror yields an image of the scene around its axis. We can observe the images generated using a TV camera with its optical axis aligned with the hyperboloidal mirror axis, as shown in Figure 6 (c).

Rees and Mich outlined a panoramic viewing device for use in a vehicle utilizing a hyperboloidal mirror (28). Their sensor was used only for monitoring by an operator, and did not consider image processing that used the special relation of a single center of projection.

Yamazawa et al. were the first group to show the merit of an omnidirectional image sensor with a single center of projection. The name of their sensor is HyperOmniVision (29)(31) (see Figure 7). By mounting HyperOmniVision on a robot so that the optical axis is vertical, a 360-degree view around the robot can be acquired. HyperOmni Vision maps a scene onto the image plane through a hyperboloidal mirror and a lens. This mapping is called "hyperboloidal projection." The hyperboloidal mirror has a focal point, which from an omnidirectional input makes it possible for any desired image to be generated and projected onto any designated image plane, such as a perspective image or a panoramic image. This easy generation of perspective or panoramic images allows the robot vision to use existing image processing methods for autonomous navigation and manipulation. It also allows a human user to see familiar perspective images or panorama images instead of an unfamiliar omnidirectional input image deformed by the hyperboloidal mirror.

The hyperboloid has two focal points (O_m, O_c). The camera center is fixed at one of the focal points. As shown in Figure 8 (a), a hyperboloidal mirror yields the image of a point in space on a vertical plane through the point P and its axis. Thus, the point P at (X,Y,Z) is projected onto the image point p at (x,y) such that

$$\tan\theta = \frac{Y}{X}$$

This means that the angle in the image, which can be easily calculated as y/x, shows the azimuth angle θ of the point P in space. Also, it can be clearly understood that all points with the same azimuth in space appear on a radial line through the image center. This useful feature is the same as in conic or spherical projection. Therefore, with a hyperboloidal projection, the vertical edges in the environment appear radially in the image and the azimuth angles are invariant to changes in distance and height. Here, the hyperboloidal mirror has a special feature. Let's consider the straight line that connects the point P in the real world and the point that the ray from the object point reflects at the hyperboloidal mirror. This line passes through the focal point O_m regardless of the location of the object point P. It means that the image taken by HyperOmni Vision can easily be transformed to a cylindrical panorama image or a common perspective image, where the origin of the transformed coordinate systems is fixed at the focal point O_m. Therefore, one can obtain an omnidirectional image of the scene on the image plane with a single center of projection O_m. Figure 8 (c) and (d) shows examples of a panorama image and a perspective image, respectively, converted from the input omnidirectional image (b).

Let us consider the view field of the hyperboloidal mirror and the resolution of the omnidirectional image obtained. The image obtained using a spherical mirror has rather good resolution in the central region (down view), but it has poor resolution in the peripheral region (side view). Figure 6 (b) shows an example of the view field of the image sensor with a spherical mirror. The view field of HyperOmni Vision is the same as that of a spherical mirror, however, the resolution of HyperOmni Vision is limited by the asymptote of the hyperboloid, thus the upper angle of HyperOmni Vision is the same as that of a conical mirror. The lower angle of conical mirror is limited by the vertex angle of the conic mirror, thus a conic mirror can not look at its foot as shown in Figure 6 (c). However, the lower angle of HyperOmni Vision is not limited. Thus, HyperOmni Vision has advantages of both conic and spherical mirrors.

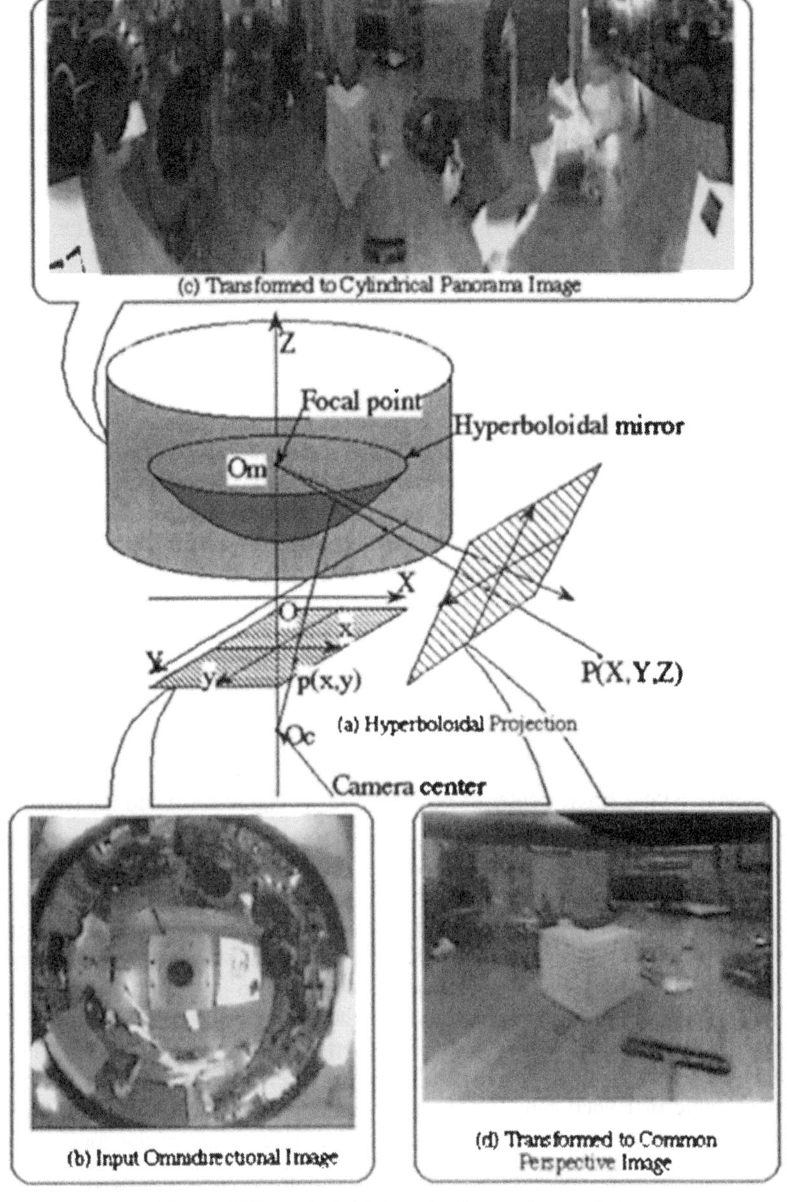

Figure 8 Hyperboloidal projection and sample images.

Paraboloidal Mirror

Another optic system, using a paraboloidal mirror, having the relation of a single center of projection, was proposed by Peri and Nayer (30)(32). Concave paraboloidal optics are frequently used to converge an incoming set of parallel rays such as a paraboloidal antenna. Basically their idea is same as that of the paraboloidal antenna. In the case of a convex paraboloidal mirror, all rays from the environment reflect at the mirror and run parallel to the rotating axis of the paraboloidal mirror. A straight line that extends the ray from the object point in the environment passes through the focal point regardless of the location of the object point. Therefore omnidirectional imaging with a single center of projection can be done by setting a camera with orthographic projection such as a telecentric lens or zoom lens in front of the paraboloidal mirror.

Peri and Nayer made a video conference system, named OmniVideo, that generated the perspective image of a desired viewpoint. The view field of the paraboloidal mirror and the resolution of the omnidirectional image were just between the hyperboloidal and the spherical mirrors (see Figure 6 (d)). Mathematically, the curvature of these mirrors are represented by

$$Z = \frac{1 + \sqrt{1 - (1 + k)c^2 r^2}}{c \cdot r^2}$$

where the z-axis is the rotating axis of the convex mirror, c is the curvature, r is the radial coordinate in lens units, and k is the conic constant. The type of the mirror is represented by the conic constant; $k < -1$ for hyperbolas, $k = -1$ for parabolas, and $k = 0$ for spheres. Comparative analysis of omnidirectional cameras with convex mirrors was done by Baker and Nayer (33)(34).

Dual Convex Mirrors

One difficulty in the design of omnidirectional imaging using mirrors is focusing. The anisotropic property of the convex mirror, also known as spherical aberration or an astigmatism, results in much blurring in the input image of the TV camera if the optical system is not well designed. Usually, it is hard to reduce aberration by a single reflective mirror.

One good idea to miniaturize and focus is to use multiple mirrors (35)(36)(37)(38). A multiple mirror system can usually obtain more clear images than can be achieved by an optical system with a single reflective mirror. For instance, to minimize blurring influence Takeya et al. designed optics similar to a reflecting telescope, which consists of two convex mirrors (43). As shown in Figure 9, a ray from the object point is reflected at the principle mirror. Then, the reflected ray is reflected again at the secondary mirror and focused on the image plane. Curvatures of both mirrors are optimized for minimizing the blur on the image plane. A fundamental configuration of the optics is similar to panoramic annular lenses (8), but the optics do not satisfy the important relation of being distortion-free. The curvatures of both mirrors are optimized for minimizing the blur on the image plane. The fundamental configuration of the optics is similar to PAL, but do not satisfy the important relation of a single center of projection.

Optics that consist of two paraboloidal mirrors or two hyperboloidal mirrors not only satisfy the important relation of a single center of projection and minimize blurring, but also can easily obtain a wide, side-view-centered panorama image. Nayer investigated such optics with two paraboloidal mirrors and developed a prototype of the omnidirectional camera (39). However, the camera did not in fact use two paraboloidal mirrors, but rather a paraboloidal and a spherical mirror because manufacturing a spherical mirror is easier than that of paraboloidal mirror. At almost the same time, Yagi proposed tiny omnidirectional image sensors called TOMs (twin reflective omnidirectional image sensors), which consist of two paraboloidal mirrors or two hyperboloidal mirrors (40)(41). Bruckstein and Richardson also demonstrated the concept of two paraboloidal mirrors; however, a real sensor has yet to be reported (42).

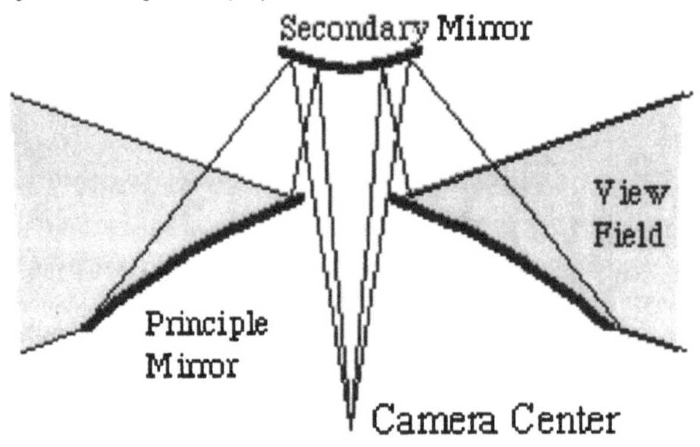

Figure 9 Optical relation of omnidirectional image sensor developed by Mitsubishi Electric Corp.

Omnidirectional Stereo

Ishiguro and Tsuji have proposed an omnidirectional stereo system that uses two panoramic views (6)(7). The two panoramic images are obtained by arranging vertical lines through the slits at the sides of the image sensor. Coarse range data can be acquired by matching similar lines between each of the panoramic images. Murray generalized the above approach according to a principle of structure from known motion (47). Kang recovered the 3-dimensional data of the scene from panoramic images at different points using the techniques of "8 points structure from motion" and multi-baseline stereo (48).

A rotating 1D CCD is a good method to obtain precise range data (50)(51). Godber developed a panoramic line-scanning range system by using two vertical line-scan devices mounted parallel to a vertical axis at a certain distance (8). Benosman proposed a different type of panoramic line-scanning range system, with two line-scan devices having the same rotational axis and arranged one on top of the other. A single point lies on the same vertical line in both high and low panoramic images, therefore, it reduces the 2-D matching problem to a 1D one (9). One of the earliest works to measure a panoramic ranging system by setting two cameras, one up and one down, was done by Saraclik (1) who used it to determine the size and shape of a room and for estimating the location of a robot.

Researchers at the Nara Institute of Science and Technology developed a high-resolution omnidirectional stereo imaging system that can take images at video-rate as mentioned above (12). The system can acquire a high resolution and single viewpoint image at video-rate, however, as they pointed out, alignment and calibration among each of the six cameras is difficult and not yet achieved. Further, it is difficult to build a compact system because acquisition from twelve cameras needs an equal number of AD converters or recorders.

A unique panoramic stereo system was investigated by Bogner (26). He set a bi-convex lobed mirror in front of the camera. A point in the environment is reflected at both lobes and focused at different positions on the image plane. Southwell et al developed a similar panoramic stereo system. They also made real-time hardware for re-mapping an input image to a panoramic image (46). A disadvantage of these systems is that ranging precision is not good because of the short length of the stereo baseline and the poor resolution of the inner omnidirectional image.

2.2. Multiple Sensing Cameras with Omnidirectional and Local Views

The tasks of an autonomous mobile robot can be classified into two categories: navigation to a goal, and manipulation of objects or communication with other robots and people at the goal. For navigation, the robot needs to estimate its own location relative to the environment, find unknown obstacles, and generate a spatial map of its environment for path planning and collision avoidance. For this, a detailed analysis is not necessary, but a high-speed rough understanding of the environment around the robot is required. Omnidirectional sensing is suitable for this purpose as it can view 360 degrees in real time, while for object manipulation or human communication, a limited field of view is sufficient. High resolution is required (namely local view) for detailed analysis of an object of interest.

Although a local perspective image can be generated from an omnidirectional camera, resolution of the omnidirectional camera system with a single camera is not sufficient for this purpose since an omnidirectional view is projected onto an image with a limited resolution. An omnidirectional camera with multiple cameras (12) can meet both requirements, however, it has the difficulty of alignment and calibration among the multiple cameras.

To perform each requirement effectively, different sensing functions are required for the vision system because the specifications such as image resolution, visual field of the sensor, and speed of data acquisition differ. Thus, as with human perception, multiple sensors and hybrid sensing are necessary for performing more tasks, more efficiently and more reliably.

A practical sensor, named the multiple image sensing system (MISS) was developed by researchers at Osaka University (31)(44). They combined two different types of optics in a single conventional TV camera; an omnidirectional camera and binocular vision. The omnidirectional camera, by using a conic mirror, can observe a 360-degree view around the robot in real time. It is suitable for observing global information around the robot, but the geometric property of the conic mirror results in much distortion in the input image of the camera. On the other hand, although the view field of binocular vision is restricted by the visual angle of the lens, it can acquire fine stereo images, and it is useful for understanding the spatial configuration of the environment. Such multiple image sensing systems can be easily reduced in size and weight. Compactness and high performance are also important factors for a mobile robot. Figure 10 (a) and (b) shows a prototype of the MISS and an example input image, respectively. There are three components: the optics of an omnidirectional imaging system, a binocular vision adapter, and a tilting optical unit for changing the viewing direction of the binocular vision. MISS has a high speed tilting unit for viewing interesting objects, but not a panning unit since panning can be done by rotating the robot itself. The omnidirectional image obtained by the conic mirror has a better resolution of the azimuth angle in the peripheral area than that in the center region. For example, the resolution of azimuth angle is approximately 0.25 degrees in the peripheral area of the image whereas it is approximately 1 degree at the circumference of a 60 pixels radius from the image center. In particular, the central region of the omnidirectional image has poor resolution of the azimuth angle. The shape of the conic mirror is improved from a circular cone to a frustum of a cone. The hole in the center region of the frustum is utilized for binocular vision imaging. Binocular vision produces an image, including a stereo one, on the center region of a single image plane. The peripheral region of the image plane is used for omnidirectional sensing.

Researchers at Mitsubishi Electric Corp. have proposed a similar multiple imaging sensor for a television communication system, which combines an omnidirectional camera with not binocular vision but a single view camera as shown in Figure 11 (a) and (b)(43).

Researchers at Osaka University have also proposed a practical multiple imaging system that consists of a hyperboloidal omnidirectional camera and active binocular vision that has a great degree of freedom for observation (45). Actually, the system consists of two robots; a main robot with an omnidirectional image sensor and a small robot with an active binocular vision as shown in Figure 12. The small robot is set on a stage of the main robot and can freely change its viewing direction because degree of freedom is an important factor for flexible human-robot interaction. The sensor is used for finding and tracking persons around the robot and focusing attention on a person of interest.

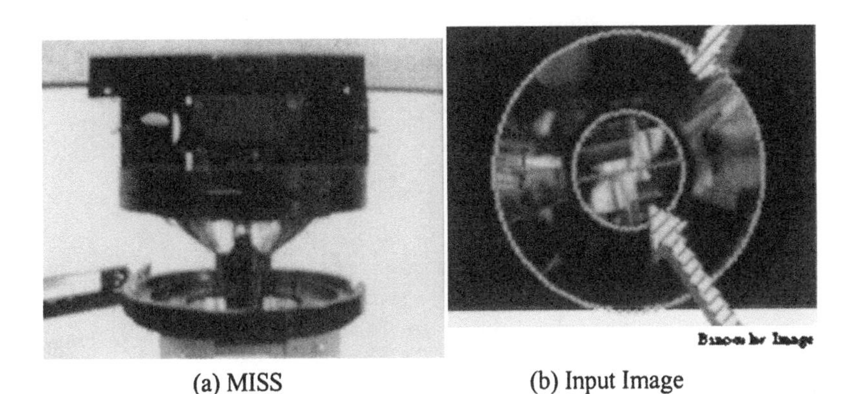

(a) MISS (b) Input Image

Figure 10 Multiple image sensing system (MISS).

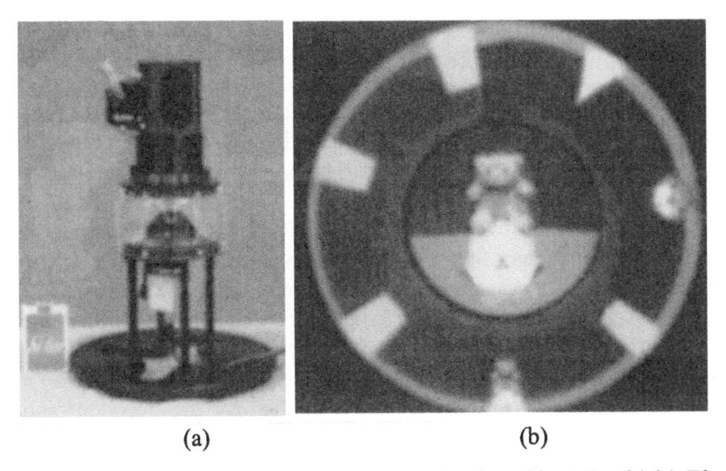

(a) (b)

Figure 11 Omnidirectional image sensor developed by Mitsubishi Electric Corp.

3. APPLICATIONS

Real-time omnidirectional cameras can be applied to a variety of fields such as autonomous navigation, telepresence, virtual reality, and remote monitoring. Examples of applications for robot navigation, remote viewing, and surveillance systems are described and discussed next.

Figure 12 Multiple imaging system.

3.1. Autonomous Robots

Advantages of omnidirectional sensing are not only wide view angles but also its special properties such as the invariability of azimuths, circumferential continuity, periodic characteristic, symmetric characteristic, and rotational invariant. Most work to date uses these special properties.

Since vertical lines in space appear as radial lines through an image center in an omnidirectional image, they can be robustly detected. The characteristic that the angle in the image coincides with the azimuth angle of the point in space is also used for efficient estimation of the 3-D position of vertical lines. Yagi proposed a method for navigating a robot by detecting the azimuth of each object in the omnidirectional image (52). The azimuth is matched with a given environmental map. The robot can precisely estimate its own location and motion (the velocity of the robot) because the omnidirectional camera can observe a 360 degrees view around the robot, even when all edges are not extracted correctly from the omnidirectional image. Under the assumption of the motion of the robot being known, environmental maps of the real scene are successfully generated by monitoring azimuth changes in the image. Several researchers have used this property for robot navigation (53). Delahoche et al. proposed an incremental map building method based on the exploitation of the azimuths data given by an omnidirectional vision and by an odometer (54). The robot position estimation and map updating are based on the use of an extended Kalman filter.

Forward and backward features along the moving direction of the robot show constant azimuth angles while the robot moves straight. By using this property, Ishiguro proposed a strategy based on path-centered representation for map generation (56)(57). Further, an environmental representation for robot navigation called T-Net, was proposed with a T-Net consisting of paths representing a rigid structure and keeping the relations between the local area (environmental map) (58).

The other important task for navigation is obstacle avoidance. Yamazawa estimated motion of a robot and detected obstacles using on the floor inverse perspective mapping (55). An input image is transformed to a perspective image, whose optical axis is perpendicular to the floor, the same as that of on the floor inverse perspective mapping. As apparent shapes of features on the floor are constant in the image, it is easy to establish correspondences among the features in consecutive images by a simple matching method. Thus, motion estimation of the robot and obstacle detection can be done robustly. Yagi proposed a method where objects moving along collision paths are detected by monitoring azimuth changes of vertical edges. Confronted with such objects, the robot changes velocity to avoid collision and determines locations and velocities (24).

While navigating to a goal, omnidirectional views are used for localization, collision avoidance and finding candidate objects of interest such as landmarks. Generally however, the resolution of the omnidirectional image is not sufficient for further analysis of the object. Effective observation has been done using the MISS (44). Since vertical lines can be robustly detected in the omnidirectional image and their 3-D locations can be obtained by motion stereo of the omnidirectional image sequence, the 3-D location of vertical edges of the unknown object are used for attention control of the binocular vision. Further, by integrating the merits of both omnidirectional and binocular vision, MISS can generate an environmental map and estimate egomotion of the robot from both sensory information sources without any assumptions of the exact motion of the robot (59).

A general technique for estimating egomotion was presented by Nelson and Aloimonous (60). To estimate egomotion with 6 DOF, their technique requires optical flows on three orthogonal great circles on spherical coordinates, however, egomotion is determined by iterative calculation, so the computational cost is high. Gluckman and Nayer estimated 6 DOF of the egomotion of the camera by using their omnidirectional camera, but the estimation error was significant (61). In the case of mobile robot navigation, the egomotion of the robot is usually caused by jogging due to unevenness in the ground plane. Under an assumption of the ground plane motion of the robot, Yagi robustly estimated the rolling and swaying motions of the robot from omnidirectional optical flows by using the special characteristics of 1) the radial component of optical flow in an omnidirectional camera has a periodic characteristic and is invariant to the ground plane motion of the robot, and 2) the circumferential component of optical flow has a symmetric characteristic (62). Additionally, the image interpolation technique was used for computation of egomotion from two panoramic images (63).

Memory-based navigation is a common approach for visual navigation. The basic operation is a comparison between present sensorial inputs and previous memorized patterns. It is easy to directly relate the robot action and sensory data without the geometrical model. Zheng's robot memorized the side scene of a route by a panoramic view while it moved along the route (64)(65). Further, range data of each object in a scene also added to this panoramic representation (66). Matsumoto et al. had the robot memorize the whole front view image at reference points along the route for visual navigation (67). The correspondence between the present input images and previous memorized images were established using DP matching and correlation methods, respectively. However, these methods need large amounts of memory for route memorization. Therefore, to reduce memorizing data, Ishiguro proposed a compact representation by expanding it into a Fourier series (68), where each input image is memorized by the coefficients of the low frequency components. His approach is simple and useful for navigation in a real environment. Aihara et al. compressed memorized data by KL transformation (69). Most of these existing memory-based approaches memorize relations between the apparent features (image) at reference points, which are useful for finding correspondences between the current position and memorized ones, and for estimating the orientation of the robot. However, apparent features at the reference point do not directly represent the robot's action against the environment and the spatial relation between the environment and the robot. Yagi proposed an approach that memorized by utilizing a series of two-dimensional Fourier power spectrums of consecutive omnidirectional images at the horizon while the robot moved to the goal position (70). This method can directly represent the temporal and spatial relations between the environment and the robot. Neural networks have also been used for scene understanding (71)(72).

From the standpoint of sensor fusion, several researchers have integrated sensor input from omnidirectional vision and ultrasonic sensors for visual navigation. For Bang, the ultrasonic sensor gave the range, and the vision sensor gave the azimuth of the edge feature (73); for Yagi, the vision sensor gave the azimuth of the edge, and the ultrasonic sensor the confirmation of free space between two vertical edges (52). Wei fused color and edge information from the omnidirectional vision sensor with range data from an ultrasonic sensor on a metric grid-based representation (74). Drocount et al. proposed sensor fusion between omnidirectional vision and range data by using the theory of Dempster-Shafer (75)(76).

As an omnidirectional view around a robot can be acquired in real-time and it is easy to communicate among multiple robots. Several groups have investigated multi-agent robot systems using omnidirectional cameras (77)(78)(79). RoboCup is a competitive test bed for evaluating such multi-robot systems (80). Recently, many groups have used omnidirectional cameras for localization and for ball and goal detection (81). Basu and Greguss applied an omnidirectional camera system for pipe inspection and endoscopy, respectively (82)(20).

3.2. Communication, Telepresence, and Surveillance

Real-time omnidirectional cameras can also be applied in other fields such as telepresence, virtual reality, and remote monitoring. Yagi et al. described an autonomous guidance robot system with an omnidirectional camera (HyperOmni Vision) and ultrasonic range sensors (83). The system can follow and navigate the observer to the desired position in the environment such as a museum or exhibition. The omnidirectional camera is used for localization of the robot and for human tracking.

Peri et al. developed a video conference system, OmniVideo. They used an omnidirectional camera with a paraboloidal mirror that has the property of a single center of projection, and generated the perspective image of the desired viewpoint (30).

Conventional virtual reality systems based on real images need to memorize panoramic or omnidirectional images at different view points with such images acquired by integrating overlapping images taken by a rotating camera (84)(85). But these can not be applied to dynamic environments. Onoe et al. developed a telepresence system that realizes virtual tours into a visualized dynamical real world without significant time delay(86)(87). Omnidirectional video is recorded in advance by a HyperOmni Vision mounted on the roof of a car or the top of a remote controlled robot that is manually driven on a public road and in a hallway, respectively. The viewer's head motion is measured by a 3D magnetic tracker attached to a head mounted display. Then the perspective image is transformed from an omnidirectional input image and displayed onto the head-mounted display corresponding to the viewer's head motion in real time. Authors have investigated how multiple users can look around from a single viewpoint in a visualized dynamic real world in different directions at the same time. However, the resolution of the transformed perspective image is too poor for monitoring details because resolution is restricted by that of the CCD camera used. Therefore, development of a high-resolution omnidirectional camera is required for such applications. For instance, in the futures, high-resolution omnidirectional video surveillance systems may wipe the smiles off the faces of bank robbers.

4. CONCLUSIONS

This comprehensive survey of omnidirectional sensing with its various applications has focused on cameras with panoramic or omnidirectional imaging capabilities, and on the application of vision to autonomous robot navigation, telepresence, and virtual reality. Below is an ordered summary.

Omnidirectional sensing: Approaches for obtaining omnidirectional images can be classified into three types according to their structure: use of multiple images, use of special lenses, and use of convex mirrors. Another criterion of classification of omnidirectional sensing is whether it has a single center of projection or not. A single center of projection is an important characteristic of omnidirectional sensing, as from an omnidirectional input image it allows any desired image to be generated and projected onto any designated image plane, such as a distortion-free perspective image or a panoramic image. This easy generation of a distortion-free perspective or of panoramic images allows a robot's vision to use existing image processing methods for autonomous navigation and manipulation. It also allows a human user to see familiar perspective images or panorama images instead of an unfamiliar omnidirectional input image. Existing systems that can satisfy this property are rotating or multiple camera systems, and the use of panoramic annular lenses, and hyperboloidal and paraboloidal mirrors.

Applications: Real-time omnidirectional cameras can be applied to a variety of fields such as autonomous navigation, telepresence, virtual reality, and remote monitoring. Examples included applications for map generation, map or landmark-based navigation, collision avoidance, egomotion estimation, memory-based navigation, robot navigation, pipe inspection, guidance robot, video conference system, remote viewing, and surveillance systems.

Acknowledgments

I am grateful to Mitsubishi Electric Corp. for providing Figure 11(a) and (b), and to Mr. Yamauchi, Managing Director of Tateyama R&D, for providing Figure 3(c) and (d).

REFERENCES

(1) K. B. Saraclik. Characterizing An Indoor Environment with A Mobile Rrobot and Uncalibrated Stereo. Proceedings of IEEE Int. Conf. Robotics and Automation, 1989, pp.984-989.

(2) J. Y. Zheng and S. Tsuji. Panoramic Representation of Scenes for Route Understanding. Proceedings of Int. Conf. Pattern Recognition, 1990, pp.161-167.

(3) J. Y. Zheng and S. Tsuji. From Anorthoscope Perception to Dynamic Vision, Proceedings of IEEE Int. Conf. Robotics and Automation, 1990, pp.1154-1160.

(4) M. Barth and C. Barrows. A Fast Panoramic Imaging System and Intelligent Imaging Technique for Mobile Robots. Proceedings of IEEE/RSJ Int. Conf. Intelligent. Robots and Systems, 1996, no.2, pp.626-633.

(5) A. Krishnan and N. Ahuja. Panoramic Image Acquisition. Proceedings of IEEE Computer Vision and Pattern Recognition, 1996, pp.379-384.

(6) H. Ishiguro, M. Yamamoto and S. Tsuji. Analysis of Omni-Directional Views at Different Location. Proceedings of IEEE/RSJ Int. Conf. Intelligent Robots and Systems. 1990, pp.659-664.

(7) H. Ishiguro and S. Tsuji. Omni-Directional Stereo. IEEE Trans. Pattern Analysis and Machine Intelligence. Vol.14, no.2, 257-262, 1992.

(8) S. X. Godber, R. Petty, M. Robinson and P. Evans. Panoramic line-scan Imaging System for Teleoperator Control. Proceedings of SPIE Stereoscopic Displays and Virtual Reality Systems. 1994, Vol.2177, pp.247-257.

(9) R. Benosman, T. Maniere and J. Devars. Multidirectional Stereovision Sensor, Calibration and Scenes Reconstruction. Proceedings of Int. Conf. Pattern Recognition. 1996, pp.161-165.
(10) H. Ishiguro and S. Tsuji. Active Vision by Multiple Visual Agents. Proceedings of IEEE/RSJ Int. Conf. Intelligent Robots and Systems. 1992, pp.2195-2202.
(11) V. Nalwa. A True Omnidirectional Viewer. Tech. Report. Bell Laboratories, Holmdel, NJ. 1996.
(12) T. Kawanishi, K. Yamazawa, H. Iwasa, H. Takemura and N. Yokoya. Generation of High-resolution Stereo Panoramic Images by Omnidirectional Imaging Sensor Using Hexagonal Pyramidal Mirrors. Proceedings of Int. Conf. Pattern Recognition. 1998, pp.485-489.
(13) T. Morita, Y. Yasukawa, Y. Inamoto, T. Uchiyama and S. Kawakami. Measurement in Three Dimensions by Motion Stereo and Spherical Mapping. Proceedings of IEEE Computer Vision and Pattern Recognition. 1989, pp.422-434.
(14) Z. L. Cao, S. J. Oh and E. L. Hall. Dynamic Omnidirectional Vision for Mobile Robots. Proceedings of SPIE Intelligent Robots and Computer Vision. 1985, Vol.579, pp.405-414.
(15) Z. L. Cao, S. J. Oh and E. L. Hall. Dynamic Omnidirectional Vision for Mobile Robots, Journal of Robotic Systems 3(1), 5-17, 1986.
(16) J. J. Roning, Z. L. Cao and E. L. Hall. Color Target Recognition Using Omnidirectional Vision. Proceedings of SPIE Optics, Illumination, and Image Sensing for Machine Vision. 1987, Vol.728, pp.57-63.
(17) Z. L. Cao and E. L. Hall. Beacon Recognition in Omni-vision Guidance. Proceedings of Int. Conf. Optoelectronic Science and Engineering,. 1990, VO- 1230, pp.788-790.
(18) R.T. Elkins and E. L. Hall. Three-dimensional Line Following Using Omnidirectional Vision. Proceedings of SPIE Intelligent Robots and Computer Vision XIII: 3D Vision, Product Inspection, and Active Vision. 1994, Vol.2354, pp.130-144.
(19) B. O., Matthews, D. Perdue and E. L. Hall. Omnidirectional Vision Applications for Line Following. Proceedings of SPIE Intelligent Robots and Computer Vision XIV: Algorithms, Techniques, Active Vision, and Materials Handling. 1995, Vol.2588, pp.438-449.
(20) P. Greguss. The Tube Peeper : A New Concept in Endoscopy. Optics and Laser Technology. pp.41-45, 1985
(21) P. Greguss. PAL-optic based Instruments for Space Research and Robotics. Laser and Optoelektronik. 28, pp.43-49, 1996.
(22) B. Rossi. Optics. Addison-Wesley Publishing Co. Inc., 1962.
(23) Y. Yagi, and S. Kawato. Panorama Scene Analysis with Conic Projection. Proceedings of IEEE/RSJ Int. Workshop on Intelligent Robots and Systems. 1990, pp.181-187.
(24) Y. Yagi, S. Kawato and S. Tsuji. Real-time Omnidirectional Image Sensor (COPIS) for Vision-Guided Navigation. IEEE Trans. on Robotics and Automation. vol.10, no.1, 11-22 1994.
(25) J. Hong, X. Tan, B. Pinette, R. Weiss and E. M. Riseman. Image-based Navigation Using 360 Views. Proceedings of Image Understanding Workshop. 1990, pp.782-791.
(26) S. L. Bogner. An Introduction to Panaspheric Imaging. Proceedings of IEEE Int. Conf. Systems, Man and Cybernetics. 1995, 4, pp.3099-3106.
(27) J. Hong, X. Tan, B. Pinette, R. Weiss and E. M. Riseman. Image-based Homing. Proceedings of IEEE Int. Conf. Robotics and Automation. 1991, pp.620-625.

(28) W. D. Rees and W. Mich.. Panoramic Television Viewing System. 1970. US Patent 3505465.
(29) K. Yamazawa, Y. Yagii and M. Yachida. Omnidirectional Imaging with Hyperboloidal Projection. Proceedings of IEEE/RSJ Int. Conf. Intelligent Robots and Systems. 1993, no.2, pp.1029-1034.
(30) V. Peri, and S. K. Nayar. Omnidirectional Video System. Proceedings of U.S-Japan Graduate Student Forum in Robotics, 1996, pp.28-31.
(31) M. Yachida. Omnidirectional Sensing and Combined Multiple Sensing. Proceedings of IEEE&ATR Workshop on Computer Vision for Virtual Reality based Human Communications. 1998, pp.20-27.
(32) S. K. Nayar. Catadioptric Omnidirectional Camera. Proceedings of IEEE Conf. Computer Vision and Pattern Recognition. 1997, pp.482-488.
(33) S. Baker and S. K. Nayer. A Theory of Catadioptric Image Formation. Proceedings of Int Conf. Computer Vision. 1998
(34) S. Baker and S. K. Nayer. A Theory of Single-Viewpoint Catadioptric Image Formation. Int. J. Computer Vision. Vol.35, No.2, pp.1-22, 1999.
(35) J. S. Chahl, and M. V. Srinivasan. Reflective Surfaces for Panoramic Imaging. Applied Optics. 1997, Vol. 36, No. 31, pp.8275-8285.
(36) I. Powell. Panoramic Lens, US Patent 5473474. 1995.
(37) Rosendahi, G. R. and Dykes, W. V. : Lens System for Panoramic Imagery, US Patent 4395093 (1983).
(38) J. E. Davis, M. N. Todd, M. Ruda, T. W. Stuhlinger, and K. R. Castle. Optics Assembly for Observing a Panoramic Scene, US Patent 5627675, 1997.
(39) S. K. Nayar and V. Peri. Folded Catadioptric Cameras. Proceedings of IEEE Conf. on Computer Vision and Pattern Recognition. 1999, Vol.II, pp.217-223.
(40) Y. Yagi and M. Yachida. Development of A Tiny Omnidirectional Image Sensor. Proceedings of JSME Conf. on Robotics and Mechatronics. 1999, No.99-9, 2 A1-66-060, pp.1-2.
(41) Y. Yagi and M. Yachida. Development of a Tiny Omnidirectional Image Sensor. Proceedings of Asian Conf. on Computer Vision. 2000, pp.23-28.
(42) A. Bruckstein and T. Richardson. Omniview Cameras with Curved Surface Mirrors. Proceedings of IEEE workshop on Omnidirectional Vision. 2000, pp.79-84.
(43) A. Takeya, T. Kuroda, K. Nishiguchi and A. Ichikawa. Omnidirectional Vision System using Two Mirrors. Proceedings of SPIE. 1998, Vol.3430, pp.50-60.
(44) Y. Yagi,, H. Okumura and M. Yachida. Multiple Visual Sensing System for Mobile Robot. Proceedings of IEEE Int. Conf. Robotics and Automation. 1994, no.2, pp.1679-1684.
(45) T. Konparu, Y. Yagi and M. Yachida. Finding and Tracking a Person by Cooperation between an Omnidirectional Sensor Robot and a Binocular Vision Robot. Proceedings of Meeting on Image Recognition and Understanding. 1998, II, pp.7-12 (in Japanese).
(46) D. Southwell, A. Basu, M. Fiala and J. Reyda. Panoramic Stereo, Proceedings of Int. Conf. Pattern Recognition. 1996, pp.378-382.
(47) D. W. Murray. Recovering Range Using Virtual Multicamera Stereo. Computer Vision and Image Understanding. vol.61, no.2, pp.285-291, 1995.
(48) S. B. Kang and R. Szeliski. 3-D Scene Data Recovery Using Omnidirectional Multibaseline Stereo. Proceedings of Computer Vision and Pattern Recognition. 1996, pp.364-370.

(49) S. B. Kang and R. Szeliski. 3-D Scene Data Recovery Using Omnidirectional Multi-baseline Stereo. Journal of Computer Vision. pp.167-183, 1997.
(50) T. Maniere, R. Benesman, C. Gastanda. and J. Devars. Vision System Dedicated to Panoramic Three-dimensional Scene Reconstruction. J. of Electronic Imaging. Vol.7, No.3, pp.672-676, 1998.
(51) R. S. Petty, M. Robinson, and J. Evans. 3D Measurement Using Rotating Line-scan Sensors. Measurement Science and Technology. Vol.9, No.3, pp.339-346, 1998.
(52) Y. Yagi, Y. Nishizawa and M. Yachida. Map-based Navigation for a Mobile Robot with Omnidirectional Image Sensor COPIS. IEEE Trans. on Robotics and Automation. vol.11, no.5, pp.634-648, 1995.
(53) C. Pegard, and E. M. Mouaddib. Mobile Robot Using a Panoramic View. Proceedings of IEEE Int. Conf. Robotics and Automation. 1996, vol.1, pp.89-94.
(54) L. Delahoche, C. Pegard, E. M. Mouaddib and P. Vasseur. Incremental Map Building for Mobile Robot Navigation in an Indoor Environment. Proceedings of IEEE Int. Conf. Robotics and Automation. 1998, pp.2560-2565.
(55) K. Yamazawa, Y. Yagi and M. Yachida. Obstacle Detection with Omnidirectional Image Sensor HyperOmni Vision. Proceedings of IEEE Int. Conf. Robotics and Automation. 1995, pp.1062-1067.
(56) H. Ishiguro, T. Maeda, T. Miyashita and S. Tsuji. Strategy for acquiring an environmental model with panoramic sensing by a mobile robot, Proceedings of Int. Conf. Robotics and Automation. 1994, no.1, pp.724-729.
(57) H. Ishiguro, T. Maeda, T. Miyashita and S. Tsuji. Building environmental models of man-made environments by panoramic sensing. Advanced Robotics. Vol.9, no.4, 399-416, 1995.
(58) H. Ishiguro, T. Miyashita and S. Tsuji. T-Net for navigating a vision-guided robot in a real world. Proceedings of Int. Conf. Robotics and Automation. 1995, Vol.1 pp.1068-1073.
(59) Y. Yagi, K. Egami and M. Yachida. Map Generation of Multiple Image Sensing Sensor MISS under Unknown Mobile Robots. Proceedings of Int. Conf. Intelligent Robots and Systems. 1997, 2, pp.1024-1029.
(60) R.C. Nelson and J. Aloinomous. Finding motion parameters from spherical motion fields. Biological Cybernetics. 58, pp.261-273, 1988.
(61) J. Gluckman and S. K. Nayer. Ego-motion and Omnidirectional Cameras. IEEE Proceedings of Int. Conf. Computer Vision. 1998, pp.999-1005.
(62) Y. Yagi, W. Nishii, K. Yamazawa and M. Yachida. Stabilization for Mobile Robot by using Omnidirectional Optical Flow. Proceedings of Int. Conf. Intelligent Robots and Systems. 1996, pp.618-625.
(63) J. S. Chahl, and M. V. Srinivasan. Visual Computation of Egomotion Using an Image Interpolation Technique. Biological Cybernetics. 74, 405-411, 1996
(64) J. Y. Zheng, and S. Tsuji. Panoramic Representation for Route Recognition by a Mobile Robot. Int. Journal of Computer Vision. 9, 1, 55-76, 1992.
(65) J. Y. Zheng, and S. Tsuji. Generating Dynamic Projection Images for Science Representation and Understanding. Computer Vision and Image Understanding. Vol.72, No.3, pp.237-256, 1998.
(66) S. Li and S. Tsuji. Qualitative Representation of Scenes Along Route. Image and Vision Computing. Vol.17, No.9, pp.685-700, 1999.
(67) Y. Matsumoto, M. Inaba, and H. Inoue. Memory-based Navigation Using Omni-View Sequence. Proceedings of Int. Conf. Field and Service Robotics. 1997, pp.184-191.

(68) H. Ishiguro and S. Tsuji. Image-Based Memory of Environment. Proceedings of Int. Conf. Intelligent Robots and Systems. 1996, pp.634-639.
(69) N. Aihara, H. Iwasa, N. Yokoya and H. Takemura. Memory-based Self-localization Using Omnidirectional Images. Proceedings of Int. Conf. Pattern Recognition. 1998, pp.1799-1803.
(70) Y. Yagi, S. Fujimura and M. Yachida. Route Representation for Mobile Robot Navigation by Omnidirectional Route Panorama Fourier Transformation. Proceedings of IEEE Int. Conf. Robotics and Automation. 1998, pp.1250-1255.
(71) E. Bideaux and P. Baptiste. Omnidirectional Vision System : Image Processing and Data Extraction. Proceedings of Japan-France Congress Mechatronics. 1994, pp.697-700.
(72) Z. Zhu, H. Xi and G. Xu. Combining Rotation-invariance Images and Neural Networks for Road Scene Understanding. Proceedings of Int. Conf. Neural Network. 1996, vol.3, pp.1732-1737.
(73) S. W. Bang,, W. Yu and M. J. Chung. Sensor-based Local Homing Using Omnidirectional Range and Intensity Sensing System for Indoor Mobile Robot Navigation. Proceedings of Int. Conf. Intelligent Robots and Systems. 1995, vol.2, pp.542-548.
(74) S. C. Wei, Y. Yagi and M. Yachida. Building Local Floor Map by Use of Ultrasonic and Omnidirectional Vision Sensor. Proceedings of IEEE Int. Conf. Robotics and Automation. 1998, pp.2548-2553.
(75) C. Drocourt, L. Delahoche, C. Pegard and C. Cauchois,. Localization Method Based on Omnidirectional Stereoscopic Vision and Dead-reckoning. Proceedings of IEEE/RSJ Int. Conf. on Intelligent Robots and Systems. 1999, Vol.2, pp.960-965.
(76) C. Drocourt, L. Delahoche, C. Pegard and A. Clerentin. Mobile Robot Localization Based on an Omnidirectional Stereoscopic Vision Perception System. Proceedings of IEEE Int. Conf. on Robotics and Automation. 1999, Vol.2 pp.1329-1334.
(77) M. J. Barth and H. Ishiguro. Distributed Panoramic Sensing in Multiagent Robotics. Proceedings of Int. Conf. Multisensor Fusion and Integration for Intelligent Systems. 1994, pp.739-746.
(78) Y. Yagi,, Y. Lin, and M. Yachida. Detection of Unknown Moving Objects by Reciprocation of Observed Information between Mobile Robot. Proceedings of IEEE/RSJ Int. Conf. Intelligent Robots and Systems, vol.2, pp.996-1003, 1994.
(79) Y. Yagi, S. Izuhara, and M. Yachida. The Integration of an Environmental Map Observed by Multiple Mobile Robots with Omnidirectional Image Sensor COPIS. Proceedings of IEEE/RSJ Int. Conf. Intelligent Robots and Systems. 1996, vol.2, pp.640-647.
(80) M. Asada, S. Suzuki, Y. Takahashi, E. Uchibe, M. Nakamura, C. Mishima, H. Ishizuka, and T. Kato. TRACKIES: RoboCup-97 Middle-size League World Cochampion. AI Magazine. Vol.19, No.3, pp.71-78, 1998.
(81) A. Bonarini, P. Aliverti, and M. Lucioni. Omnidirectional Vision Sensor for Fast Tracking for Mobile Robots, Proceedings of IEEE Instrumentation and Measurement Technology Conference. 1999. Vol. 1, pp.151-155.
(82) A. Basu and D. Southwell. Omni-directional Sensors for Pipe Inspection. Proceedings of IEEE Int. Conf. Systems, Man and Cybernetics. 1995, 4, pp.3107-3112.
(83) Y. Yagi, K. Sato, K. Yamazawa and M. Yachida. Autonomous Guidance Robot System with Omnidirectional Image Sensor. Proceedings of Int. Conf. Quality Control by Artificial Vision. 1998, pp.385-390.

(84) S. E. Chen. QuickTime VR - An Image-Based Approach to Virtual Environment Navigation. Proceedings of ACM SIGGRAPH Conf. Computer Graphics. 1995, pp.29-38.

(85) L. A. Teodosio and M. Mills. Panoramic Overviews for Navigating Real-Rorld Scenes. Proceedings of ACM Int. Conf. Multimedia. 1993, pp.359-364.

(86) Y. Onoe, N. Yokoya, K. Yamazawa and H. Takemura. Visual Surveillance and Monitoring System Using an Omnidirectional Video Camera. Proceedings of IAPR Int. Conf. Pattern Recognition. 1998, vol.I, pp.588-592.

(87) Y. Onoe, K. Yamazawa, H. Takemura, and N. Yokoya. Telepresence by Real-Time View-Dependent Image Generation from Omnidirectional Video Streams, Computer Vision and Image Understanding, 71(2):154-165, 1998.

6

Computational Sensors : Vision VLSI

Kiyoharu Aizawa
University of Tokyo, Tokyo, Japan

SUMMARY

Computational sensors (smart sensors, a vision chip in other words) are a very small integrated system, in which processing and sensing are unified on a single VLSI chip. It is designed for a specific targeted application. Research activities of computational sensors are described in this chapter. There have been quite a few proposals and implementations in computational sensors. First, their approaches are summarized from several points of view, such as advantages vs. disadvantages, neural vs. functional, architecture, analog vs. digital, local vs. global processing, imaging vs. processing, and new processing paradigms. Then, several examples are introduced, which are spatial processings, temporal processings, A/D conversions, and programmable computational sensors.

1. INTRODUCTION

In biological vision, the retina is the imaging sensor. The retina not only detects image signals but also processes limited, low-level tasks, in order to enhance imaging performance and to preprocess tasks for later stages. Thus, given the precedent in biological vision, integration of processing and sensing is reasonable in order to augment sensing and processing performance.

Traditionally, a visual information processing system is comprised of discrete system modules; an image sensor (normally CCD camera), and an A/D conversion and a digital processing system. The combined system tends to be large and the image acquisition is limited only to that of NTSC standard spatio-temporal resolution. Development of VLSI technologies led to producing a new generation of image sensors that integrate sensing and processing on a single chip, we called a computational sensor, or a smart sensor, an intelligent sensor, or a vision chip.

Computational sensors are designed for specific targeted applications. Circuits for both photo-detection and processing co-exist on a single chip. Sensing and processing have interactions in a low level. The major advantages of computational sensors are "size" and "speed". Differing from the conventional visual processing systems that are comprised of module systems and follow the sense-read-processing paradigm, tight integration of sensing and processing realizes a very small integrated system on a chip. The speed is not restricted by the video rate any more.

The freedom from the restriction of a video standard has a very significant impact on real-time image processing. For example, let's take a look at motion estimation that requires heavy processing in the conventional framework in which the rate is limited by the video rate (30Hz), which is the communication bottleneck between the sensor and the processor. Because the complexity of motion estimation is determined by the dimensions of the search space, a faster frame rate can reduce the difficulty of the task. If the frame rate can be much faster than the video rate, the search area can be very small and the motion estimation can be much easier.

Fundamental device technology of image sensors is changing, too. So far, CCD sensors have been dominating, but CMOS active pixel sensors (CMOS sensors in short) are emerging as a competitive new image sensor technology. Although CMOS passive pixel sensors have existed, CMOS active pixel sensors are relatively new (38, 39). CMOS active pixel sensors have several transistors as an active amplifier in a pixel. Because of the CMOS process compatibility, the CMOS sensor is easier to integrate with additional circuits, and it can achieve functionality with lower power dissipation. The CMOS sensor was integrated with A/D conversion and timing control etc., which resulted in the on-chip-digital-camera. The major advantage of the CMOS sensor is the functionality.

Most computational sensors have been made by an ordinary CMOS process, and it can be viewed as a functional CMOS sensor. At present, CMOS sensors are being improved in imaging quality and products employing CMOS sensors have been recently introduced in the market. (For example, digital still cameras using CMOS sensors and CMOS sensor modules for cellular phones appeared.) In the near future, its functionality will be more and more important. Then, interest in the sensor device technology in CMOS sensors and interest in the processing technology for computational sensors will overlap. The sensor device technology and the processing applications share a common field of research and development as shown in Fig.1.

In this chapter, computational sensors are described from several points of view. Some examples are described, and then the chapter is concluded.

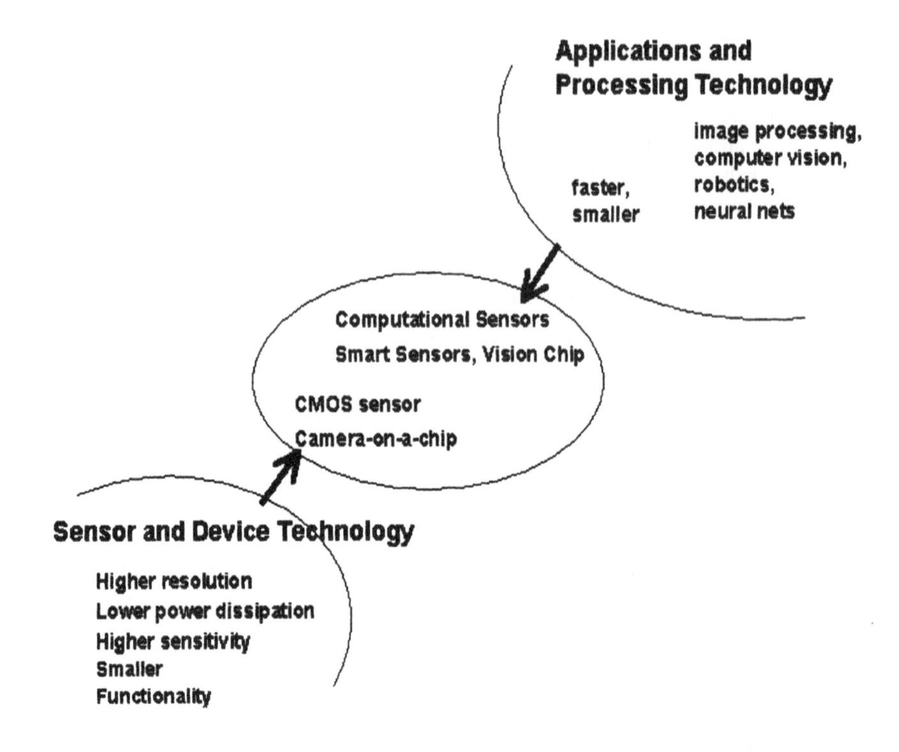

Figure 1 Computational sensors and related fields.

2. COMPUTATIONAL SENSORS

Computational sensors (or smart sensors, or vision chip) are a small integrated system, in which processing and sensing are unified on a VLSI chip by co-existing circuits of photo-detection and processing. Computational sensors are designed for specific target applications.

In the early days, in order to make a fast image processor, CCD technology was adopted for image processing. The convolution operation was implemented by CCD, which divided and shifted charge packets (6). A pixel parallel focal plane processor was even made under CCD technology, and a generic focal plane architecture based on CCD was discussed in (7). However, CCD is not easily integrated with CMOS computational circuits. Now, most of the computational sensors are made by an ordinary CMOS process.

Various computational sensors have been made, and in this section they are surveyed from several points of view.

2.1 Advantages vs. Disadvantages

When compared to the traditional module system based visual information processing systems, computational sensors have advantages and disadvantages, which are listed below.

- Advantages
    1. Size
    2. Speed
    3. Power dissipation
    4. Application specific image format
    5. Mutual interaction between sensing and processing.

As noted in the introduction, small size and fast speed are the most advantageous characteristics in a computational sensor. Compared to the conventional module processing system, power dissipation is also reduced by orders of magnitude. Image format is able to be optimized for the target application. An extreme example is a polar coordinate sensor (8, 9), in which the pixels are arranged in such a way that the density of the pixel changes from high to low from the center to the peripheral (Fig.2). Last but not the least, the mutual interaction between sensing and processing is the main advantage over the traditional systems. Computational sensors can handle analog data of the photo-detector so that it can feedback the results of the processing to parameters of sensing, or it can finish processing during integration.

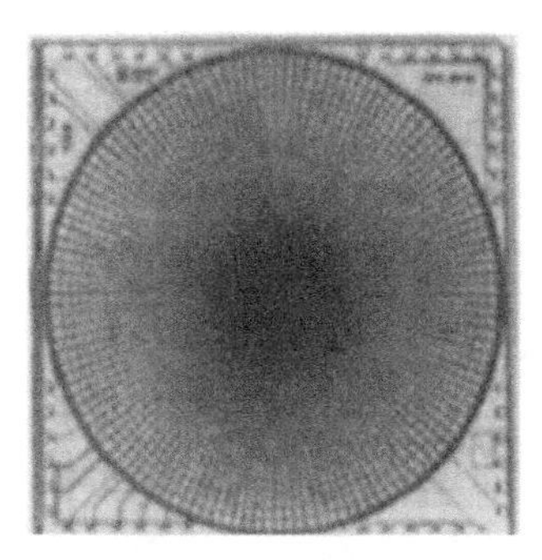

Figure 2 Polar coordinate sensor: photo detectors are arranged in concentric rings and with a size varying linearly by the distance from the geometric center (9).

- Disadvantages
    1. Precision
    2. Resolution and fill factor
    3. Design difficulty
    4. Programmability.

Computational sensors have disadvantages too, and they are listed above. The majority of computational sensors utilize analog processing circuits, and the precision should be chosen for the target applications. Because they use many analog circuits, the design is not easy. When each pixel contains processing circuits, the resolution and the fill factor is low, which results in poor imaging capability. Computational sensors are not for general purpose. Architecture as well as processing schemes must be optimized to reduce the effects of those disadvantages.

2.2 Neural vs. Functional

Computational sensors became widely known by the work of silicon retina (3), which computes spatial and temporal derivatives of an image projected onto its focal plane. In the human retina, communication between neurons is based on the analog and graded potentials (not spikes). Thus, the human retina is inherently well-modeled by electronic circuits and has been attempted to be modeled by a resistive network and logarithmic photo-detector (Fig.3). Some early vision tasks have been implemented. Using resistive networks, optical flow field computation and edge detection, etc. were implemented as a computational sensor (3, 4, 5, 25). These works were rather focused on modeling biological neural systems and suffered from low resolution and low precision. One of the interesting research activities developed from the neural network-based research is an artificial eye or visual prosthesis that could help a blind person obtain visual information (10).

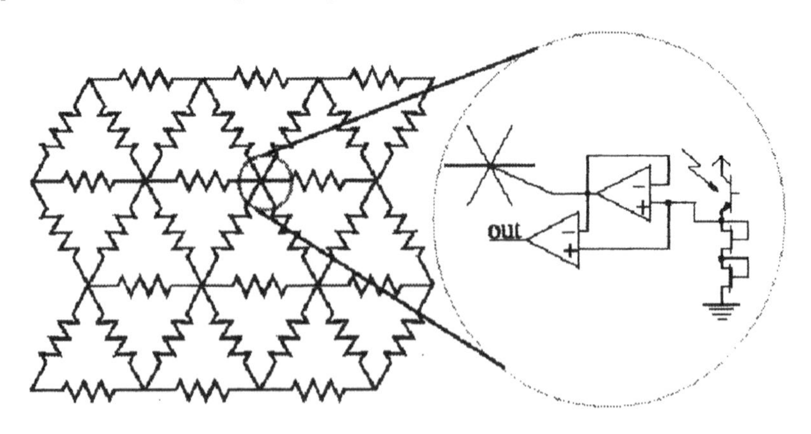

Figure 3 Silicon retina by resistive network (3, 1).

On the other hand, many computational sensors are designed for specific target tasks such as compression (18, 19, 22), enhancement (23,27), multi-resolution (14,43), moment computation (42), A/D conversion (33, 34, 35, 36, 37), range finding (40, 41,44), etc. Although some tasks are functionally similar to those of the human retina, they are designed by special analog processing schemes that are very different from that of biological neural networks. Computational sensors, which makes use of very limited resources (area, power), do not necessarily follow biological computational principles. In the rest of the chapter, we leave from the neural network-based approach and focus on the functional approach. In Section 3, some examples will be introduced.

2.3 Architectures: Pixel Parallel vs. Column Parallel

There are two approaches to parallel processing architecture on the focal plane, from the point of view of how to locate processing elements together with the photo-detector array. They are, as shown in Fig.4, (1) pixel parallel architecture which has a processing element in each pixel, and (2) column parallel architecture which shares a processing element for pixels in a column.

The two have advantages and disadvantages. Parallelism is higher in pixel parallel architecture, and the processing speed is fast, but it has lower fill factor, lower resolution, and higher power dissipation. Column parallel architecture has lower parallelism, but the photo-detector array can be almost the same as that of an ordinary CMOS sensor, and it does not sacrifice the fill factor and resolution. The architecture should be chosen for the target application.

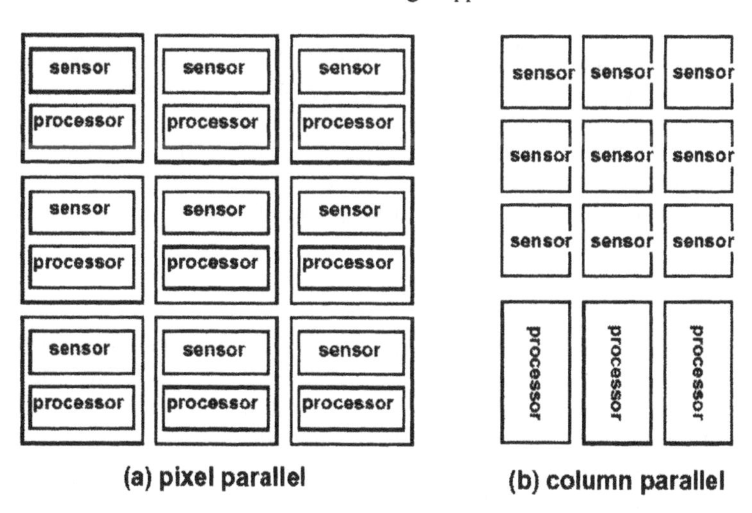

Figure 4 Pixel parallel and column parallel architectures for focal plane processing.

2.4 Analog vs. Digital

Computational sensors are designed for specific applications and most of them make use of analog processing which effectively use a physical principle of a device to perform a computation. Analog processing is fast and compact, and analog signal representation is efficient, but its precision is not high. Analog processing also enables processing during integration.

However, there have been attempts toward programmable computational sensors based on digital processing, which enables several simple image processing operations. Implementations using both column parallel architecture (28,29) and pixel parallel architecture (30, 31, 32,45) have been investigated, and are introduced in the next section.

2.5 Local Processing vs. Global Processing

Spatial processing on computational sensors requires a connection between pixels. Local connection is easy, but global connection is not. Then, most of the tasks that were achieved by computational sensors are spatially local processes such as edge detection, filtering, or temporal processing that does not need any inter-pixel connections. A global processing which cannot be achieved by spatially local processing (for example, finding the largest intensity in the image), need to handle the outputs of the local processing for the entire image. A few global processing tasks (such as a sorting operation) have been made on the sensor plane (27).

2.6 Imaging vs. Processing

The resource of the focal plane is very limited. There is generally a trade-off between "imaging capability" and "processing complexity." Processing complexity is related, to some extent, to the complexity of circuits. Higher complex processing needs higher complex circuitry, which results in a lower fill factor, lower resolution, higher power dissipation, and higher noise. It may lead to significant degradation of imaging capability. Imaging capability of computational sensors should be of major concern.

2.7 New Processing Paradigm

Traditional digital image processing systems are comprised of a camera, A/D conversion, and digital processing modules. It always uses normal NTSC video camera and their algorithms and performances for real time processing are rather optimized for the NTSC spatio-temporal resolution. Because of this fact, traditional image processing methods tend to get more complex in order to improve reliability. For example, in the case of motion estimation, motion of only 30 frames per second of NTSC temporal resolution easily results in large displacements between frames which makes motion estimation harder. However, for example, if the temporal resolution is much higher, motion between frames is much smaller and its estimation is much easier. Thus, advances in the sensor significantly change the difficulties of the processing tasks.

Computational sensors also enable mutual interaction between the photodetectors and the processors. The processing results can be fed back to photodetectors to change imaging conditions. The processing can also be done during integration. This is not possible in traditional digital processing systems.

3. EXAMPLES FOR COMPUTATIONAL SENSORS

Recent examples of computational sensors are introduced in this section. They were chosen from the categories of (1) spatial processing, (2) temporal processing, (3) programmable computational sensors and (4) A/D conversion on the focal plane. (1) and (2) have features distinctive of analog processing on the sensor focal plane.

3.1 Spatial Processing

3.1.1 Analog filtering by variable sensitivity photo-detector (VSPD) cells

Variable sensitivity photo-detector (VSPD) cell was developed using special GaAs photo-detector pixel (11). The interesting feature is that the output of the pixel depends on a sensitivity control voltage both in sign and magnitude. The control line is shared by pixels in a column. Thus, 1D analog filtering is easily achieved by applying the control signals to the pixels in a column and summing their outputs in the current mode on an output line for the pixels in the column. A 128x128 array was prototyped. The filtering operation was extended 2 dimensionally by adding a multiple output in a 1-D memory array (12) (Fig.5).

A n-MOS version of VSPD array was also developed by the same group (13). The n-MOS VSPD can control only the sign of the output of the pixel. Each pixel contains a PD and current direction conversion circuit of several transistors. A 256x256 array with 35x26μm^2 pixels was developed by a 2μm n-MOS process.

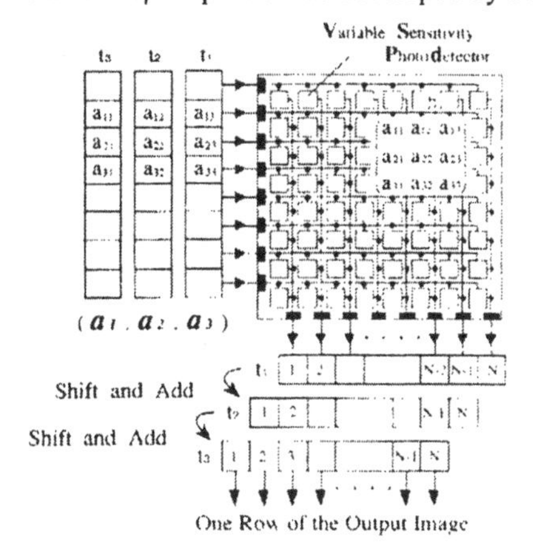

Figure 5 Diagram of analog processing of VSPD pixel array (12).

3.1.2 Multiresolution processing on a focal plane

For a variety of processing tasks, it is desirable to have image data available at varying resolutions to increase processing speed and efficiency. Multiresolution is a standard image processing methodology and it is usually generated by a software-based image pyramid approach.

A multiresolution CMOS sensor was developed (14). It is programmable to average and read out any size n x n block kernel region of pixels. The architecture is shown in Fig.6. The prototyped sensor has a 128x128 pixel array. At the bottom of each column, it has a network of switched capacitors to store signal levels of pixels. The column circuitry contains additional capacitors to perform averaging on any size kernel. Resolution is determined by the kernel. X-Y addressability of the sensor enables zoom into areas of interest.

A multiresolution CMOS sensor with local control of 2D moving average was also developed (46). Multiple shift registers were utilized to locally control the size of the moving average, and the size of the kernel can be changed up to 7x7 pixels.

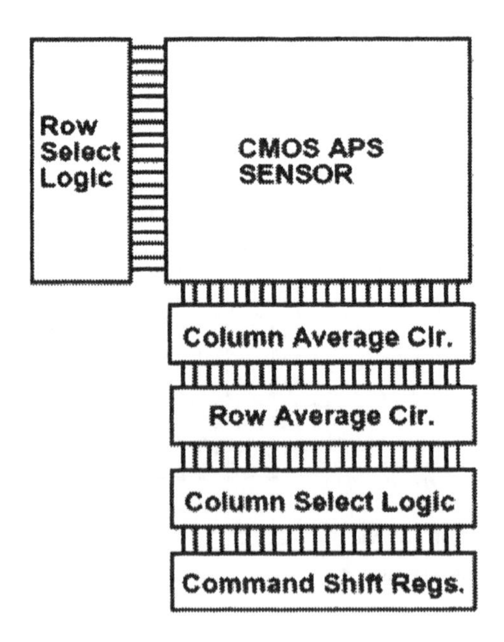

Figure 6 Architecture of a multiresolution CMOS sensor.

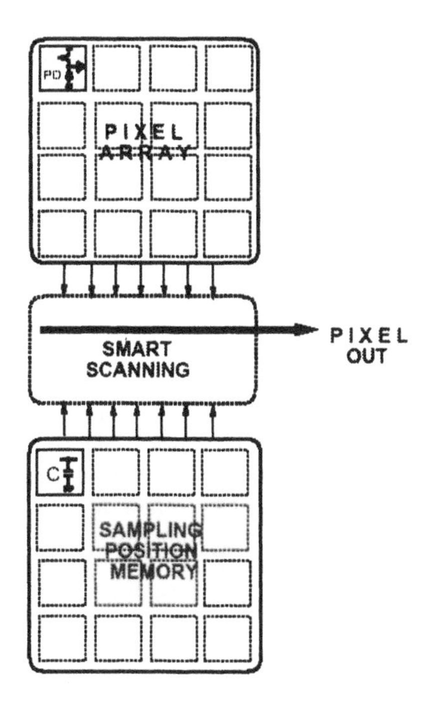

Figure 7 Architecture of a variable spatially sampling sensor.

3.1.3 Spatially variant sampling sensor

Biological vision has spatially variant sampling structure, which has high resolution in the fovea and sparse sampling in the periphery. Because of this structure, the region of interest is densely analyzed while sparsely keeping information outside of the region. It is desirable to have spatially variant sampling on the focal plane. It significantly reduces data and speeds up the entire processing.

One spatially variant sampling sensor is the polar coordinate sensor in Fig.2, which has a fixed sampling structure. A spatially variant sampling sensor, in which sampling positions are programmable, was developed (15). The architecture of the sensor is shown in Fig.7. It has a CMOS pixel array and a memory array, where each memory element corresponds to each pixel and contains a binary control data to determine the corresponding pixel to be read out. By using a smart scanning shift register (18), selected pixels are read out in a compact sequence. Contrary to the random access CMOS sensor (17), the spatially variant sampling sensor reads out at high speed without the need of a pixel address for every access. Access to the memory array is separated, and rewriting the memory dynamically changes the sampling positions. A prototype of 64 x 64 pixels was developed.

A different approach was taken in (16), where pixels are scanned by quad-tree, and one quad-tree code determines a block of pixels to be read out. The sensor was applied to an active range finder (47).

3.2 Temporal Processing

Temporal processing can be very different between the computational sensor and the traditional digital processing system. Three examples are described in this section.

3.2.1 Compression

One inherent difficulty of very fast image processing is transferring very high bandwidth image signals out of the sensor. Then, compression is one of the most desirable processing tasks on the focal plane. Compression integrated on the focal plane overcomes the problem of high bandwidth, that is, the fundamental limitation to the feasibility of high pixel-rate imaging such as high frame rate imaging.

Compression sensors integrated with analog compression using conditional replenishment have been developed (18, 19, 20, 21). Figure 8 illustrates the scheme of conditional replenishment. The current pixel value is compared to the last replenished value stored in the memory, and the value and address of an activated pixel, for which the magnitude of the difference is greater than the threshold, are extracted and output from the sensor. The value of the memory of the activated pixel is replenished by the current input value, and that of a non-activated pixel is kept unchanged. The conditional replenishment algorithm is not complex and lends itself easily to highly parallel processing. This compression scheme becomes more efficient at higher frame rates because motion is less at higher frame rates.

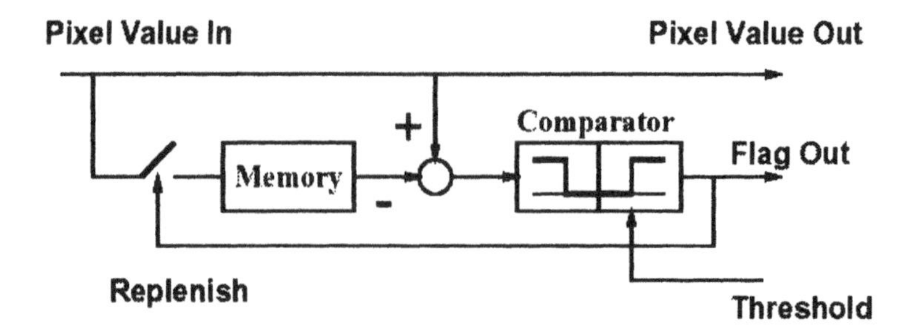

Figure 8 Compression scheme of conditional replenishment.

Compression sensors of both pixel parallel and column parallel architecture were developed. As shown in Fig.9, in the pixel parallel approach (18, 19), each pixel contains a photo-diode (PD), a comparator and a memory. Rate control for the entire frame is also implemented on the focal plane in the pixel parallel prototype. The drawback is fill factor and power dissipation. The first prototype of pixel parallel architecture, under a 2 μm CMOS process, has 32 x 32 pixels of 160 x 160 μm^2 pixels, and the fill factor is 1.9 %.

In order to improve imaging performance, column parallel architecture was also developed (20, 21). Column parallel architecture has PD array, memory array, and comparators shared by pixels in a column. Then, pixels in a row are processed in parallel. The PD array is an almost ordinary CMOS sensor and the fill factor is very much improved (38.5 % in the 32x32 pixel prototype under a 1 μm CMOS process). Because each comparator is shared by the pixels in a column, the total number of comparators is reduced and power dissipation is reduced. The operation speed is more than 15,000 frames per sec even in the column parallel prototype. The resolution of the column parallel prototype was recently improved to 128x128 pixels (21).

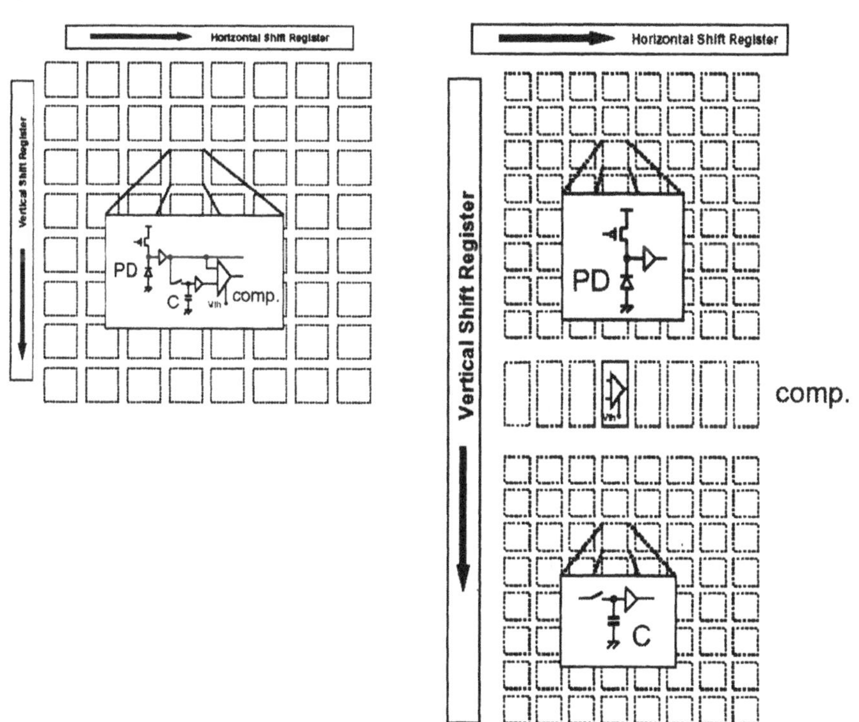

Figure 9 Illustrations of compression sensors of pixel parallel architecture (left) and column parallel architecture (right).

As for compression, it is notable that a compression sensor with analog DCT transform on the focal plane was also developed (22).

3.2.2 Adaptive integration sensor: enhancement and wide dynamic range

A computational sensor has been investigated, each pixel of which adapts its integration time to motion and light in such a way that the integration is finished if motion is detected or pixel intensity is saturated (23). The pixel value is normalized after it is read out. Then, moving areas of pixels have shorter integration time with little motion blur, brighter areas have shorter integration without saturation, and darker areas have longer integration with better Signal-to-Noise-Ratio. The pixel-wise adaptivity to motion and light improves intrascene temporal resolution and intrascene dynamic range. The processing scheme is illustrated in Fig.10. A prototype of 32 x 32 pixels was developed, which uses column parallel processing architecture. In the second prototype, the resolution was improved to 128x128 pixels.

3.2.3 Motion detection and estimation

A motion detection scheme is rather different in a computational sensor compared to an ordinary digital processing scheme (24). Motion detection on the sensor focal plane ranges between change detection in a pixel, 1D motion direction

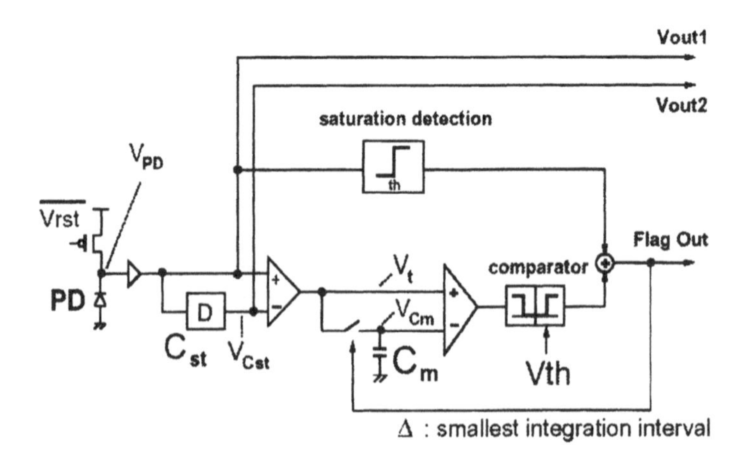

Figure 10 Processing scheme for each pixel of an adaptive integration sensor, in which integration time is adapted to motion and light

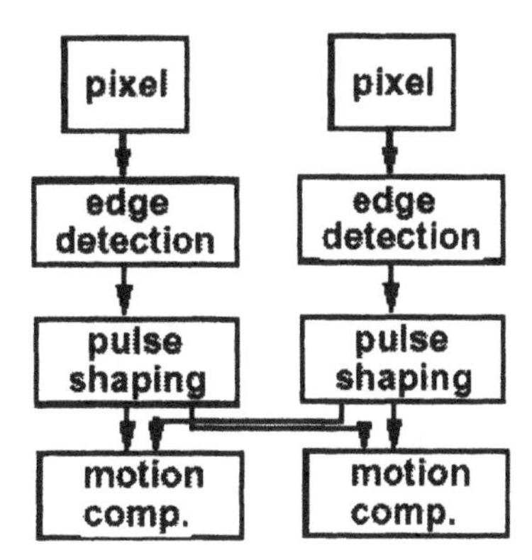

Figure 11 Motion computation using correlation between adjacent pixels.

detection, 1D motion vector (direction and magnitude) detection, and 2D motion vector detection. Because of the limitation of spatial processing, a typical approach to the motion detection on the focal plane uses the correlation between only adjacent pixels, as illustrated in Fig.11. Temporal edge in each pixel or spatial edge is detected and the motion of the edge is measured using correlated information between adjacent pixels. For example, the time delay between the pulses is measured. In most cases, the motion detection of the computational sensor is limited to one dimensional case. A review about motion processing is found in (24).

Differing from the above approach, an optic flow scheme (3, 25) and block matching scheme (26) have been also investigated. In the former, the motion field is computed by the spatial gradient technique. In (25), formulation of optic flow is simplified and 2D motion vectors are computed by pixel parallel processing. Block matching motion estimation was also attempted in such a way that edges are detected first, and block matching of edges is performed, at very fast frame rates, in a search-area of ±1 pixels in both x and y directions. Matching is done in the column parallel way (26).

3.2.4 Sorting Sensor

A computational sensor that sorts pixels by the magnitudes of their intensities and produces cumulative histogram was developed (27). Finding the order is one of the global operations that is not easy for computational sensors. In the sorting sensor shown in Fig.12, each pixel contains an inverter, and intensity of pixels is observed by their response time. Temporal processing is efficiently applied to the task. The principle is as follows: when the inputs have different intensities, the responses are separated in time. A global processor (a single operational amplifier) surveys the pixel array, aggregates triggers of pixels, and the voltage provided by the global processor to all pixels changes from high to low linearly to the number of accumulated triggered pixels. Each pixel has a memory that records the voltage of the global processor at the instance when the pixel is triggered. Then, the sorting order is in the memory of each pixel.

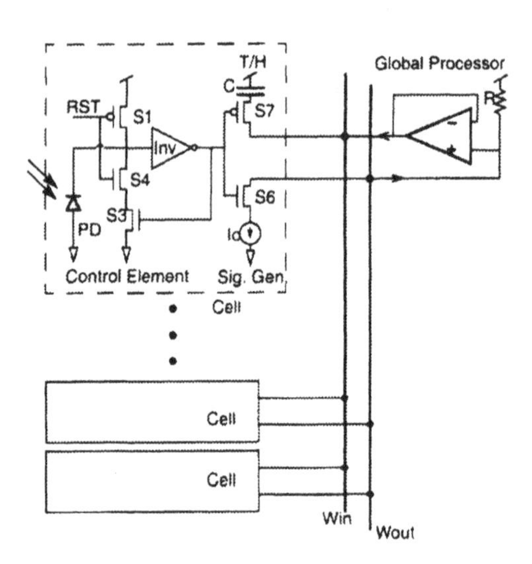

Figure 12 Architecture of a sorting sensor (27).

3.3 Programmable Computational Sensors

There have been attempts toward programmable computational sensors based on digital processing, which enables several simple image processing operations. Implementations using both column parallel architecture (28, 29) and pixel parallel architecture (30, 31, 32) have been investigated.

The column parallel architecture (28) has pixel array, and A/D conversion, register, ALU, an memory for each column, which is shown in Fig.13. Following this architecture, IVP developed and commercialized 512x512 pixel array with 8-bit A/D conversion (29). This vision chip, for example, executes binary edge detection in 0.5 msec/frame and gray scale edge detection in 10msec/frame.

The pixel parallel digital approach was also investigated. A simple 2D boolean operation was achieved by pixel parallel fashion in (30); the prototype contains 65 x 76 pixels with 1-bit pixel level digitization by thresholding, and executes, for example, edge detection in 5 μsec. A digital processing scheme based on efficient A/D conversion (named near sensor image processing, NSIP) was developed, in which the pixel intensity is represented by the time that it takes for the PD

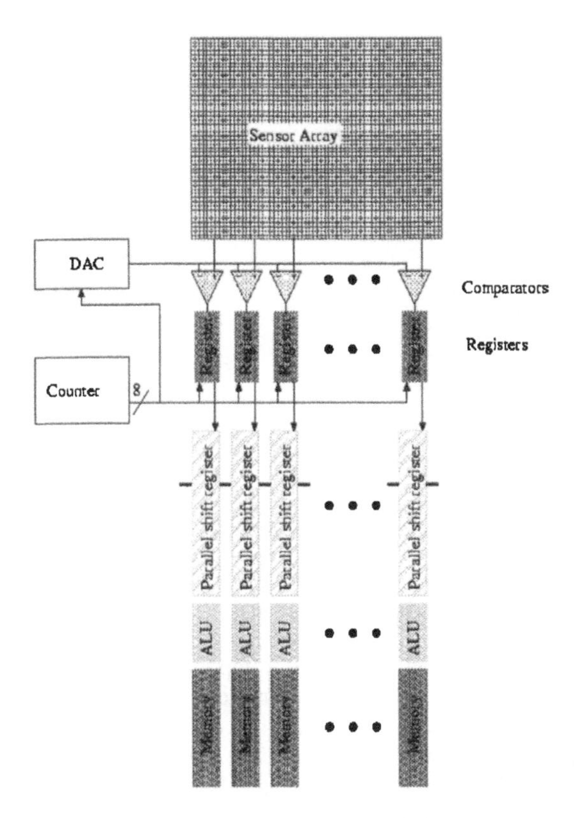

Figure 13 Architecture of a digital programmable computational sensor based on column parallel architecture (28,1).

voltage from the precharged voltage to the reference voltage (31). The prototype of 32x32 pixel arrays with a processing circuit and A/D conversion for each pixel were implemented. The most straightforward approach in the digital-based pixel parallel scheme is (32), in which each pixel contains A/D conversion, ALU, and local memory as shown in Fig.14. The processing speed is very fast, for example, edge detection is done in 1 μsec. The prototype of 64x64 pixel arrays was recently developed. As mentioned in the introduction, the pixel parallel approach has disadvantages in fill factor and resolution. The fill factors for the above three are very small.

Apart from the specific computational sensor, CMOS sensor which is directed toward "camera-on-a-chip" also needs A/D conversion, timing control, gain control, white balance, interfaces etc. integrated together with pixel array.

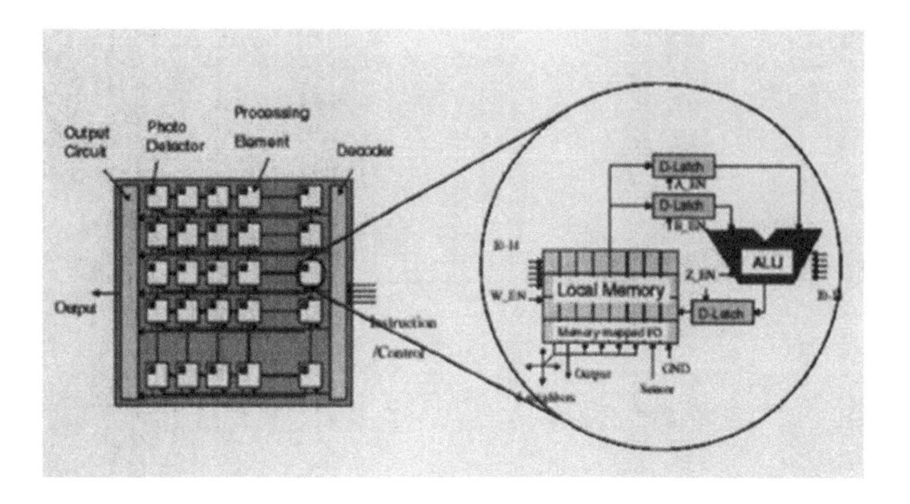

Figure 14 Architecture of computational sensor based on pixel parallel digital approach (32)

3.4 A/D Conversion (ADC)

A/D conversion (ADC) on focal plane is important especially for the "camera-on-a-chip" direction, which requires higher precision for imaging quality. Various approaches were proposed, and parallelism differs among them. Single ADC integration (no parallelism), column parallel ADC, and pixel parallel ADC have been investigated. Quite a few works have been made on the focal plane ADC, and in this section early works among them are introduced.

In the case of single ADC integration, a color CMOS sensors of 352x288 (33) and 306x244 pixels (34) integrated with flash ADC were developed. In the latter prototype, smoothing, edge detection, gain control, and γ correction were carried out for the digitized image by digital processing, and finally DAC was performed, and analog video signal was output from the chip.

In the case of the column parallel ADC integration, in which ADC can be much slower compared to the single ADC approach, single slope ADCs (28, 29), as shown in Figure 13, and successive approximation ADC (35) have been implemented on the focal plane. A commercial product of a 512 x 384 pixels color CMOS sensor integrated with 8-bit column parallel ADC was in market (36). Pixel parallel ADC was also investigated (37), in which Σ-Δ ADC was contained in each pixel. Multi-channel bit serial ADC was developed for pixel-level ADC (48).

4. CONCLUSION

Computational sensors are a very small integrated system, in which processing and sensing are unified on a single VLSI chip. It is designed for a specific targeted application. Research activities of computational sensors are summarized in this chapter, along with examples of computational sensors of spatial processing, temporal processing, programmable computational sensors, and A/D conversion.

Integration of sensing and processing (or a part of processing) on a single chip is very challenging. The integrated system is quite different from the traditional module system which is limited to NTSC spatio-temporal resolution. The processing paradigm can be quite different from the digital processing framework.

REFERENCES

(1) A.Moini, Vision Chips, Kluwer Academic, 1999.

(2) C.Koch and H.Li, Vision Chips, IEEE Computer Society Press, 1995.

(3) C.Mead, Analog VLSI and neural systems, Addison-Wisley, 1989.

(4) C.Koch, Implementing early vision algorithms in analog hardware, SPIE vol. 1473, Visual Information Processing From Neurons to Chips, 1991.

(5) S.Kameda, A.Honda, T.Yagi, A one-dimensional analog vision chip system with adaptive mechanisms, Proc. Scientific Research on Priority Areas, Ultimate Integration of Intelligence of Silicon Electronic Systems, pp.245-252, Mar. 1998.

(6) G.R.Nudd et al, A charge coupled device image processor for smart sensor applications, Image Understanding Systems and Industrial Applications, SPIE Proc., Vol.155, pp. 15-22, 1978.

(7) E.R.Fossum, Architectures for focal image processing, Optical Engineering, Vol.28, pp.866-871, 1989.

(8) J. van der Spiegel et al., A foveated retina-like sensor using CCD technology, Analog VLSI implementation of neural systems, chap.8 pp.189-212, Kluwer Academic, 1989.

(9) F.Pardo, B.Dierickx, D.Scheffer, CMOS Foveated image sensor: Signal scaling and small geometry effects, IEEE Trans. ED, Vol.44, No.10, pp.1731-1737, Oct. 1997.

(10) IEEE Spectrum, Towards Artificial Eye, May 1996.

(11) K.Kyuma et al., Artificial retinas - Fast, versatile image processors, Nature, vol.372, pp.197-198, 1994.

(12) E.Funatsu et al., An artificial retina chip made of a 128x128 pn-np variable-sensitivity photodetector array, IEEE Photonics Technology Letters, Vol.7, No.2, pp.188-190, 1995.

(13) E.Funatsu et al., An artificial retinal chip with current-mode focal plane image processing functions, IEEE Trans. ED, vol.44, No.10, pp.1977-1982, Oct. 1997.

(14) S.E.Kemeny, et al. Multiresolition image sensor, IEEE Trans. CSVT, Vol.7, No.4, pp.575-583, Aug. 1997.

(15) Y.Ohtsuka, T.Hamamoto, K.Aizawa, A new image sensor with space variant sampling control on a focal plane, IEICE Trans. IEICE Trans. Information and Systems, Vol.E83-D No.7 pp.1331-1337 July 2000.

(16) J.Akita, K.Asada, An image scanning method with selective activation of tree structure, IEICE Trans. on Electronics, Vol.E80-C, No.7, pp.956-961, 1997.

(17) O.Yadid-Pecht, R.Ginosar, A Random Access Photodiode Array for Intelligent Image Capture, IEEE Trans. ED, pp.1772-1780, Aug. 1991.
(18) K.Aizawa et al., On sensor compression, IEEE Workshop on CCD and Advanced Image Sensor, Apr. 1995.
(19) K.Aizawa, et al., Computational image sensor for on sensor compression, IEEE Trans. ED, Vol.44, No.10, pp.1724-1730, Oct. 1997.
(20) K.Aizawa, T.Hamamoto, Y.Ohtsuka, M.Hatori, M.Abe, Pixel parallel and column parallel architectures and their implementation of on sensor image compression, IEEE Workshop on CCD and Advanced Image Sensor, June 1997.
(21) T. Hamamoto, Y. Otuska, K.Aizawa, 128x128 pixels image sensor for on-sensor compression, IEEE Int. Conf. Image Processing (ICIP98), Oct. 1998.
(22) S.Kawahito et al., A CMOS image sensor with analog two dimensional DCT based compression circuits for one-chip cameras, IEEE J.SSC, Vol.32, No.12, pp.2030-2041, Dec. 1997.
(23) T.Hamamoto, K.Aizawa, A computational image sensor with adaptive pixel-based integration time, IEEE Journal of Solid State Circuits, Vol.36 No.4 pp.580-585 April 2001.
(24) R. Sarpeshkar et al., Analog VLSI architectures for motion processing: From fundamental limits to system applications, Proc. IEEE, Vol.84, No.7, pp.969-987, July 1996.
(25) R.A.Deutschmann and C.Koch, Compact real-time 2-D gradient-based analog VLSI motion sensor, SPIE Vol.3410, Int. Conf. Advanced Focal Plane Arrays and Electronic Cameras (AFPAEC), pp.98-108, May, 1998.
(26) Z.Li, K.Aizawa, M.Hatori, Motion vector estimation on focal plane by block matching, SPIE Vol.3410, Int. Conf. Advanced Focal Plane Arrays and Electronic Cameras (AFPAEC), pp.98-108, May, 1998.
(27) V. Brajovic and T. Kanade , A VLSI Sorting Image Sensor: Global Massively Parallel Intensity-to-Time Processing for Low-Latency, Adaptive Vision, IEEE Transactions on Robotics and Automation, Vol. 15, No. 1, February, 1999.
(28) K.Chen et al., PASIC: A processor - A/D converter - sensor integrated circuits, IEEE ISCAS90, Vol.3, pp.1705-1708, 1990.
(29) IVP product notes (smart vision sensor).
(30) T.M.Bernard, Y.Zavidovique, F.J.Devos, A programmable artificial retina, IEEE J. SSC, Vol.28, No.7, pp.789-798, July 1993.
(31) VLSI implementation of a Focal Plane Image Processor - A realization of the near sensor image processing concept, IEEE Trans. VLSI systems, Vol.4, No.3, Sept. 1996.
(32) T.Komuro, S. Suzuki, I. Ishii, M. Ishikawa, Design of massively parallel vision chip using general purpose processing elements, IEICE Trans., Vol.J81-D-I, No.2, pp.70-76, 1998.
(33) M.Loinaz et al., A 200mW 3.3V CMOS color camera IC producing 352x288 24b video at 30 frames/s, ISSCC Dig. Tech. Papers, pp.168-169, Feb. 1998.
(34) S.Smith et al., A single chip 306x244 pixel CMOS NTSC video camera, ISSCC Dig. Tech. Papers, pp.170-171, Feb. 1998.
(35) Z.Zhou, B.Pain, E.Fossum, CMOS active pixel sensor with on-chip successive approximation analog-to-digital converter, IEEE Trans. ED, Vol.44, No.10, pp. 1759-1763, 1997.
(36) http://www.micron.com/imaging
(37) B.Fowler, et al., CMOS FPA with multiplexed pixel level ADC, IEEE Int. Workshop on CCD and Advanced image sensors, 1995.

(38) F.Ando et al., A 250,000 pixel image sensor with FET amplification at each pixel for high speed television cameras, ISSCC Dig. Tch. Papers, 1990, pp.212-213.
(39) E.R.Fossum, CMOS image sensors: Electronic camera on a chip, IEEE Trans. ED, Vol.44, No.10, pp.1689-1698, Oct. 1997.
(40) A. Gruss, L.R.Carely, T.Kanade, Integrated sensor and range finding analog signal processor, IEEE J. SSC, vol.26, pp.184-190, 1991.
(41) A.Yokoyama, K.Sato, T.Ashigaya, S.Inokuchi, Realtime range imaging using an adjustment-free photoVLSI: Silicon range finder, IEICE Trans.D-II, Vol.J79-D-II, No.9, pp.1492-1500, Sep.1996.
(42) J.L.Wyatt, Jr., D.L.Standley, W.Yang, The MIT vision chip project: analog VLSI system for fast image acquisition and early vision processing, IEEE Int. Conf. Robotics and Automation, pp. 1330-1335, Apr.1991.
(43) Y.Ohtsuka, I.Ohta, K.Aizawa, Programmable spatially variant multiresolution readout on a sensor focal plane, IEEE Int. Symp. Circuit and Systems (ISCAS2001), pp.III 632-635, 2001.
(44) Y.Oike, M. Ikeda, and K. Asada, A CMOS Image Sensor for High-Speed Active Range Finding Using Column-Parallel Time-Domain ADC and Position Encoder, IEEE Trans. on Electron Devices, Vol. 50, No. 1, Jan. 2003.
(45) A. Namiki, T. Komuro, M. Ishikawa, High Speed Sensory-Motor Fusion Based on Dynamics, Proc. of the IEEE, Vol.90, No.7, pp.1178-1187 July 2002.
(46) Y.Ohtsuka, I.Ohta, K.Aizawa, Programmable spatially variant multiresolution readout capability on a sensor focal plane,IEEE Int. Symp. Circuit and Systems (ISCAS2001), pp.III 632-635, May 2001.
(47) Y. Oike, M. Ikeda, K. Asada, A CMOS Image Sensor for High-Speed Active Range Finding Using Column-Parallel Time-Domain ADC and Position Encoder, *IEEE* Trans. on Electron Devices, Vol. 50, No. 1, Jan. 2003.
(48) D. Yang, B. Fowler, A. El Gamal, and H. Tian, A 640x512 CMOS Image Sensor with Ultrawide Dynamic Range Floating-Point Pixel-Level ADC, In IEEE Journal of Solid State Circuits, Vol.34, No.12, Pages 1821-1834, Dec. 1999.

7

Processing and Recognition of Face Images and Its Applications

Masahide Kaneko

The University of Electro-Communications, Chofu, Tokyo, Japan

Osamu Hasegawa

Tokyo Institute of Technology, Nagatsuta, Yokohama, and

National Institute of Advanced Industrial Science and Technology, Tsukuba, Ibaraki, and

PRESTO, Japan Science and Technology Corp. (JST), Kawaguchi, Saitama, Japan

SUMMARY

Human faces convey various information, including that which is specific to each individual person and that which reflects some parts of the mutual communication among persons. Information exhibited by a face is what is called "non-verbal information" and usually verbal media cannot easily describe such information appropriately. Recently, detailed studies on the processing of face images by a computer were carried out in the engineering field for applications to communication media and human computer interaction as well as automatic identification of human faces. Two main technical topics are the recognition of human faces and the synthesis of face images. The objective of the former is to enable a computer to detect and identify humans and further to recognize their facial expressions, while that of the latter is to provide a natural and impressive user interface on a computer in the form of a "face." These studies have also been found to be useful in various non-engineering fields related to a face, such as psychology, anthropology, cosmetology, and dentistry. Most of the studies in these different fields have been carried out independently before, although all of them deal with a "face." Now in virtue of the progress in the above engineering technologies, common study tools and databases for facial information have become available.

On the basis of these backgrounds, this chapter surveys recent research trends in the processing of face images by a computer and its typical applications. Firstly, the various characteristics of faces are considered. Secondly, recent research activities in the recognition and synthesis of face images are outlined. Thirdly, the applications of digital processing methods of facial information are discussed from several standpoints: intelligent image coding, media handling, human computer interaction, facial impression, and psychological and medical applications. The common tools and databases used in the studies of processing of face images and some related topics are also described.

1. INTRODUCTION

A "face" is very familiar to us, and at the same time it plays very important roles in various aspects of our daily lives. The importance of a "face" exists in the fact that it conveys a wide variety of information relating to each person. Information expressed by a face is generally classified into two categories. One is that of an individual person, that is, information utilized to identify an individual person from others. For example, who he/she is, gender, age, and so on. The other is that reflects the physical and psychological conditions and emotions in each person. Each type of information appears on a face in different manners and works as a unique message differing from verbal messages in the communication with other persons. Logical messages are transmitted by languages. On the other hand, a face can express a lot of messages that cannot be expressed appropriately by words and this helps to make our communication richer and often more complicated.

From the early stage of digital image processing technology, a face image was picked up as an interesting subject to process. Studies have been mainly carried out from the viewpoint of the automatic recognition of face. Another important approach to handle face images was activated in the middle of the 1980's in Japan and has spread to other countries. This approach was called "intelligent image coding" or "model-based image coding." It was originally studied for the very low bit-rate coding of moving face images, however, its unique idea has influenced various fields related to faces and has made studies in these fields more active. Technologies developed in the study of intelligent image coding can be applied not only to the compression of image data, but also to the improvement of human interface between a user and a computer, synthesis of an anthropomorphic agent, psychology, and so on.

Interests to realize the novel human interface have also increased very much recently in the field of human-computer interaction. The target is to materialize a new type of interface that understands the user's face and gestures through input images and furthermore possesses its own face. This trend has come from the progress in multimedia technologies, rapid and remarkable increase in computing power, and the strong desire to provide a more natural and human-friendly computer interface.

A "face" is an important research subject not only in the engineering field, but studies have also been carried out in numerous other fields so far, such as psychology, anthropology, cosmetology, dentistry (especially orthodontics), and so on. These studies have advanced individually until now, and engineering approaches were not necessarily utilized. However in recent days, results of engineering studies on face images, especially the results obtained in the field of intelligent image coding, have stimulated studies in other non-engineering fields.

On the basis of these backgrounds this chapter surveys recent research activities in the digital processing of face images and its applications. First, in Section **2** characteristics of faces and possibilities to deal with them in the engineering field are briefly considered. In Sections **3** and **4**, research trends in two major component technologies are introduced, that is, recognition of face images and synthesis of face images with natural expressions. In Section **5** applications of processing of facial information are discussed from several standpoints. In Section **6** common tools and databases that are useful for the research of facial information processing and for the development of related systems are introduced. In Section **7** academic activities related to a "face" are introduced. Lastly, Section **8** summarizes this chapter.

2. CHARACTERISTICS OF FACES AND APPLICATIONS IN THE ENGINEERING FIELD

Before describing recent research trends, let us briefly consider general characteristics of faces and discuss the possibilities to utilize them in actual engineering applications.

2.1 Expressiveness

A face is one of the main output channels by which a person expresses his/her messages in a non-verbal manner, and its utility depends on an abundance of expressiveness in line of sight, facial expressions, and movement of the head. It is said that there are more than 20 expression muscles in a face and more than 60 kinds of expressions that one can intentionally manifest. However, it has been found that most messages that expressions show in a conversation are included in their dynamic change process and timing of expression and utterance. On the other hand, it has been reported that the line of sight can express a human will and feelings together with expressions and at the same time it can express a transfer of the priority for utterance during dialogue and the spatial location of points of attention. This facial expressiveness is innate to some extent in its subtle and sophisticated control, but it is said that humans learn and grasp the particular facial expressive rules of their society during their growing process.

In the future, by utilizing the above-mentioned characteristics in an engineering form, human-computer interaction will be improved much more and further computers will help to support face-to-face communication among humans. In order to realize them, it is indispensable to establish methodologies such as a computational estimation method of the partner's intentions based on both the contents of utterance and the timing between utterances and facial expressions. Those will be carried out based on the technical background like real-time measurement of three-dimensional (3-D) facial motions and line of sight and the knowledge of the social rules of facial expressions.

2.2 Individuality and Similarity

Each face possesses individuality because facial components, such as eyes, eyebrows, nose and mouth, have individual differences in size, shape, color, and arrangement. On the other hand, a face possesses similarity or commonality originated from multiple attributes, such as blood relationship, race, gender, age, and occupation. By using individuality, it will be possible to construct a system that identifies an individual from others and further estimates his/her age and/or gender. In order to realize such a system, it is necessary to establish technologies for extraction of individual facial characteristics from face images and estimation of their similarities.

2.3 Variety of External Appearance

Even for the same person, external appearances are not steady and constant. They change according to various factors. While short-term changes include hairstyle, suntan, make-up, wearing glasses, and so on, long-term changes include color of skin and hair, amount of hair, wrinkles of skin, facial shape, and so forth. Since these changes are inevitable in the handling of face images, the technologies related to facial image processing need to possesses functions to treat them appropriately.

3. DETECTION AND RECOGNITION OF FACES

3.1 Overall Framework of Detection and Recognition of Faces

Detection and recognition of faces in images is important in various applications such as intelligent human computer interaction, person identification or

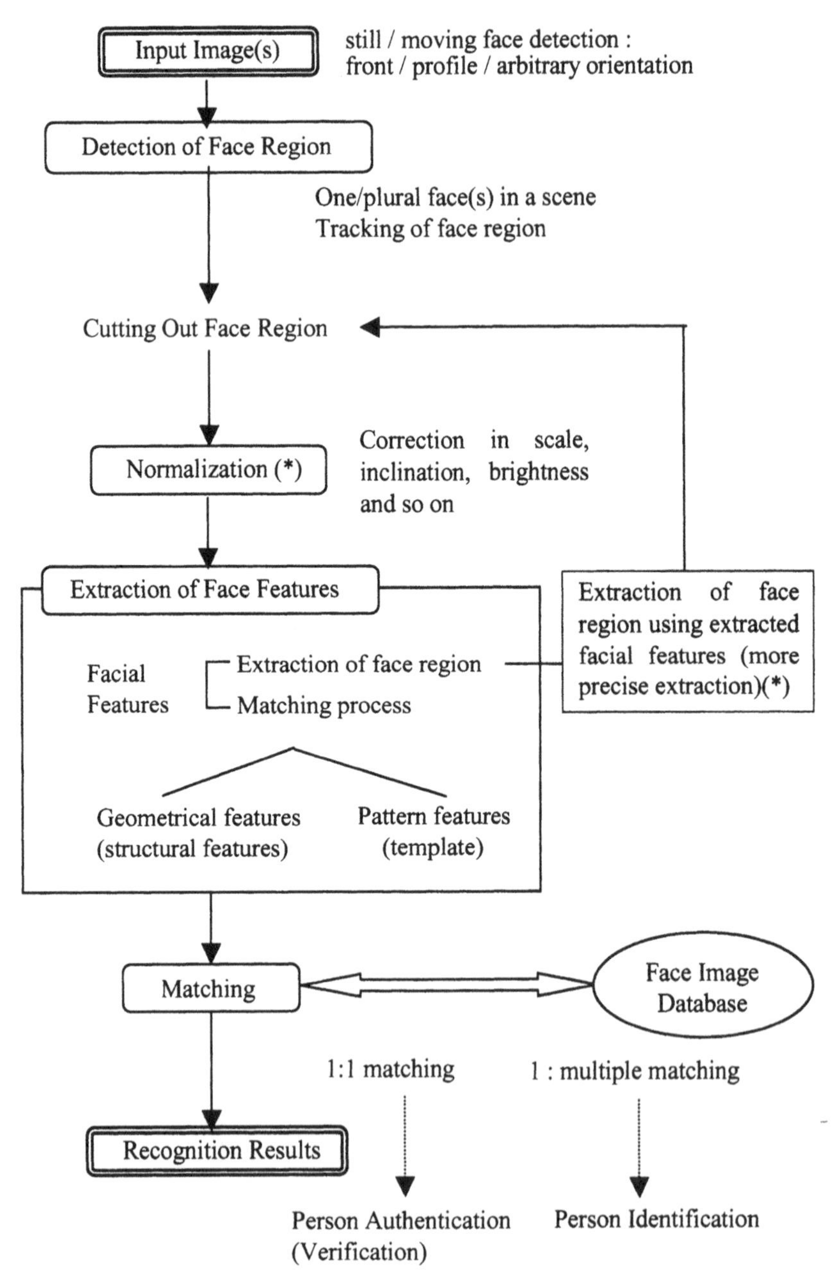

Figure 1 Fundamental flow in automatic detection and recognition of a face. (Item marked with (*) will be carried out, if necessary.)

verification by face, classification of photographs, and so on. The fundamental flow in automatic detection and recognition of a face is shown in Fig.1. Input face images are classified into three categories: frontal face, profile, and face with arbitrary orientation. Most studies have treated frontal face images, and a rather high face recognition rate has been achieved.

General processing steps in detection and recognition of faces are summarized as follows.

3.1.1 Detection / finding of face region

Except for a mug shot type of photograph in which a face occupies a large space in an image and a shot in which a person faces to a camera at the predefined position, the location and orientation of a face in an image are unknown in general. Therefore, it is required to detect a face region in an input image automatically. Here we need to consider two cases: one face in a scene and more than one face in a scene. When we treat moving images, the tracking of the facial region, considering the changes in mutual positions between a TV camera and a person, is required.

Actual face detection methods in a still image can be classified into four categories [169]: knowledge-based methods, feature invariant approaches (facial features, skin color, texture and so on), template matching methods (predefined face templates or deformable templates), and appearance-based methods (eigenface, neural networks, support vector machine, and so on).

3.1.2 Cutting out face region and normalization

The detected face region is cut out from an input image. Normalization is carried out on the face region to correct dispersion in scale, inclination, and brightness contained in each of the face images. If the relative position between a camera and a person is unknown, the normalization is carried out based on facial features such as the position of eyes extracted from an input image.

3.1.3 Extraction of facial features

Feature parameters are calculated for the face region. Two types of facial features can be considered: facial features used in normalization and those used in a feature matching process. The same features can be used for these two purposes. However, different features are normally used for each purpose. Extracted facial features can be used to evaluate the result of cutting out the face region. If the evaluation result is not good, the detection of the face region is carried out again to improve the performance in extracting the face region.

3.1.4 Matching with face image database

Extracted facial features are matched with facial features stored in a database. For person authentication, facial features of an input face image are compared with those of a specific person that are stored in a database. On the other hand, for person identification, face images having facial features that most resemble those of an input facial image are searched in a database.

3.2 Merit and Demerit of Face from the Viewpoint of Biometrics

A face has unique features as biometrics among other biological features such as fingerprints, iris pattern, retina pattern, shape of palm, voiceprint, and so on. Before discussing the details in trends of face recognition, it would be useful to summarize the merit and demerit of a "face" in the light of applications to person identification and person verification.

3.2.1 Merits

1. Since humans identify others by their faces, a face is the most familiar among various biometrics. In the identification by fingerprint, a person needs to put his/her finger onto the input device. In identification by iris pattern, a person needs to look closely to a camera. Therefore these often accompany uncomfortable feelings, but identification by face gives less reluctant feelings. Moreover, human can easily understand the recognition processes for face images, whereas it is difficult to understand them for other biometrics.
2. In face recognition, a person does not need to contact with any input hardware because a camera is used as the input device. This imposes less constraints on him/her. In order to make the identification process easier and to improve the identification ratio, it is desirable to capture a full-sized still image. An identification system for a walking person has also been studied. In this system a user is not required to assume a specific pose in front of a camera.
3. For image acquisition, a standard camera is required, and no special input devices are necessary. Small charge-coupled device (CCD) cameras are inexpensive and are widely used in our daily lives now. Thus we can capture face images easily by using them.
4. Identification procedures in the system can be confirmed by storing facial images in an access log.
5. Reliability of the face identification system can be improved by using security tools such as ID cards together with other biometrics.

3.2.2 Demerits

1. Capturing face images may invade personal privacy. (This problem is more or less common to other biometrics.)
2. Variations in face images are unavoidable. Typical factors causing variation include hair, suntan, glasses, beard, make-up, aging, differences in capturing conditions (relative position between a camera and a face, lighting, and motion) and so on. Countermeasures against these problems are the development of more robust recognition methods, the control of image capturing conditions, and so on.
3. The system may be deceived by photographs or dolls. 3-D motion information may be a good key to solving this problem.
4. Reliability of the identified result by face images is often lower than that obtained by fingerprints or by iris patterns. For a security system, other security tools should be used together with face recognition. Of course, there are many applications that do not require such a high identification ratio as a security system.
5. Because the history of face recognition technology as a practical means of person recognition is short, it is technically unstable compared with other biometrics such as fingerprint matching technology.

3.3 Detection of Faces in Images

3.3.1 Detection of faces in still images

Detection of faces in input images will serve as an essential initial step toward building a fully automated face recognition system. It also finds applications in human-computer interfaces, surveillance, and census systems. In [169], face detection methods in a single image are classified into four categories: knowledge-based methods (the rules about the relationships between facial features), feature invariant approaches (edge features, texture, skin color, and multiple features), template matching methods (predefined face templates or deformable templates), and appearance-based methods (models or templates of faces are learned from a set of training images). These categories are not necessarily rigid and some methods overlap category boundaries.

Here, several face detection methods will be briefly introduced. Turk and Pentland presented the Karhunen-Loeve transform (KLT)-based approach for locating faces by using "eigenfaces" obtained from a set of training images [158]. An input image is projected onto the eigenspace (here "face space") consisted of eigenvectors (here "eigenfaces"). The distance between an image region and the face space is calculated for all locations in the image. The distance from the face space is used as a measure of "faceness." The localization performance of this method, however, is degraded with changes in scale and orientation. The idea of "eigenface" has been adopted in many works on face detection and recognition.

Osuna et al. developed support vector machines (SVMs)-based approach for detecting faces in still images with complex backgrounds [111]. This system detects faces by exhaustively scanning an image for face-like patterns at many possible scales, by dividing the original image into overlapping sub-images and classifying them using SVMs to determine the appropriate class (face/non-face). Multiple scales are handled by examining windows taken from scaled versions of the original image. In the experiment, a database of face and non-face 19×19 pixel patterns and SVMs with the second-order polynomial as a kernel function were used. The detection rate was 97.1% for 313 high-quality images with one face per image. However, in general, training an SVM and scanning an image at multiple scales are computationally expensive, thus the realization of real-time processing is difficult.

Schneiderman et al. proposed a trainable object detector for detecting faces and cars of any size, location, and pose [140]. To cope with variations in object orientation, the detector uses multiple classifiers, each spanning a different range of orientation. Each of these classifiers determines whether an object is present in a specified size within a fixed-size image window. To find the object at any location and size, these classifiers scan the image exhaustively. Each classifier is based on the statistics of localized parts. Each part is a transform from a subset of wavelet coefficients to a discrete set of values. Such parts are designed to capture various combinations of locality in space, frequency, and orientation. In building each classifier, they gathered the class-conditional statistics of these part values from representative samples of object and non-object images. They trained each classifier to minimize classification error in the training set by using the AdaBoost (Adaptive Boosting) technique with confidence-weighted predictions. For details of the AdaBoost, see [141]. In detection, each classifier computes the part values within the image window and looks up their associated class-conditional probabilities. The classifier then makes a decision by applying a likelihood ratio test. Their trainable object detector achieves reliable and efficient detection of human faces and passenger cars with out-of-plane rotation.

Keren et al. [71] developed the face detection algorithm which shows the favorably comparable performance to the eigenface and support vector machines-based algorithms, and can be calculated substantially faster. Very simple filters are applied sequentially to yield small results on the multitemplate (hence, "anti-faces"), and large results on "random" natural images.

Approaches presented by Sung and Poggio [148] and Feraud et al. [28] aimed to detect faces in images with complex backgrounds. Sung and Poggio proposed an example-based learning approach for locating vertical frontal views of human faces. This approach models the distribution of human face patterns by means of a few view-based "face" and "nonface" model clusters. At each image location, a different feature vector is calculated between the local image pattern and the distribution-based model. A trained classifier determines the existence of the human face based on the difference feature vector measurements. Feraud et al. presented the approach based on a neural network model: the constrained generative model (CGM). A conditional mixture of networks is used to detect side-view faces and to decrease the number of false detection.

Hotta et al. developed a scale-invariant face detection and classification method that uses shift-invariant features extracted from a log-polar image (see Fig.2) [54]. Scale changes of a face in an image are represented as a shift along the horizontal axis in the log-polar image. In order to obtain scale-invariant features, shift-invariant features are extracted from each row of the log-polar image. Autocorrelations, Fourier spectrum, and PARCOR coefficients are used as shift-invariant features. These features are then combined with simple classification methods based on linear discriminant analysis to realize scale-invariant face detection and classification. The effectiveness of this face detection method was confirmed by experiments using face images captured under different scales, backgrounds, illuminations, and dates. To evaluate this face classification method, they performed experiments using 2,800 face images with 7 scales under 2 different backgrounds and face images of 52 persons.

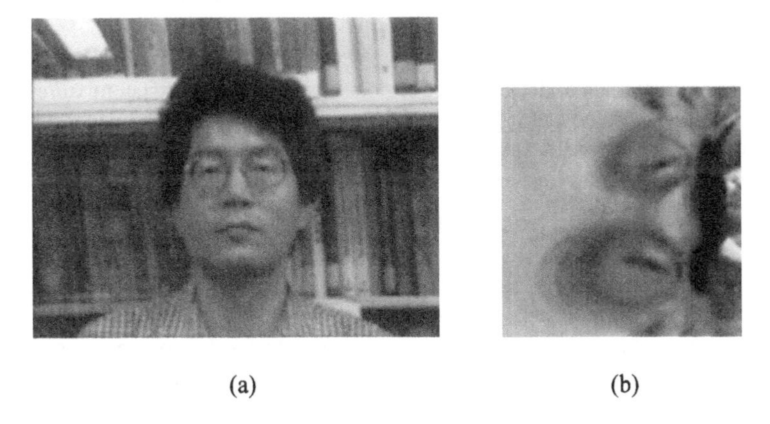

(a) (b)

Figure 2 Example of a log-polar image. (a) An input image and (b) the corresponding log-polar image.

Skin color was proved to be an effective feature to detect face and hand regions. However skin color is affected by various factors such as differences in races, illuminating conditions, suntan, physical conditions, and so on. Quan et al. [55] developed a robust skin color extraction method based on both a Gaussian mixture model of human skin color and image segmentation using an automatic and adaptive multi-thresholding technique. Since skin color extraction alone is not sufficient to extract face regions, the separation of hair region using the color distribution of skin and hair in the Cr Cb space and structural features of a face are employed to select a face region among skin color regions. Figure 3 shows examples of detected face regions by this method.

Wu et al. proposed the face detection method in color images based on the fuzzy theory [166]. Two models to extract the skin color region and the hair color region were used, and then they are compared with the prebuilt head-shape models by using a fuzzy-theory-based pattern matching method to detect face candidates.

As mentioned above, many face detection methods have been presented and they are progressing markedly. There is presumably a high possibility that these technologies will be progressively unified in the framework of the future face finding system.

(a) (b)

Figure 3 Examples of detected face regions based on the Gaussian mixture model of human skin color [55]. For each of (a) and (b), left: input image, right: detected face regions.

3.3.2 Detecting and tracking faces from moving images

For the detection of faces, not only colors but also motions of faces are regarded as promising features. Hasegawa et al. developed a face detection and tracking system on a Transputer-based parallel computer network (48 processor elements) [47]. This system extracts some visual features such as motions, colors, edges, and others from input images and integrates them on a "visual feature (attention) map" as a bottom-up process. On the other hand, this system searches the map by using face templates as a top-down process in parallel. It was reported that the system can detect and track one primal face (a user of the system) in real-time that is nearest and faces to the camera.

Viola el al. presented a machine learning approach for visual object detection that is capable of processing images very rapidly and achieving high detection rate [163]. Their work was distinguished by three key contributions. The first is the introduction of a new image representation named "integral image", which allows the features used by their detector to be computed very quickly. The second is a learning algorithm, based on AdaBoost, which selects a small number of critical visual features from a larger set and yields efficient classifiers. The third is a method for combining increasingly more complex classifiers in a "cascade" that allows background regions of the image to be quickly discarded while spending more computation on promising object-like regions. The cascade can be viewed as an object-specific focus-of-attention mechanism. Unlike previous approaches it provides statistical guarantees that discarded regions are unlikely to contain the object of interest. In the domain of face detection, the system yields detection rate comparable to that of the best previous systems. For real-time applications, the detector runs at 15 frames per second. S.Z. Li inspired and extended the above-mentioned Viola's approach and proposed an approach called "FloatBoost learning" for classification [87].

3.4 Face Recognition

3.4.1 General trends in the study of face recognition

There are many applications that involve face recognition: for example, an electric mug-book search system for identifying suspects, systems for controlling access to specific rooms or computers, and systems used at immigration controls and entrances for identifying passport holders. Recently face recognition has become also important to realize a new type of interface between humans and computers/robots. A computer or robot will recognize each user and give appropriate responses depending on the user.

Either profiles or frontal views are used for face recognition, although frontal faces are more common. Many methods have been proposed for recognizing frontal faces, which are broadly classified into the following two groups:

1. Geometrical feature-matching approaches measure the distances among face features, such as the eyes, nose, and mouth, and recognize these measurements as the characteristics of the face.
2. Template pattern matching approaches recognize faces as two-dimensionally distributed light and shade patterns [11].

Historically, the geometrical feature matching approaches, which use a lesser amount of data than the latter method, were developed first [65]. Around 1990, as computing power rapidly enlarged, the development of template pattern matching approaches began and the number of researchers studying face recognition increased. As shown in Section **3.3.1**, Yang et al. classified the face detection methods into four categories and drew the line between template-matching methods and appearance-based methods [169]. However, here the "template pattern matching approach" means both template matching methods (in a narrow sense) and appearance-based methods.

Because so many face recognition methods were proposed, they should be mutually compared and evaluated in terms of algorithm and performance. The Counterdrug Technology Development Program Office of the United States Department of Defense started the Face Recognition Technology (FERET) program in 1993 and continued to sponsor it [122, 130]. The main objective of this in 1993 and continued to sponsor it [122, 130]. The main objective of this program was to identify the most useful face recognition algorithm and/or approach by testing and comparing existing methods. The tests were conducted by, in principle, locating one face at the center of an image. Three tests have been conducted so far: in August 1994, March 1995, and September 1996, and the results were made public. Most of the methods that obtained high scores in these tests were based on template pattern matching approaches.

In 2000 and 2002, the face recognition vendor test (FRVT) was done in the United States to provide performance measures for assessing the ability of automatic face recognition systems that are commercially available to meet real-world requirements [33, 34].

3.4.2 Geometrical feature matching approach

The basic idea of this approach is the extraction and matching of geometrical facial features, such as the mouth position, chin shape, and nose width and length. Adequate scale and normalization should be used for each face image, and feature points, such as the corners of the eye, should be appropriately identified to measure the distances among them.

Brunelli and Poggio proposed a method that estimates the feature points in a face from the local projection information of horizontal and vertical edge components, measures the distances among these feature points, and recognizes a face from these measurements and chin shape data [11]. However, faces are three-dimensional, rotate, and show various expressions. The geometrical feature matching approach is useless unless such factors are adequately corrected. Brunelli and Poggio reported that the template pattern matching approach, which is described in the next section, recognized faces better than the geometrical feature matching approach even when face images taken under uniform conditions were used.

3.4.3 Template pattern matching approach

The template pattern matching approach regards face images as spatially distributed patterns of light and shade. A recognition system may recognize faces as different patterns when the position of a face shifts or when the lighting conditions change. Therefore, the system should be robust against changes in face position, scale, rotation, and lighting conditions. Usually, pre-processings are carried out such as searching and positioning of faces within images, scaling, and contrast normalization. Pre-learning based on face images with various face angles and lighting conditions taken in advance is also effective.

The typical approach based on the idea of template pattern matching is the "eigenface" method. In this method, principal component analysis (PCA) is applied to a set of face images and eigenvectors (here "eigenfaces") representing input face images are obtained. Around one hundred eigenvectors are sufficient to represent various face images. Each face can be reconstructed by the weighted sum of eigenvectors. Therefore, each face can be described by a set of weights for eigenfaces. Face recognition is carried out by firstly projecting an input face image onto the eigenspace and obtaining the corresponding set of weights for eigenfaces. Then the similarity between the obtained set of weights and those in a database is measured to recognize or identify the input face. The eigenface method was examined by Kirby and Sirovich in Brown University [72] and Turk and Pentland in MIT [158], and many works on face detection and recognition have adopted the idea of eigenfaces. For example, Pentland et al. improved the original eigenface method and developed the "modular eigenface" method, which constructs an eigenspace for each facial feature [119]. Swets et al. proposed a method that uses multidimensional discriminant analysis instead of PCA [149]. Martinez and Kak compared the performance of PCA and LDA (linear discriminant analysis) using a face database. Their conclusion was that when the training data set is small, PCA can outperform LDA and, also, that PCA is less sensitive to different training data sets.

Moghaddam et al. developed a probabilistic face-recognition method [98] that constructs an eigenspace from both the difference between the normalized face images of many persons and the difference between images of one person, which are caused by expressions and lighting conditions. Bayesian matching based on the spatial distribution of data is used to neglect changes caused by expression and/or lighting conditions.

Examples of other methods based on the idea of "eigenfaces" are the LFA (local feature analysis) method done at Rockefeller University and the discriminate analysis method done at University of Maryland. The method developed by Rockefeller University employs LFA to extract eigen features of a face [118]. Face recognition by this method is robust to the differences in color information and to aging. LFA carries out the principal component analysis on local features such as the nose, eyebrows, mouth and cheeks (the contour where the curvature of bone changes). Local features are represented by 32-50 small blocks. The combination of these small blocks represents the features of a whole face. The result of LFA is described by face code (or face print) using, at the minimum, 84 bytes.

Various methods that employ the ideas differing from "eigenfaces" have been also studied intensively. The University of Southern California (USC) developed an approach using the Gabor jets and graph matching method. This approach describes facial features by coefficients obtained by projecting a facial image onto Gabor jet bases [165]. Gabor jet bases are Gabor wavelet bases with different scales and directions. The position of each jet constitutes a vertex of a 2-D graph representing a geometrical model of the face. Face recognition is carried out by comparing the coefficients and distances between vertexes in an input image with those obtained from facial images in the database using an elastic graph matching. A user is not required to stand still in front of a camera. Even when a user is walking or coming toward a camera, this method can recognize a user's face.

A similar method was proposed by Lyons et al. [91]. Their method was based on labeled elastic graph matching, a 2-D Gabor wavelet representation, and linear discriminate analysis. Experimental results were presented for the classification of gender, race, and expression. Tefas et al. enhanced the performance of elastic graph matching in frontal face authentication [154]. The starting point is to weigh the local similarity values at the nodes of an elastic graph according to their discriminatory power. Fisher's discriminate ratio was then reformulated to a quadratic optimization problem subject to a set of inequality constraints by combining statistical pattern recognition and support vector machines.

Robustness to variations in lighting conditions, expressions, face posture, scaling and perspective is crucial to realize a practical face recognition system and many works have picked up this problem. Adini et al. conducted an empirical study to evaluate the sensitivity of several image representations to changes in illumination, viewpoint, and expression [1]. Belhumeur et al. developed a face recognition algorithm that is insensitive to large variations in lighting direction and expression [7]. They linearly projected an image onto a subspace based on Fisher's linear discriminant and produced well-separated classes in a low-dimensional subspace. In their experiments, the proposed "Fisherface" method showed lower error rates than the eigenface method. Ben-Arie and Nandy presented the pose-invariant face recognition method based on a three-dimensional (3-D) object representation that unifies a viewer and a model-centered object representation [8]. A unified 3-D frequency domain representation called volumetric frequency representation (VFR) encapsulates both the spatial structure of an object and a continuum of its views in the same data structure. Vel and Aeberhard proposed a line-based face recognition algorithm that uses a set of random line segments of a 2-D face image as the underlying image representation, together with the nearest neighbor classifier as the line matching scheme [159]. Lam and Yan presented an analytic-to-holistic approach that can identify faces at different perspective variations. Fifteen feature points are located on a face. Features windows are set up for the eyes, nose, and mouth and are compared with those in the database by correlation [81].

Liu and Wechsler introduced evolutionary pursuit (EP) as an adaptive representation method for image coding and classification [88]. EP starts by projecting the original image into a lower dimensional whitened PCA space. Directed but random rotations of the basis vectors in this space are then searched by GAs (genetic algorithms) where evolution is driven by a fitness function defined in terms of performance accuracy and class separation. EP improves the face recognition performance when compared against PCA ("eigenfaces") and shows better generalization abilities than the Fisher linear discriminant

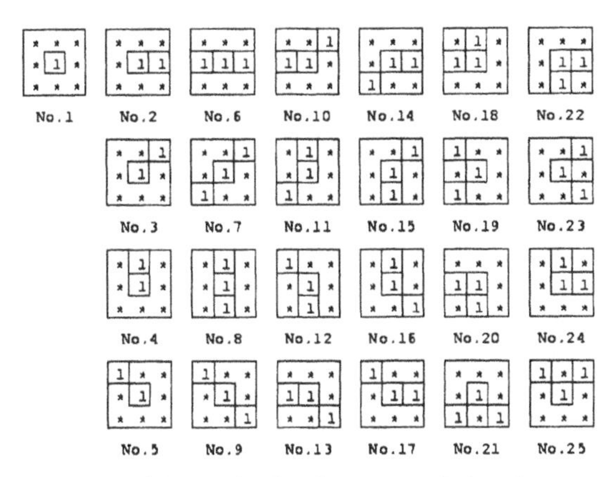

Figure 4 Higher-order local auto-correlation features.

("Fisherfaces"). Salah and Alpaydin developed a serial model for visual pattern recognition based on the selective attention mechanism [135]. The idea in selective attention is that not all parts of an image give us information. The model was applied to a more complex face recognition problem. Lanitis et al. presented a compact parameterized model of facial appearance which takes into account sources of variability including individual appearance, 3-D pose, facial expression, and lighting [82]. The model was created by performing a statistical analysis over a training set of face images and represents both shape and gray-level appearance. Even less than 100 of their parameters can still give a good reconstructed face to a given face.

Wiskott and Malsburg integrated template pattern matching with geometrical feature matching to attain robustness against changes caused by face angles and expressions [165].

Kurita et al. developed a unique method for recognizing faces by extracting the higher-order local auto-correlation features from face images, using these features as vectors peculiar to each face, and conducting a linear discriminant analysis [80]. This method uses vectors of limited orders (at most several kilobytes per person) and is shift-invariant in principle. Figure 4 shows 25 masks of 2-D local auto-correlation features. Goudali et al. expanded this method [37], conducted discriminant analysis on higher-order local auto-correlation features extracted from pyramidal images of various resolutions, and identified 11,600 face images of 116 persons with 99.9% accuracy.

3.5 Analysis and Recognition of Facial Expressions

Facial expressions reflect the internal condition of human and play an essential role in interpersonal communication. In the near future, automatic analysis of facial expressions will also become important in man-machine interaction. Studies of facial expressions were originally carried out in the field of psychology. By virtue of advances in computer technology and in image processing, automatic analysis and recognition of facial expressions were studied intensively in the 1990's. The problems of facial expression analysis consists of three main steps: face detection in a facial image or image sequence, facial expression data extraction, and facial expression classification. Pantic et al. surveyed the past work in solving these problems [116]. Facial expressions are analyzed by using either static facial images or facial image sequences, both of which require, at present, frontal face images and accurate estimation of face features. Here, several recent methods to analyze and recognize facial expressions are briefly introduced.

Matsuno et al. presented a method to recognize expressions of a person using static images [94]. Two- or three-dimensional facial elastic models are fitted on the face and the differences in shape between the expressionless and expressive faces are detected. Their system, called potential net, which uses 2-D elastic models, gave 80% correct results in recognizing the expressions of 45 face images in front of complicated backgrounds. Kobayashi et al. developed another method [75], which analyzes the gray levels of 234 points on 13 lines that are established over the eyebrows, eyes, and mouth. For face images with various expressions, obtained gray level data is inputted into a neural network as pattern vectors, and is used to recognize facial expressions.

Most systems that recognize expressions from dynamic images detect the parts of the face (muscles) that moved, convert the changes into optical flow [24, 92], and estimate the expression from the distribution of flow vectors. These methods have difficulty in accurately detecting flow vectors if the head moves. Black et al. proposed a method that minimizes the effects of head movement [9]. The shapes of the eyebrows, eyes and mouth in an image are roughly simulated with elastic, local parametric models. The models are deformed and normalized as the face moves (changes in scale, position, and rotation), and the changes in shape parameters are successively calculated. Donato et al. explored and compared techniques for automatic classification of facial actions in image sequences [20]. The best performance was obtained using the Gabor wavelet representation and the independent component representation. Both achieved 96% accuracy for classifying 12 facial actions of the upper and lower face.

Essa and Pentland derived a new, more accurate, representation of human facial expressions called "FACS+" [24]. They used a computer vision system, for observing facial motion by using an optimal estimation on object flow method coupled with geometric, physical, and motion-based dynamic models describing the facial structure. They probabilistically characterized facial motion and muscle activation in an experimental population. Tian et al. developed an automatic face analysis (AFA) system to analyze facial expressions based on both permanent facial features (brows, eyes, mouth) and transient facial features (deepening of facial furrows) in a nearly frontal-view face image sequence [156]. This system recognizes fine-grained changes in facial expression into action units of the facial action coding system, instead of a few prototypic expressions.

3.6 Other Approaches

Recently, other approaches that can simultaneously recognize faces and facial expressions have been studied. Cootes et al. developed an active contour model (an active appearance model (AAM) by a one-dimensional elastic model) of face (eyebrows, eyes, nose, mouth, and chin) [18, 21, 83]. The AAM contains a statistical, photo-realistic model of the shape and gray-level appearance of faces, and it is used as a basis for face recognition. They demonstrated the use of the AAM's efficient iterative matching scheme for image interpretation, and obtained interesting results from difficult images. More concretely, as a part of the face moves (such as the eyebrows), the model tracks them and changes the shape. This method, which sequentially records dynamic changes in facial shape, can be used for recognizing facial expressions. Furthermore the fitting result of an active shape model to a different person yields an individual shape, and it can be also used for identifying faces. By analyzing the characteristics such as gender and age, this system can be used for identifying them from face images.

A similar approach was presented by von der Maluberg et al. They demonstrated a system capable of tracking, in real world image sequences, landmarks such as the eyes, mouth, or chin on a face [95]. In the standard version, knowledge previously collected about faces is used for finding the landmarks in the first frame. In the second version, the system is able to track the face without any prior knowledge about faces and is thus applicable to other object classes. By using Gabor filters as visual features, and by both avoiding limiting assumptions and many parameters, their tracking tool is simple and easy to use. This method was used to estimate the pose of a face as one of the applications of the tracking results.

From the viewpoint of performance improvement in biometric applications, integration of the face recognition system and fingerprint verification system was implemented for more reliable personal identification [53]. The integrated system overcame the limitations of both systems. The decision fusion scheme was also proposed to enable performance improvement by integrating multiple cues with different confidence measures.

3.7 Practical Face Recognition Systems

Practical face recognition systems have been developed through activities such as the FERET Program and so on [29]. In one Web page on face recognition [132], 21 companies are listed as vendors of practical face recognition systems. The appendix in Ref. [36] studied several commercial systems and compared their recognition algorithms, input methods, countermeasures to disguise by photographs, and constraint conditions. Moreover, other systems such as "Smartface" [35] and "NeoFace" [171] in Japan have also been developed. In 2000 and 2002, the face recognition vendor test (FRVT) was carried out in the United States to provide performance measures for assessing the ability of automatic face recognition systems to meet real-world requirements [33, 34]. Ten commercial firms participated in FRVT2002. An extremely large dataset – 121,589 operational facial images of 37,437 individuals – was used to compute performance statistics.

Table 1 summarizes some typical commercial systems for practical face recognition. Face recognition system by Viisage Technology [162] and that by Visionics Co. have been developed based on the research activities at MIT (Massachusetts Institute of Technology) [98, 119, 120, 158] and USC (University of Southern California) [165], respectively, relating to the FERET Program. A method used at Rockefeller University [118] also entered the FERET Program at the beginning. Face recognition software by Cognitec Systems GmbH in Germany [17], which was formerly started at Siemens Nixdorf in 1995, was not tested in the FERET Program but was tested in the FRVT.

Evaluation for current mature face recognition systems including some of above systems was conduced in FRVT2002, which addressed several important topics in face recognition. As for the verification performance on the HCInt visa dataset, where a false accept rate is 1%, systems by Cognitec, Identix, and Eyematic showed the better results among ten systems. Detailed evaluation results are reported in [34].

The method proposed by MIT is based on eigenfaces. A commercial system developed by Viisage Technology was used together with driving licenses and ID cards issued by companies and civil services in the United States. Eigenspaces for a whole face are sensitive to changes in facial expressions, lighting conditions, and the differences in face posture. Toshiba Inc. developed a method to cope with this problem [35]. A facial image sequence containing various face postures and different facial expressions is used as inputs instead of a single still facial image. Matching processes with the distribution of input patterns will lead to face recognition, which is robust to the face postures and facial expressions. Furthermore, the constrained subspace consisting of components that are free from lighting conditions is calculated for the subspace to represent facial features. The constrained subspaces are used for matching with the database. This method can be carried out in real-time using a 450MHz Pentium II personal computer. This method was named "Smartface" and was bundled with a mini notebook type PC, Toshiba Libretto. Toshiba Co. also developed a security access control system named "FacePass."

A commercial version based on the LFA method at Rockefeller University was named "FaceIt" and is supported by Identix Inc. [56] (formerly by Visionics Co., which was merged with Identix Inc. in June 2002). Its recognition accuracy is more than 99% and its memory-based processing speed for one to multiple matchings is 60 million faces per minute using a 733MHz Pentium PC. The FaceIt

Table 1 : Typical face recognition systems commercially available.

Product Name	Vendor	Original Idea	Method
FacePASS, FaceFINDER, FaceEXPLORER FacePIN	Viisage Tech.	MIT Media Lab.	Eigenfaces (principal component analysis)
Smartface	Toshiba Co.	Toshiba Co.	Based on moving image sequence (matching with distribution of input patterns), constrained mutual subspace method
FaceIt	Identix Inc. (previously, Visionics Co.)	Rockefeller Univ., Computer Neuroscience Lab.	LFA (local feature analysis), faceprint (face code)
-----	Eyematic Interfaces Inc.	Univ. of Southern California, Computational and Biological Vision Lab.	Projection onto the set of Gabor jets, elastic graph matching
Face Key	Omron Inc.	ibid.	ibid.
TrueFace	eTrue.com Inc. (previously, Miros Inc.)	eTrue.com Inc.(previously, Miros Inc.)	Neural networks, template code
FaceVACS	Cognitec Systems GmbH	Cognitec Systems GmbH	Combination of local image transform and global transform
NeoFace	NEC	NEC	Generalized learning vector quantization, perturbation space method, adaptive regional blend matching

system has been adopted in a criminal surveillance system using street cameras in Newham, London. This system has contributed to reduce the occurrence of crimes. As an actual result, the reduction rate of total number of crimes is 34%. House robberies and crime damages have been reduced 72% and 56%, respectively. This system has also been utilized in cash dispensers (ATMs), immigration office, the Pentagon, state police offices in the United States, and so on.

The Gabor wavelets-based method by USC was employed in the commercial product by Eyematic Interfaces Inc. [25]. The Omron Co. in Japan innovated this method and developed an access control system named "FaceKey" and the supporting system for roaming patients [110].

"TrueFace," developed by eTrue.com Co. (previously Miros Co.), utilizes neural networks. It carries out the extraction of the facial region from complex backgrounds and matches extracted regions with a facial feature database. Facial features extracted from an input facial image are parts with high contrast, such as eyes, sides of nose, mouth, eyebrows, and cheeks. These features are represented as normalized numerals and described by template codes with about 1500 bytes. A matching process with a database is carried out for the template code by using a feature value neural network. By changing weights by learning, it can adapt to various lighting conditions and improve recognition accuracy. Matching speed is 500 faces per second with a Pentium 200MHz PC. This system can adapt to variations in face orientation within ±15 degrees. If multiple data with different face orientations are preregistered in the database, this method can adapt to large variations in face orientation. To cope with the disguise of a 2-D picture, a stereo pair is used as input information. A stereo pair can be captured by a stereo camera or by shooting a pair of pictures with different face orientations in front of a single camera. TrueFace has been installed in check cashing machines by Mr. Payroll Co. in the United States. It has helped to realize no forgeries for more than one-half million users. TrueFace has been also utilized in the airplane passengers control system of Exigent International Co. and has been authorized as a security product by ICSA (International Computer Security Association) in the United States. All of the intellectual property including TrueFace and related assets of the Miros division of eTrue.com were acquired by Viisage Technology, Inc. in June 2002.

Cognitec Systems GmbH has developed face recognition software named "FaceVACS" [17]. Face and eye localization is performed first. After the quality of the face image is checked to see whether it is sufficient for the following processing steps, only the face region is cut out as a rectangle image and is normalized. Facial features are extracted from normalized face images by the combination of local image transforms applied at fixed image locations and a global transform applied to the result of local image transforms. Local image transforms capture local information relevant to distinguishing a person. The parameters of a global transform were chosen to maximize the ratio of the interperson variance to the intraperson variance in the space of the transformed vectors using a large face image database.

"NeoFace," developed by NEC, utilizes a generalized learning vector quantization to discriminate face and non-face images. For the matching process, a perturbation space method and an adaptive regional blend matching method have been developed to reduce the effects caused by both global and local variations in face images [171].

4. SYNTHESIS OF FACE IMAGES

4.1 Three-Dimensional Approaches

In the field of face image synthesis by using a 3-D shape model [27], two major approaches have been historically developed, in the following order:

1. 3-D shape model + rendering approach. This is the conventional approach in computer graphics [113].
2. 3-D shape model + texture mapping approach [40, 41, 46, 66, 155]. Here, a texture mapping method maps texture information obtained from actual photographs onto the surface of 3-D shape model.

Recently several different approaches to the latter method have been studied intensively to yield photorealistic face images [38, 89, 124, 125]. Pighin et al. presented a well-rounded method to give photorealistic and textured 3-D facial models and to synthesize animation of facial expressions [124]. They first took photographs of eight different expressions, including a neutral or expressionless face, from five angles, on which they manually selected feature points. A 3-D shape model of the head was created for each expression by estimating the 3-D positions of these feature points. They translated these facial expressions from one to another by using 3-D shape morphing, and blended textures for the images. This method was also used for creating view-dependent images and for editing face animation. They acquired a set of face models that can be linearly combined to express a wide range of expressions [125]. They used these models for a given video sequence to estimate the face position, orientation, and facial expression at each frame.

Guenter et al. reported another method with a similar effect [38]. They posted 182 markers on the face of an actor, who was asked to show various expressions. They used six calibrated video cameras to record the facial movement. The 3-D position of each marker was then estimated. Based on this estimation, they deformed a 3-D polygonal model of actor's face, which had been created in advance, and merged the texture recorded by the six cameras. The markers on the face were then removed from the texture images by digital processing.

Comparing these two methods, the former method requires more labor since face features are manually extracted, and so much labor would be required for treating many expressions. The latter method, on the other hand, automatically extracts, follows, and erases feature points, and can relatively easily synthesize face images with various expressions. However, many markers must be placed on the subject's face and the subject's head must be fixed to extract only the facial movements caused by changes in expression. It is expected that these methods will be integrated in order to utilize the advantages of each.

Liu et al. developed a system that constructs a textured 3-D face model and realistic animation from videos with only five manual clicks on two images to tell the system where the eye corners, nose top, and mouth corners are [89]. Vetter proposed a technique to synthesize face images from new viewpoints using only a single 2-D image [161]. This technique draws on a single generic 3-D model of a human head and on prior knowledge of faces based on example images of other faces seen in different poses.

As for the construction of a 3-D individual head model, methods based on two orthogonal views, that is frontal and side views, have been studied [57, 137]. Ip et al. modified a generic 3-D head model by a set of the individual's facial features extracted by a local maximum curvature tracking (LMCT) algorithm [57]. Sarris et al. employed a combination of rigid and nonrigid transformations to adapt 3-D model feature nodes to the calculated 3-D coordinates of the 2-D feature points [137]. Automatic adaptation of a generic 3-D face model to an individual face image is also important to realize a practical model-based image coding, which will be discussed in Section **5.1**. Kampmann estimated facial features from the videophone sequences: the eye center, eyebrows, nose, chin, and cheek contours, and the 3-D face model is adapted to the individual face using these estimated facial features [64]. Zhang presented the automatic method to adapt the face model to the real face and also to determine initial values of action units [170]. Kuo et al. modified several existing methods to automatically locate the feature point positions from a single front-view facial image [79].

Modeling of hair and synthesis of realistic 3-D hair images have been also studied. For example, Chen et al. used a trigonal prism wisp model to represent each hair [14].

4.2 Two-Dimensional Approaches

The image-based rendering approach is one of the major techniques for facial image synthesis from 2-D face images. This approach has been studied to generate realistic facial animation with an arbitrary head pose and/or facial expression from a limited number of real images, to generate speech animation, and to render the face under arbitrary changes in lighting.

Mukaigawa et al. presented a method for synthesis of arbitrary head poses and/or facial expressions by combining multiple 2-D face images [99]. In their method, arbitrary views of a 3-D face can be generated without explicit reconstruction of its 3-D model. The 2-D coordinate values of a set of feature points in an arbitrary facial pose and expression can be represented as a linear combination of those in the input images. Face images are synthesized by mapping the blended texture from multiple input images. By using the input images with facial expressions, realistic face views can be synthesized. Moiza et al. presented a technique to create realistic facial animation given a small number of real images and a few parameters for the in-between images [97]. These image patterns in facial motion were revealed by an empirical study.

Ezzat et al. proposed a refined approach to create a generative, videorealistic, speech animation module with machine learning techniques [26]. In their approach, a human subject is first recorded using a video camera as he/she utters a predetermined speech corpus. After processing the corpus automatically, a visual speech module is learned from the data that is capable of synthesizing the human subject's mouth uttering entirely novel utterances that were not recorded in the original video. The synthesized utterance is recomposited onto a background sequence that contains natural head and eye movements. The final output is videorealistic in the sense that it looks like as if a video camera recorded the subject. At run time, the input to the system can be either real audio sequences or synthetic audio produced by a text-to-speech system, as long as they have been phonetically aligned. Samples of synthesized sequences are downloadable from their web site.

Shashua et al. developed an approach for "class-based" re-rendering and recognition with varying illumination [142]. They input a face image and sample face images with varying illumination conditions, and then re-rendered the input image to simulate new illumination conditions. They assumed human faces have a Lambertian surface. The key result in their approach is that by defining an illumination-invariant signature image they enabled an analytic generation of face images under varying illumination conditions.

Debevec et al. presented a method to acquire the reflectance field of a face and used it to render the face under arbitrary changes in lighting and viewpoints [19].

4.3 Manipulation of Facial Appearance

Manipulation of facial appearance is one of the interesting approaches in the synthesis of facial images. Rowland et al. developed shape and color transformations that allow individual face images to be moved in appearance along quantifiable dimensions in "face space" [131]. For example, it is possible to reconstruct a plausibly colored image of an individual's face from a black-and-white image, to change the apparent age, ethnic appearance and gender of a face, and so on. Tiddeman et al. created realistic moving facial image sequences that retain an actor's personality (individuality, expression, and characteristic movements) while changing the facial appearance along a certain specified facial dimension [157]. Lanitis et al. picked up the topic of aging effects on face images and carried out simulation of aging effects [83]. They described how the effects of aging on facial appearance can be explained using learned age transformations. Their proposed framework can be used to predict how an individual may look in the future or how he/she used to look in the past. It can also be used to eliminate the effect of aging variation in a face recognition system.

Kalra discussed the modeling of vascular expression, although most of the existing facial animations consider only muscular expressions [62]. This model enables visual characteristics such as skin color to change with time and provide visual clues for emotions like paleness and blushing.

4.4 Virtual Humans

Synthesis of animation of virtual humans including faces has been studied [10, 12, 63]. Cavazza discussed motion control to mimick behavior in the virtual human in response to some external stimulus or the presence of another virtual human in the environment [12]. Kalra developed an interactive system for building a virtual human, fitting texture to the body and head, and controlling skeleton motion [63]. Case studies were presented for tennis and dance performance. Brogan considered the behaviors of virtual human responding to the user's action [10]. Character motion is dynamically simulated to populate a virtual environment.

4.5 Automatic Drawing of Facial Caricatures

Facial caricatures represent features of individual face compactly and effectively. By extracting and emphasizing facial features of an individual person from an input photograph, it is possible to draw facial caricatures automatically by a computer. Most conventional approaches are based on a method that calculates the differences between an input face and an average face, and exaggerates them by the extrapolation technique. The PICASSO system [78] applied this method to extracted edge features from an input image.

In conventional approaches, the shapes of facial parts and their arrangement are treated together. On the other hand, a unique method to synthesize a facial caricatures using eigenspaces has been developed [68, 69]. In this method, eigenspaces are calculated for shape of each facial part and arrangement of facial parts, respectively. The exaggeration process is performed through each eigenspace independently. Thus, this method possesses the high flexibility in drawing facial caricatures. Figure 5 shows an example of synthesized facial caricatures by this method. Each of the principal components in each eigenspace represents a specific facial feature. Figure 6 shows typical features in facial contour and arrangement of facial parts represented by the four major principal components for eigenspace of facial contour and arrangement of facial parts, respectively [70]. By giving suitable weights to these principal components, it is possible to synthesize an arbitrary facial caricature having specific facial features without using an input photograph.

Another approach, which reflects the drawing style of each human caricaturist by making a correspondence between the eigenspace of input face images and that of caricatures, was also presented [39].

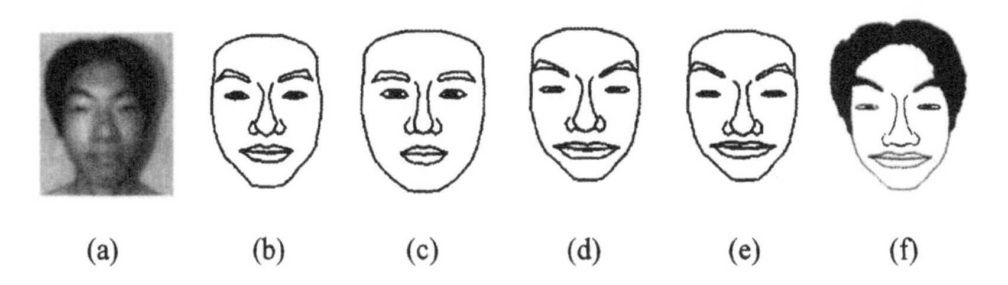

Figure 5 Synthesis of facial caricatures using eigenspaces [68]. (a) Original image, (b) initial contour shape, (c) average face, (d) only shapes are emphasized, (e) both shapes and arrangement are emphasized, and (f) final facial caricature with hair.

5. APPLICATIONS OF PROCESSING FACE IMAGES

In Sections **3** and **4**, trends in the studies of recognition and synthesis of face images are surveyed. As already mentioned in the introduction, applications of processing face images have been considered from a wide variety of standpoints. This section describes several typical and interesting applications, which will demonstrate the possibility of future development in the fields related to face images.

5.1 Application to Image Coding

The idea of "intelligent image coding" [40], which was originally studied for the very low bitrate coding of a moving face image sequence, has made a great impact on the various fields related to facial information. The basic framework of intelligent image coding can be briefly summarized as follows [41, 46, 66, 67]. A 3-D shape model of head is provided at both transmitting and receiving sides. The coding process is realized by transmitting only the motion and shape information of the face in the input images so that the face images, which are the same as those observed on the transmitting side, can be reconstructed on the receiving side. Since the knowledge about objects, in this case "faces," is described as models including a 3-D shape model, this coding method is also called "model-based image coding." "Semantic coding" is another technical term for this type of coding method, since it treats meanings of image contents.

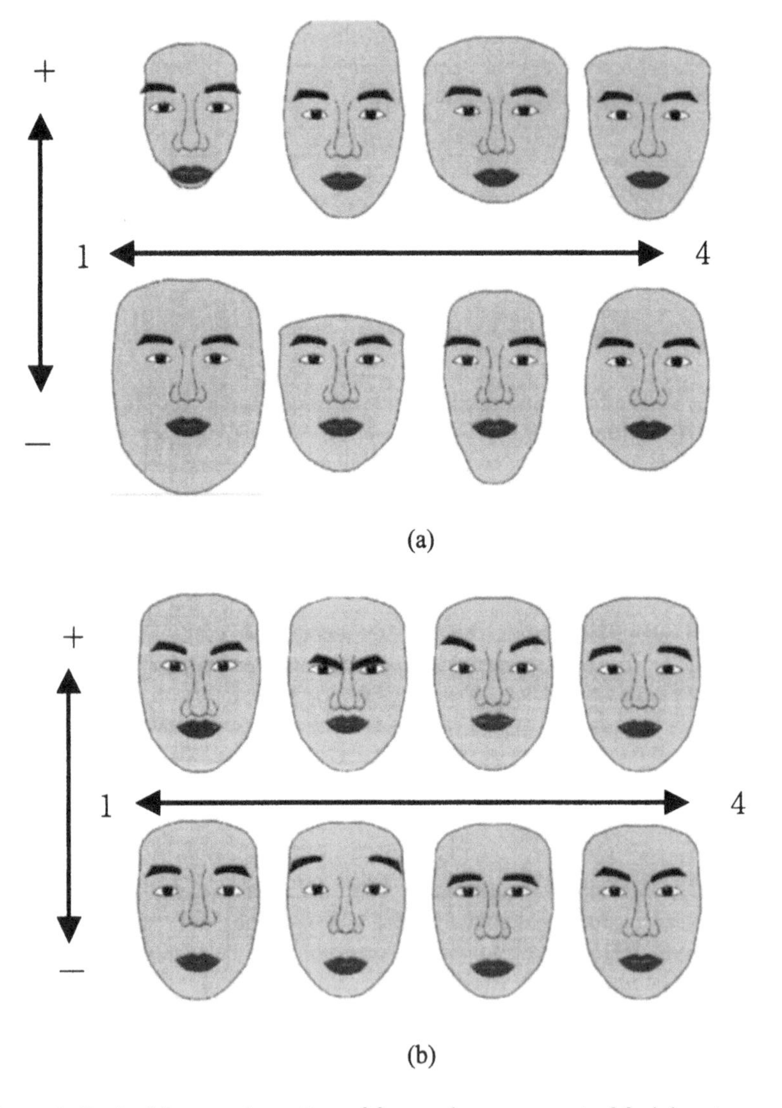

Figure 6 Typical features in outline of face and arrangement of facial parts represented by each of the four major principal components [70]. (a) Outline of face and (b) arrangement of facial parts.

In an intelligent image coding method, 2-D and 3-D analysis of motion and shape of each facial part in an input image, and synthesis of a natural face image sequence using a small number of parameters representing motions and changes in shape, are two key technologies. As for the analysis of face images, more robust methods have been studied toward the construction of a practical system. For example, Park et al. devised the GRIP (gradient, referential frame, iteration and pixel selection) method which can analyze the 3-D motion of head even in the case that motion is rather large or some part of face is occluded [113]. Pei et al. proposed a more robust and accurate method to estimate the global motion of head [114]. They extracted 3-D contour feature points and used a chamfer distance matching to estimate the global motion of head.

Li et al. presented the two-view facial movement estimation method [86]. Both the motion of head and the facial expressions are estimated. To eliminate the effect of the facial expressions on the estimation of the head movement, an M-estimation technique is used to obtain more reliable results. Yokoyama et al. proposed a method using the active contour model under consideration with an axis symmetry of the face for a stable detection of facial features [167].

As for the synthesis of natural face images, a texture mapping of facial texture obtained from the real photograph onto a 3-D shape model was introduced [41, 46, 66] in the latter half of the 1980's. This idea improved the quality of reconstructed face images drastically compared to images generated by the shading methods in those days. Since facial expressions are important in face-to-face communication, synthesis of expression images has been intensively studied. AUs (action units) used in the FACS (facial action coding system) in the psychological field [22] have been quantified and implemented as computer software [3].

The idea of intelligent image coding has been adopted in MPEG-4 (MPEG: Moving Picture coding Experts Group; international standard for audio visual object coding) as the synthetic / natural hybrid coding (SNHC). MPEG-4 consists of various profiles covering a wide range of audio and visual codings. Simple facial animation profile treats facial definition parameters (FDPs) which constitute a face model, and facial animation parameters (FAPs) which describe movement of each of characteristic points in FDPs. FAPs are typically correlated and also constrained in their motion due to the physiology of the human face. Ahlberg et al. utilized this spatial correlation to express FAP data efficiently [2]. To realize SNHC as a practical coding method, automatic estimation of FAPs is essential. Sarris et al. presented a determination method of FAPs by using 3-D rigid and nonrigid motion of human facial features found from 2-D image sequences [138]. Smolic et al. presented the recursive and predictive method for the real-time estimation of long-term 3-D motion parameters from monocular image sequences[144]. This method solves the problem of error accumulation in long-term motion estimation and gives stable estimation results. Lavagetto et al. proposed the facial animation engine implementing a high-level interface for the synthesis and animation of faces that is in full compliance with MPEG-4 specification [84].

Handling of face images are also utilized to improve the quality of decoded images in conventional waveform coding of videophone images. Nakaya et al. [105] and Kaneko et al. [67] discussed hybrid methods of model-based coding and waveform coding to improve the subjective quality of videophone images at a very low bitrate. There are several variations in a hybrid coding method. Three typical types are: (1) to apply model-based coding and waveform coding to the facial area and background, respectively, (2) to use the model-based synthesized face image as the prediction for a real face image, and (3) to apply finer quantization to the facial area estimated by a head model. Chai et al. proposed a face segmentation method using a skin color map and applied it to improve the perceptual quality of a videophone sequence encoded by the H.261-compliant coder [13]. Luo et al. presented a face location technique based on contour extraction within the framework of a wavelet-based video compression [90]. The quantization is controlled depending on the face location.

5.2 Media Handling

Recent years have seen an increased demand for multimedia applications, including video on demand, digital libraries, and video editing/authoring. For handling these large pools of information, one of the top requirements is to find key persons and their names in multimedia data, such as still and/or moving images, natural languages, transcripts, and so on.

Sato and Kanade developed "Name-It" [139], a system that associates names and faces in given news videos. An important basic function of Name-It is to guess "which face corresponds to which name" in given news videos. First, the system finds faces in videos under the following conditions: frontal, centered, long duration, and frequently. Second, the system tracks them by following skin color region and extracts frontal face images from tracked image sequences. Third, the system extracts candidates of names from transcripts in videos and finds an association between a face and a name by calculating their co-occurrence using their timing information. For the face identification, they adopted the "eigenface" method. This kind of approach, which employs both image and natural language information, is one of the most promising approaches for multimedia handling. It is expected to be able to access not only news videos but also "real" contents of multimedia data.

Recently digitization of broadcasting, in which MPEG-2 is used as the video compression method, has become widely discussed in many countries. A huge amount of MPEG coded video data will be produced and stored every day in the near future. From this point of view, a highly efficient system that can rapidly detect human face regions in an MPEG coded video sequence was proposed [164]. Since detection of the face is carried out directly in the compressed domain, there is no need to carry out the decoding process for image retrieval.

5.3 Human-Computer Interaction

5.3.1 Multimodal interface

Up to now, although computing power has progressed remarkably, the standard interfaces for human-computer interaction (such as a keyboard and a mouse) have not progressed much over the past 30 years. These interfaces are quite different from the style of communication between humans.

Studies on multimodal interface [100] have begun to improve this situation. Instead of the conventional keyboard and mouse, a user can operate a computer through natural spoken language, facial expressions, head movements, line of sight, and gestures. To realize such an interface, various methods to recognize facial expressions, line of sight, and gestures from images have been studied [48]. With the focus of attention on the complemental nature of facial expressions and languages, trials that simultaneously recognize both face images and speech sounds (utterances) have been carried out [146].

"Share attention" between a user and a robot is also becoming one of the important research topics now. A method to realize share attention is discussed by Chen et al. in the realization of an automatic detection of the attended object by a user using image information [15]. The user's gaze direction is the key to detect the currently concerned object. The face posture is also a useful key, when the quality of image is not sufficient to distinguish the gaze direction. The saliency map is calculated using the detected face posture and acuity distribution, and it is used with the distance image to identify the concerned object. Furthermore, the cross-modal attention using both visual and auditory information has been studied [16].

5.3.2 Interactive anthropomorphic agents

The progress in the analysis and synthesis of facial images has made it possible to provide an anthropomorphic (human-like) agent in a computer as a human-friendly and intelligent interface between a user and a computer [5]. To realize such an agent, various elemental technologies are required: image analysis and synthesis, speech analysis and synthesis, natural language processing, machine learning and others. Among them, the role of the face of the agent is basically equal to that of the human face.

As for the analysis of face images, there are two important meanings in the study of an anthropomorphic agent. One is to give a visual function to the agent to identify an individual user by his/her face, and to recognize the user's head motion and facial expressions. The other is to detect facial features of a specific person so as to synthesize facial images resembling him/her. As for facial image synthesis, it is very important to synthesize a realistic face on a computer display, with which a user can make an interaction.

The fundamental idea of the anthropomorphic agent was impressively proposed in a concept video entitled "Knowledge Navigator" produced by Apple Co. in the late 1980's. Actual synthesis of agents began in the 1990's when the performance of computer hardware advanced much more. Since then, many kinds of agents with various characteristics have been reported.

In 1995, scientists at the Electrotechnical Laboratory (ETL, currently Advanced Industrial Science and Industry: AIST) in Japan developed a multimodal anthropomorphic agent [49] with seeing, hearing, speaking, and smiling functions. By the seeing function, the agent could not only learn and identify individual users, but also recognize objects in real time. The learning process was executed in a dialogue manner. Figure 7 shows an example of interaction between a user and the agent.

Figure 7 An example of interaction between a user and the agent [49].

Another aspect of the role of an anthropomorphic agent is in the relationship with users. There are three typical types. The first one is an agent that answers questions from a user. Examples of this type of agent are TOSBURG II based on the speech dialogue system [153], WebPersona guiding an access to the World Wide Web [129], and Pedagogical Agent used as a "teacher" [85, 128]. The second type is an agent thar can interact with multiple users [101, 108]. In [101], this type of agent is named "social agent." The third type is an agent that acts as a substitution for a real person [106].

Though the anthropomorphic agents have human-like figures, interactive agents can be also represented as animals and toys. Even in such agents, the representation of the face is one of the essential factors. Examples of this type are Peedy (bird like a parrot) by Microsoft Co. [121], Silas (dog) in the ALIVE system of MIT [4], Woggles (rubber ball) by CMU [112], and so on.

5.3.3 Face robot

In addition to a human-like agent on a computer screen, mechanical face robots have been constructed by several groups [76, 77, 96, 150, 151]. The important feature of these robots is that they can change their facial expressions and express emotions.

A face robot developed in Tokyo University of Science is human-sized and has an appearance quite similar to a real human [76, 77]. Face skin is made of special silicon rubber. Facial expressions are controlled by SMA (shaped memory alloy) actuators driven by electricity. All actuators are mounted in the head part. Figure 8 (a) and (b) show the frontal and side views of the mechanics, respectively. Examples of expressions displayed on this face robot are shown in Fig. 8 (c) and (d). (c) and (d) represent smile and surprise expressions, respectively. As seen from Fig.8, this robot gives vivid lifelike impressions.

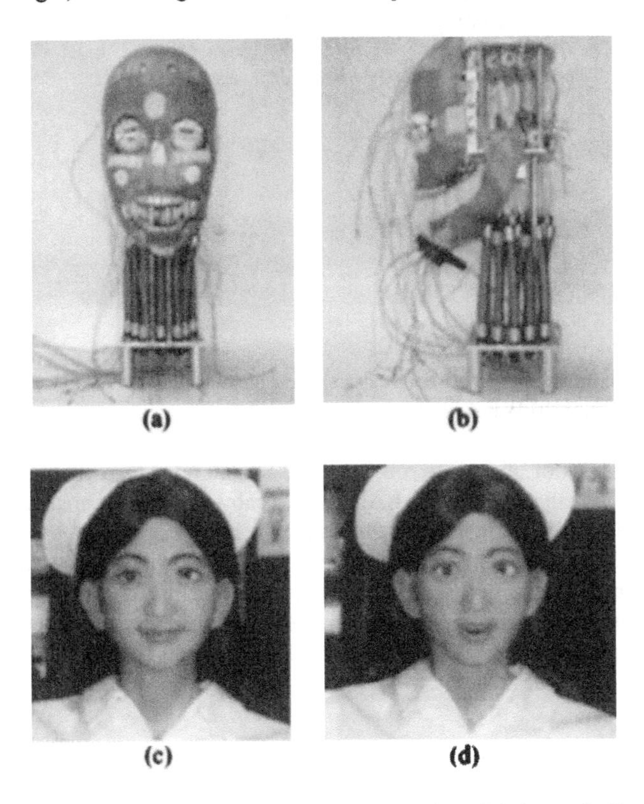

Figure 8 Face robot developed at Tokyo University of Science [76]. (a) Front view of mechanics, (b) side view of mechanics, (c) smile expression displayed on the face robot, and (d) surprise expression displayed on the face robot.

On the other hand a human-like head robot developed at Waseda University has a mechanical appearance as shown in Fig.9. Figure 9 (a) is WE-3RV (Waseda Eye No.3 Refined V) and (b) is WE-4 (Waseda Eye No.4). WE-4 is an improved version of WE-3RV. WE-4 has 29 freedoms in total and is installed with various sensors: vision, sound, touch, and smell. WE-4 can express emotions by movement of eyeballs, lips and chin, face color, and speech. In addition to the mechanism to express facial expressions, a psychological model is equipped to control emotions momentarily according to input stimuli.

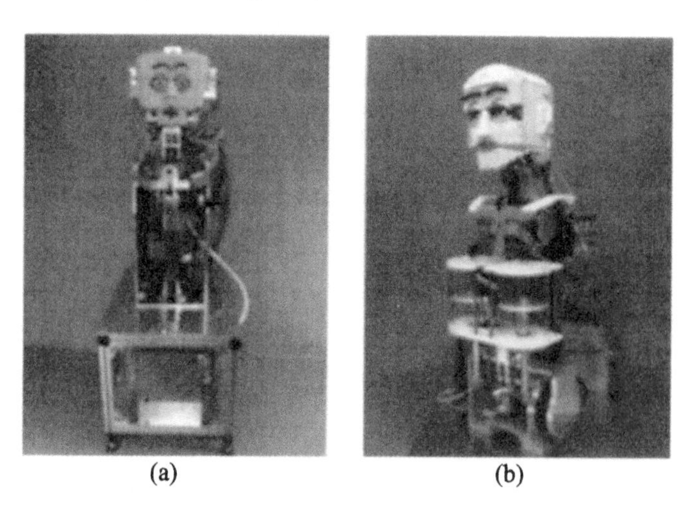

(a) (b)

Figure 9 Human-like head robot developed at Waseda University. (a) WE-3RV [150], and (b) WE-4 [151].

Examples of other face robots are a sociable humanoid robot "Kismet" by MIT [73], and the MY REAL BABY doll (baby-type robot) created by Hasbro Inc. and iROBOT Inc. in the United States, which has many facial expressions and different combinations of sounds and words [58]. "Kismet" is an expressive robotic creature with perceptual sensors and motor systems. Kismet perceives a variety of natural social cues from visual and auditory channels. It delivers social signals to humans through gaze direction of the eyes, facial expressions, orientation of the head, and vocal babbles.

5.4 Face Impression

So far, the face impression has not been dealt with systematically. In [102] this theme was discussed by using an average face image calculated for a collection of real face images, which were sampled from persons belonging to a certain group such as the same occupation, the same age, the same race, the same company or school, and so on. Examples of such average face images are shown on a Web page [42].

There are many linguistic expressions in the appearance of human faces such as a gentle face, severe face, manly face, pretty face, young face, and so on. Xu et al. and Mizuno et al. discussed the synthesis of facial caricature corresponding to each of these linguistic expressions [69, 70]. Fuzzy inference methods are applied to represent ambiguous expressions for human faces numerically.

5.5 Psychological and Medical Applications

It is said that there are three types of facial expressions of laughing, that is, the laugh of pleasantness, the laugh of unpleasantness, and the laugh of sociability. Classification of these types was tried by experiments examining temporal difference of eye and mouth movements [107]. In this experiment, synthesized face images were used as stimulus images. In relation to psychological applications, the construction of a 3-D model of emotion space was tried [134]. Transformation from this emotion space to fundamental categories of facial expressions was also examined.

The aging simulation of a face based on a skeletal model [103] and the assessment of cerebral disease by using synthesized face images [147] are examples to demonstrate the potential applicability of digital processing of face images to various non-engineering fields relating to facial information.

Takanobu et al. proposed a mathematical model that simulates the nonlinearity in the actual movement of the facial muscles for chewing food by a mastication robot [152]. When the lower jaw rapidly closes, it may come in hard contact with the upper jaw if the food is a crushable one. To clarify the mechanism of such rapid jaw motion, they focused on the nonlinearity of the facial muscles known in the field of the physiology and/or biomechanics. They proposed a nonlinear spring mechanism for the non-linear control of the muscles, and confirmed the effectiveness of this mechanism through experiments.

From the surgical point of view, modeling of not only the facial surface but also tissue data or bone data was carried out [6, 74]. In [74], using a finite element method, a prototype system was constructed for surgical planning and prediction of human facial shapes after craniofacial and maxillofacial surgery for patients with facial deformities. In [6], a physics-based 3-D head model was reconstructed from cephalograms, which are normalized X-ray images of the head shot from front and side. This model is used to simulate a surgical operation on a jaw.

6. COMMON TOOLS AND DATABASES FOR STUDYING FACIAL INFORMATION

6.1 Advanced Agent Project

Amid increasing interests in various fields dealing with facial information and the advancement of research regarding face images in the engineering field, the cooperative work toward the development of common software tools related to processing face images has been carried out by more than ten research groups in Japan. This activity was called Development of a Facial Image Processing System for Human-like "Kansei" Agent (or Advanced Agent Project) [43, 44]. This activity had been intensively carried out in a three-year plan from 1995 to 1998 as one of the development themes on fostering unique information technology in the Information-technology Promotion Agency, Japan (IPA).

The main objective in this project was the development of the processing system related to the recognition and synthesis of face images. The developed face recognition system included basic tools to detect facial portions in an input image and recognize facial shapes. The extraction of the position and the outline of each facial component from a still image was carried out by using methods such as extraction of facial area by color information and Fuzzy pattern matching, detection of center point of each pupil, extraction of outline of each facial component by a dynamic contour model (snake), and so on.

The synthesis system of face images employs a 3-D shape model of a human head and a texture mapping method to synthesize animated face images that are rich in expressive ability. Handling of facial expressions is based on the methodology of EU (expression unit) suggested as AUs in the field of psychology. Through the newly developed GUI (graphical user interface), users can interactively decide parameter values and generate expression images dynamically. Synthesis of mouth shapes accompanying utterances is also carried out according to the deformation rules defined separately from expressions.

6.2 Real World Computing Project

From 1992 to 2002, the Real World Computing (RWC) project was carried out in Japan under the Ministry of International Trade and Industry (MITI, currently Ministry of Economy Trade and Industry: METI). In this project, scientists constructed a database (called a multimodal database) of information containing human gestures synchronized with speech sounds. Since there is a close relationship between dynamic expressions and utterances, the importance of this kind of database will probably increase for the systematic study of facial expressions in the future. This database is freely available to the public [133].

In addition, a computer graphics (CG) model of the human face and body, the control programs for this model, and the manuals were constructed by the Electrotechnical Laboratory (ETL, currently Advanced Industrial Science and Technology: AIST). They are all freely available [23, 50]. The CG face and body can be controlled by commands transmitted through "sockets." For more details about the sockets, see the reference manual of UNIX and/or MS-Windows. The model and the program seem useful in various research and development areas needing the CG of face images.

6.3 Other Tools and Databases

Regarding digital processing of face images, a software system for interactive synthesis of face images using a 3-D shape model was introduced in [104]. This system includes various tools, such as a shape-fitting tool for 3-D facial models, an expression synthesis tool, a facial morphing tool, and a tool called "Fukuwarai" (a traditional Japanese game to arrange the parts of facial components, such as eyes, eyebrows, nose, and mouth, with a blindfold). The shape-fitting tool for 3-D facial models is used to adjust a standard 3-D model to the specific input face by changing the coordinates of control points. The expression synthesis tool generates facial expressions based on the concept of FACS (facial action coding system). The facial morphing tool independently morphs the shape and texture of several faces. The "Fukuwarai" tool can create face images in which the position, slant and shape of each of the eyebrows, eyes, nose, mouth, and facial outlines are changed interactively in the manner of a "Fukuwarai" game. One may not only use this tool for a "Fukuwarai" game, but also for synthesis of face images to retrieve similar face images from the database [109].

Eyematic Interfaces Inc. developed facial animation software named "FaceStation" (now the updated version "FaceStation2" is available) [25]. FaceStation integrates computer vision technology with speech analysis technique to realize real and accurate facial animation. Some other free or commercial software systems such as face recognition and face detection are available from home pages through the Internet [27, 32].

Several face image databases have been constructed to promote related studies and developments. Most of them [29, 30, 148, 168] are designed mainly to evaluate performance of face recognition methods and contain only gray scale images. The FERET database consists of a large number of monochrome images taken under several different conditions; facial expressions, illumination, pose angle, and date. The background is simple and uncluttered, and only the head and neck parts of an individual person are captured. To date, 14,126 images from 1,199 individuals are included [30]. A brief description about other monochrome face image databases such as the MIT database, the Yale database, and so on are given in [168].

Face image databases of color images have been also constructed: the XM2VTSDB by Surrey University in England, the database by Softpia Japan Foundation in Japan, and the PIE database by Carnegie Mellon University(CMU) in the United States [93, 143, 145]. To evaluate face detection methods using skin color information, such color image databases are indispensable. The XM2VTSDB is an extension of the M2VTS (multi modal verification for teleservices and security applications) multimodal database [93] from the European ACTS (advanced telecommunications technologies and services) projects [123]. It contains both frontal and side-view face images of 295 subjects. Two types of shots were recorded: one consisted of speech movements and the other consisted of a rotating head movement. The face image databases by Softopia Japan Foundation contain two types of databases [145]. One contains images captured by 15 video cameras operating at the same time. The other contains the rotated face images captured by 8 video cameras operating at the same time. The PIE (pose, illumination, and expression) database contains over 40,000 facial images of 68 people using the CMU 3-D room [143]. Each person was imaged across 13 different poses, under 43 different illumination conditions, and with 4 different expressions.

7. ACTIVITIES RELATED TO A "FACE" IN THE ACADEMIC DOMAIN

Since a "face" is not only of interest to engineering applications but more widely discussed in many fields, the unique academy named the Japan Academy of Facial Studies (abbreviated as "J-face") was established in Japan in March 1995 [61]. In this academy there are about 860 members from a wide range of fields, such as psychology, anthropology, dentistry, cosmetics, and art anatomy, as well as electronic and mechanical engineering. J-face holds the Symposium "Face" in the spring and the National Conference in autumn every year. It also publishes the journal named "Kao Gaku (in Japanese)" which means the studies on "face."

As a research group where researchers in science and engineering fields who are interested in "faces" gather, the Technical Group on Human Communication Science was organized in the Human Communication Engineering Group of the Institute of Electronics, Information and Communication Engineers in Japan [52]. Many topics related to engineering applications of face images have been discussed there.

Papers and articles on engineering applications of face images are widely published in research meetings related to human interface, national conference proceedings, journals, and transactions, in the Institute of Electronics, Information and Communication Engineers, the Institute of Image Information and Television Engineers, the Japanese Society for Artificial Intelligence, the Information Processing Society of Japan, the Society of Instrument and Control Engineers of Japan, and so on. Many papers in bulletins and journals related to vision, graphics, and human interface are presented internationally by such groups as IEEE, IAPR, and

ACM [136, 160]. References [45], [59] and [115] are typical examples of special issues on the subject of facial information processing.

In addition, at recent large international conferences related to image processing, sessions dealing only with research related to "faces" were scheduled frequently. Moreover, an international conference dealing only with "faces and gestures" has been held four times through 2002 [31]. Recently, international workshops on lifelike animated agents were also held in 2002 [126, 127].

8. CONCLUSIONS

"Faces" are very familiar to us, but it is not easy to grasp all the information expressed by a face appropriately for a computer and even for humans. Automatic recognition and synthesis of face images have been challenging topics in the field of computer vision and image processing.

Owing to the accumulation of results of intensive studies and with the remarkable progress in computing power, more refined and more robust algorithms for recognition and synthesis of face images have been devised. Recent applications of these technologies to human communication and computer interface as well as automatic identification of human faces were discussed intensively.

On the basis of these backgrounds, this chapter has surveyed recent research trends in the processing of face images by a computer and its various applications. After considering several representative characteristics of faces, this chapter reviewed recent research activities in the detection and recognition of faces in input images and synthesis of more natural face images. Synthesis of facial caricatures was also introduced. Then interesting applications of digital processing of face images were discussed from several standpoints: intelligent image coding, media handling, human-computer interaction, face robot, face impression, and psychological and medical applications. Common tools and databases used in the studies of processing of face images were also described. Finally activities in the academic fields related to facial information were introduced.

We hope the continuous and still more active development of technologies related to the processing of facial information in order to make our lives more comfortable and richer in future society which will be widely supported by computers.

REFERENCES

[1] Y.Adini, Y.Moses, and S.Ullman: "Face recognition: the problem of compensating for changes in illumination direction," IEEE Trans. Pattern Analysis and Machine Intelligence (PAMI), vol.19, no.7, pp.721-732, July 1997.

[2] J.Ahlberg and L.Haibo: "Representing and compressing facial animation parameters using facial action basis functions," IEEE Trans. Circuits and Systems for Video Technology, vol. 9, no.3, pp.405-410, April 1999.

[3] S.Akamatsu: "Computer recognition of human face - A survey," Trans. IEICE D-II, vol.J80-D-II, no.8, pp. 2031-2046, Aug. 1997 (in Japanese).
[4] ALIVE: Artificial life interactive video environment : http: //lcs.www.media.mit.edu/projects/alive
[5] E.Andr ed.: Proc. of the Workshop on "Animated interface agents: Making them intelligent," in conjunction with IJCAI-97, Nagoya, Japan, Aug. 1997.
[6] Y.Aoki, M.Terajima, A.Nakashima, and S.Hashimoto: "Physics-based 3D head model reconstruction from cephalograms for medical application," IEICE Trans. D-II, vol.J84-D-II, no.2, pp.390-399, Feb. 2001 (in Japanese).
[7] P.N.Belhumeur, J.P.Hespanha, and D.J.Kriegman: "Eigenfaces vs. Fisherfaces: recognition using class specific linear projection," IEEE Trans. Pattern Analysis and Machine Intelligence (PAMI), vol.19, no.7, pp.711-720, July 1997.
[8] J.Ben-Arie and D.Nandy: "A volumetric/iconic frequency domain representation for objects with application for pose invariant face recognition," IEEE Trans. Pattern Analysis and Machine Intelligence (PAMI), vol.20, no.5, pp.449-457, May 1998.
[9] M.J.Black and Y.Yacoob: "Tracking and recognizing rigid and non-rigid facial motions using local parametric models of image motion," Proc. of the International Conference on Computer Vision (ICCV'95), pp. 374-381, June 1995.
[10] D.C.Brogan, R.A.Metoyer, and J.K.Hodgins: "Dynamically simulated characters in virtual environments," IEEE Computer Graphics and Applications, vol.18, no.5, pp.58-69, Sep./Oct. 1998.
[11] R.Brunelli and T.Poggio: "Face recognition: features versus templates," IEEE Trans. Pattern Analysis and Machine Intelligence (PAMI), vol.15, no.10, pp. 1042-1052, Oct. 1993.
[12] M.Cavazza, R.Earnshaw, N.Magnenat-Thalmann, and D. Thalmann: "Motion control of virtual humans," IEEE Computer Graphics and Applications, vol.18, no.5, pp.24-31, Sep./Oct. 1998.
[13] D.Chai and K.N.Ngan: "Face segmentation using skin-color map in videophone applications," IEEE Trans. Circuits and Systems for Video Technology, vol.9, no.4, pp.551-564, June 1999.
[14] L.-H.Chen, S.Saeyor, H.Dohi, and M.Ishizuka: "A system of 3D hair style synthesis based on the wisp model," Visual Computer, vol.15, no.4, pp.159-170, 1999.
[15] B.Chen, M.Meguro, and M.Kaneko: "Forming share attention between user and robot based on face posture estimation," IEEJ Trans. EIS, vol. 123, no.7, pp.1243-1252, July 2003 (in Japanese).
[16] B.Chen, M.Meguro, and M.Kaneko: "Selection of attention point for a human-like robot in interaction with users," Technical Report of IEICE, HCS2002-56, Mar. 2003 (in Japanese).
[17] http: //www.cognitec-systems.de/index.html

[18] T.F.Cootes, G.J.Edwards, and C.J.Taylor: "Active appearance models," Proc. of European Conference on Computer Vision (ECCV'98), vol.2, pp. 484-498, June 1998.
[19] P.Debevec, T.Hawkins, C.Tchou, H.-P.Duiker, W.Sarokin, and M. Sagar: "Acquiring the reflectance field of a human face," Proceedings of ACM SIGGRAPH 2000, pp.145-156, 2000.
[20] G.Donato, M.S.Bartlett, J.C.Hager, P.Ekman, and T.J.Sejnowski: "Classifying facial actions," IEEE Trans. Pattern Analysis and Machine Intelligence (PAMI), vol.21, no.10, pp.974-989, Oct. 1999.
[21] G.J.Edwards, T.E.Cootes, and C.J.Taylor: "Face recognition using active appearance models," Proc. European Conference on Computer Vision, vol.2, pp. 581-595, Freiburg, Germany, June 1-5, 1998.
[22] P.Ekman and W.V.Friesen: "Facial action coding system," Consulting Psychologist Press, 1977.
[23] ETL CG Tool Home Page: http: //www.etl.go.jp/etl/gazo/CGtool
[24] I.A.Essa and A.P.Pentland: "Coding, analysis, interpretation, and recognition of facial expressions," IEEE Trans. Pattern Analysis and Machine Intelligence (PAMI), vol.19, no.7, pp.757-763, July 1997.
[25] http: //www.eyematic.com
[26] T.Ezzat, G.Geiger, and T.Poggio: "Trainable videorealistic speech animation", Proc. ACM SIGGRAPH 2002, San Antonio, Texas, July 2002.
[27] Facial Animation Home Page: http: //mambo.ucsc.edu/psl/fan.html
[28] R.Feraud, O.J.Bernier, J.-E.Viallet, and M.Collobert: "A fast and accurate face detector based on neural networks," IEEE Trans. Pattern Analysis and Machine Intelligence (PAMI), vol.23, no.1, pp.42-53, Jan. 2001.
[29] P.J.Phillips, H.Wechsler, J.Huang, and P.J.Rauss: "The FERET database and evaluation procedure for face-recognition algorithms," Image and Vision Computing, 16, pp.295-306, 1998.
[30] P.J.Phillips, H.Moon, S.A.Rizvi, and P.J.Rauss: "The FERET evaluation methodology for face recognition algorithms," IEEE Trans. on Pattern Analysis and Machine Intelligence (PAMI), vol.22, no.10, pp.1090-1104, Oct. 2000.
[31] Proc. of the IEEE International Conference on Automatic Face and Gesture Recognition, June 1995, Oct. 1996, April 1998, March 2000, and May 2002.
[32] The Face Recognition Home Page: http: //www.cs.rug.nl/~peterkr/FACE/face.html
[33] http: //www.frvt.org
[34] P.J.Phillips, P.Grother, R.J.Micheals, D. M. Blackburn, E. Tabassi, and M. Bone: "Face recognition vendor test 2002, Overview and summary," March 2003.
[35] K.Fukui, O.Yamaguchi, K.Suzuki, and K.Maeda: "Face recognition under variable lighting condition with constrained mutual subspace method -- learning of constraint subspace to reduce influence of lighting changes," IEICE Trans. D-II, vol.J82-D-II, no.4, pp.613-620, Apr. 1999 (in Japanese).

[36] S.Gong, S.J.McKenna, and A.Psarrou: "Dynamic vision : From images to face recognition," Imperial College Press, 2000.

[37] F.Goudali, E.Lange, T.Iwamoto, K.Kyuma, and N.Otsu: "Face recognition system using local autocorrelations and multiscale integration," IEEE Trans. Pattern Analysis and Machine Intelligence (PAMI), vol.18, no.10, pp. 1024-1028, Oct. 1996.

[38] B.Guenter, C.Grimm, D.Wood, H.Malvar, and F.Pighin: "Making faces," SIGGRAPH98 Conference Proc., pp.55-66 , July 1998.

[39] G.Hanaoka, M.Kaneko, and H.Harashima: "Facial caricature by computer based on the style of individual human caricaturist," Trans. IEICE D-II, vol.J80-D-II, no.8. pp.2110-2118, Aug. 1997 (in Japanese).

[40] H.Harashima: "Intelligent image coding and communication," Journal of the Institute of Television Engineers of Japan, vol.42, no.6, pp.519-525, June 1988 (in Japanese).

[41] H.Harashima, K.Aizawa, and T.Saito: "Model-based analysis and synthesis coding of videotelephone images -- conception and basic study of intelligent image coding," Trans. IEICE, vol.E72, no.5, pp.452-459, May 1989.

[42] http: //www.hc.t.u-tokyo.ac.jp/facegallery/index.html

[43] H.Harashima, et al.: "Facial image processing system for human-like 'Kansei' agent," Proc. of the 15th IPA Technical Conference, vol.15, pp.237-249, Oct. 1996 (in Japanese).

[44] H.Harashima, et al.: "Facial image processing system for human-like 'Kansei' agent," 1996 Annual General Report of Individual Theme, IPA (The Information-technology Promotion Agency, Japan), pp.99-106, March 1997 (in Japanese).

[45] H.Harashima et al. Ed.: IEICE Trans. Special issue on "Computer and cognitive scientific researches of faces," vol.J80-A, no.8, pp.1213-1336, Aug. 1997, and vol.J80-D-II, no.8, pp.2029-2258, Aug. 1997 (in Japanese).

[46] K.Aizawa, H.Harashima, and T.Saito: "Model-based analysis synthesis image coding (MBASIC) system," Signal Processing: Image Communication, vol.1, no.2, pp.139-152, Oct.1989.

[47] O.Hasegawa, K.Yokosawa, and M.Ishizuka: "Real-time parallel and cooperative recognition of facial images for an interactive visual human interface," Proc. of the 12th IAPR International Conference on Pattern Recognition (ICPR'94), pp. 384-387, Oct. 1994.

[48] O.Hasegawa, S.Morishima, and M.Kaneko: "Processing of facial information by computer," Electronics and Communications in Japan, Scripta Technica, Part 3, vol.81, no.10, pp.36-57, Oct. 1998 (translated from Trans. IEICE D-II, vol.J80-D-II, no.8, pp. 2047-2065, Aug. 1997 (in Japanese)).

[49] O.Hasegawa, K.Sakaue, K.Itou, T.Kurita, S.Hayamizu, K.Tanaka, and N.Otsu: "Agent oriented multimodal image learning system", Proc. of the Workshop on Intelligent Multimodal Systems, in conjunction with IJCAI-97, pp.29-34, Aug. 1997.

[50] O.Hasegawa and K.Sakaue: "CG tool for constructing anthropomorphic interface agents," Proc. of the Workshop on Animated Interface Agents, pp. 23-26, Aug. 1997.
[51] http://www.humanoid.waseda.ac.jp/index-j.html and http://www.takanishi.mech.waseda.ac.jp
[52] Human Communication Engineering Group, The Institute of Electronics, Information and Communication Engineers Home Page: http: //www.ieice.or.jp/hcg
[53] L.Hong and A.Jain: "Integrating faces and fingerprints for personal identification ," IEEE Trans. Pattern Analysis and Machine Intelligence (PAMI), vol.20, no.12, pp.1295-1307, Dec. 1998.
[54] K.Hotta, T.Mishima, and T.Kurita: "Scale invariant face detection and classification method using shift invariant features extracted from Log-Polar image," IEICE Trans. Information and Systems, Vol.E84-D, No.7, pp.867-878, 2001.
[55] Q.Huynth-Thu, M.Meguro, and M.Kaneko: "Skin-color extraction in images with complex background and varying illumination, " Proc. of 6th IEEE Workshop on Applications of Computer Vision: WACV 2002, pp.280-285, Florida, USA, Dec. 2002.
[56] http: //www.identix.com
[57] H.H.S.Ip and L.Yin: "Constructing a 3D individualized head model from two orthogonal views," Visual Computer, vol.12, no.5, pp.254-268, 1996.
[58] http: //www.irobot.com
[59] Journal of the Institute of Image Information and Television Engineers, Special Edition on "Image engineering of human face and body," vol.51, no.8, pp. 1169-1174, Aug. 1997 (in Japanese).
[60] S.Iwashita and T.Onisawa: "Facial caricature drawing based on the subjective impressions with linguistic expression," IEICE Trans. D-I, vol.J83-D-I, no.8, pp.891-900, Aug. 2000 (in Japanese).
[61] Japan Academy of Facial Studies Home Page: http: //www.hc.t.u-tokyo.ac.jp/jface
[62] P.Kalra and N.Magnenat-Thalmann: "Modeling of vascular expressions in facial animation," Computer Animation '94. pp.50-58, 1994.
[63] P.Kalra, N.Magnenat-Thalmann, L. Moccozet, G.Sannier, A.Aubel, and D.Thalmann: "Real-time animation of realistic virtual humans," IEEE Computer Graphics and Applications, vol.18, no.5, pp.42-56, Sep./Oct. 1998.
[64] M.Kampmann: "Automatic 3-D face model adaptation for model-based coding of videophone sequences," IEEE Trans. Circuits and Systems for Video Technology, vol.12, no.3, pp.172-182, March 2002.
[65] T.Kanade: "Computer recognition of human faces," Basel and Stuttgart, Birkhauser Verlag, 1977.
[66] M.Kaneko, Y.Hatori, and A.Koike: "Coding of facial images based on 3-D model of head and analysis of shape changes in input image sequence," Trans. IEICE B, vol.J71-B, no.12, pp.1554-1563, Dec. 1988 (in Japanese).

[67] M.Kaneko, A.Koike, and Y.Hatori: "Coding of moving facial images using combination of model-based coding and waveform coding," Proc. of 6th Picture Coding Symposium of Japan, 3-2, pp.61-64, Oct. 1991 (in Japanese).

[68] G.Xu, M.Kaneko, and A.Kurematsu: "Synthesis of facial caricatures using eigenspaces and its applications to communication," Proc. of 1999 International Workshop on Very Low Bitrate Video Coding (VLBV'99), pp.192-195, Kyoto, Japan, Oct. 1999.

[69] G.Xu, M.Kaneko, and A.Kurematsu: "Synthesis of facial caricature using eigenspaces," IEICE Trans. D-II, vol.J84-D-II, no.7, pp.1279-1288, July 2001 (in Japanese).

[70] T.Mizuno, M.Meguro, and M.Kaneko: "Analysis of facial features using eigenspaces and its application to synthesis of facial caricature," IEICE Technical Report, HCS2002-53, Mar. 2003 (in Japanese).

[71] D.Keren, M.Osadchy, and C.Gotsman: "Antifaces: A novel, fast method for image detection," IEEE Trans. Pattern Analysis and Machine Intelligence (PAMI), vol.23, no.7, pp.747-761, July 2001.

[72] M.Kirby and L.Sirovich: "Application of the Karhunen-Loeve procedure for the characterization of human faces," IEEE Trans. Pattern Analysis and Machine Intelligence (PAMI), vol.12, no.1, pp.103-108, Jan. 1990.

[73] http: //www.ai.mit.edu/projects/humanoid-robotics-group/kismet/kismet. html

[74] R.M.Koch, M.H.Gross, F.R.Carls, D.F.vonBüren, G.Fankhauser, and Y.Parish: "Simulating facial surgery using finite element methods," Proc. of SIGGRAPH 96. pp.421-428, 1996.

[75] H.Kobayashi, A.Tange, and F.Hara: "Real-time recognition of 6 basic facial expressions," Proc. of IEEE International Workshop on Robot and Human Communication (RO-MAN'95), pp.179-186, July 1995.

[76] H.Kobayashi, F.Hara, G.Uchida, and M.Ohno: "Study on face robot for active human interface: Mechanisms of face robot and facial expressions of 6 basic emotions," Journal of Robotics Society of Japan, vol.12, no.1, pp. 155-163, Jan. 1993.

[77] H.Kobayashi: "Study on face robot platform as a KANSEI medium," Journal of Robotics and Mechatronics, vol.13, no.5, pp.497-504, 2001.

[78] H.Koshimizu and K.Murakami: "Facial caricaturing based on visual illusion: A mechanism to evaluate caricature in PICASSO system," Trans. IEICE D, vol.E76-D, no.4, pp.470-478, 1993 (in Japanese).

[79] C.J.Kuo, R.-S.Huang, and T.-G.Lin: "3-D facial model estimation from single front-view facial image," IEEE Trans. Circuits and Systems for Video Technology, vol.12, no.3, pp.183-192, Mar. 2002.

[80] T.Kurita, N.Otsu, and T.Sato: "A face recognition method using higher order local autocorrelation and multivariate analysis," Proc. of the 11th IAPR International Conference on Pattern Recognition (ICPR'92), vol. II, pp.213-216, Aug. 1992.

[81] K.-M.Lam and H.Yan: "An analytic-to-holistic approach for face recognition based on a single frontal view," IEEE Trans. Pattern Analysis and Machine Intelligence (PAMI), vol.20, no.7, pp.673-686, July 1998.

[82] A.Lanitis, C.J.Taylor, and T.F.Cootes: "Automatic interpretation and coding of face images using flexible models," IEEE Trans. Pattern Analysis and Machine Intelligence (PAMI), vol.19, no.7, pp.743-756, July 1997.

[83] A.Lanitis, C.J.Taylor, and T.F.Cootes: "A unified approach to coding and interpreting face images," Proc. of the International Conference on Computer Vision (ICCV'95), pp. 368-373, June 1995.

[84] F.Lavagetto and R.Pockaj: "The facial animation engine: toward a high-level interface for the design of MPEG-4 compliant animated faces," IEEE Trans. Circuits and Systems for Video Technology, vol.9, no.2, pp.277-289, Mar. 1999.

[85] J.C.Lester, J.L.Voerman, S.G.Towns, and C.B.Callaway: " Cosmo: a life-like animated pedagogical agent with deictic believability," Proc. of the Workshop on Animated Interface Agents: Making them Intelligent, in conjunction with IJCAI-97, pp.61-69, Aug. 1997.

[86] H.Li and R.Forchheimer: "Two-view facial movement estimation," IEEE Trans. Circuits and Systems for Video Technology, vol.4, no.3, pp.276-287, June 1994.

[87] S.Z.Li, Z.Q.Zhang, H. Shum, and H.J.Zhang: "FloatBoost learning for classification," Proceedings of the 16-th Annual Conference on Neural Information Processing Systems (NIPS), Vancouver, Canada, Dec. 9-14, 2002.

[88] C.Liu and H.Wechsler: "Evolutionary pursuit and its application to face recognition," IEEE Pattern Analysis and Machine Intelligence (PAMI), vol.22, no.6, pp.570-582, June 2000.

[89] Z.Liu, Z.Zhang, C.Jacobs, and M.Cohen: "Rapid modeling of animated faces from video," Journal of Visualization and Computer Animation, vol.12, no.4, pp.227-240, 2001.

[90] J.Luo, C.W.Chen, and K.J.Parker: "Face location in wavelet-based video compression for high perceptual quality videoconferencing," IEEE Trans. Circuits and Systems for Video Technology, vol.6, no.4, pp.411-414, Aug. 1996.

[91] M.J.Lyons, J.Budynek, and S. Akamatsu: "Automatic classification of single facial images," IEEE Trans. Pattern Analysis and Machine Intelligence (PAMI), vol.21, no.12, pp.1357-1362, Dec. 1999.

[92] K.Mase: "Recognition of facial expressions by optical flow," Trans. IEICE, Special Issue on Computer Vision and Its Applications, vol.74E, no.10, pp. 3474-3483, Oct. 1991.

[93] K.Messer, J.Matas, J.Kittler, J.Luettin, and G.Maitre: "XM2VTSDB: the extended M2VTS database," Proceedings of Second International Conference on Audio and Video-Based Biometric Person Authentication, pp.72-77, 1999.

[94] K.Matsuno, C.W.Lee, S.Kimura, and S.Tsuji: "Automatic recognition of human facial expressions," Proc. of the International Conference on Computer Vision (ICCV'95), pp.352-359, June 1995.

[95] T.Maurer and Christoph von der Malsburg: "Tracking and learning graphs and pose on image sequences of faces," Proc. IEEE 2nd International Conference on Automatic Face and Gesture Recognition (FG '96) pp. 76-81, 1996.

[96] P.Menzel and F.D'Aluisio: "ROBO SAPIENS : Evolution of a new species (Material World Books)," MIT Press, Aug. 2000. (Japanese translation version was published by Kawadeshobou-Shinsya in 2001).

[97] G.Moiza, A.Tal, I.Shimshoni, D.Barnett, and Y.Moses: "Image-based animation of facial expressions," Visual Computer, vol.18, no.7, pp. 445-467, 2002.

[98] B.Moghaddam and A.Pentland: "Probabilistic visual learning for object representation," IEEE Trans. Pattern Analysis and Machine Intelligence (PAMI), vol.19, no.7, pp.696-710, July 1997.

[99] Y.Mukaigawa, Y.Nakamura, and Y.Ohta: "Face synthesis with arbitrary pose and expression from several images - An integration of image-based and model-based approach -," Proc. of Asian Conference on Computer Vision (ACCV'98), Vol.1, pp.680-687, Jan. 1998.

[100] K.Nagao: "Design of interactive environment," Kyoritsu Shuppan Co., Ltd., 1996 (in Japanese).

[101] K.Nagao and A.Takeuchi: "Social interaction: multimodal conversation with social agents," Proc. of the 12th National Conference on Artificial Intelligence (AAAI-94), vol.1, pp.22-28, 1994.

[102] A.Nagata, M.Kaneko, and H.Harashima: "Analysis of face impression using average faces," Trans. IEICE A, vol.J80-A, no.8. pp.1266-1272, Aug. 1997 (in Japanese).

[103] M.Nakagawa, T.Munetugu, Y.Kado, F.Maehara, and K.Chihara: "The facial aging simulation based on the skeletal model," Trans. IEICE A, vol.J80-A, no.8. pp.1312-1315, Aug. 1997 (in Japanese).

[104] M.Nakagawa, T.Ohba, O.Delloye, M.Kaneko, and H.Harashima: "Interactive synthesis of facial images based on 3-dimensional model," Proc. of the 1996 Engineering Sciences Society Conference of IEICE, SA-10-5, pp.342-343, Sept. 1996 (in Japanese).

[105] Y.Nakaya, Y.C.Chuah, and H.Harashima: "Model-based/waveform hybrid coding for videotelephone images," Proc. ICASSP91, Canada, M9.8, pp.2741-2744, May 1991.

[106] Y.Nakayama: " Multimodal interface agent," Proc. of 1997 ITE (Institute of Image Information and Television Engineers) Annual Convention, S5-4, pp.477-480, Sept. 1997 (in Japanese).

[107] S.Nishio and K.Koyama: "A criterion for facial expression of laugh based on temporal difference of eye and mouth movement," Trans. IEICE A, vol.J80-A, no.8. pp.1316-1318, Aug. 1997 (in Japanese).

[108] K.Nitta, O.Hasegawa, T.Akiba, T.Kamishima, T.Kurita, S.Hayamizu, K.Itoh, M.Ishizuka, H.Dohi, and M.Okamura: "An experimental multimodal disputation system" Proc. of the Workshop on Intelligent Multimodal Systems, in conjunction with IJCAI-97, pp.23-28, Aug. 1997.

[109] T.Ohba, M.Kaneko, and H.Harashima: "Interactive retrieval of facial images using 'Fukuwarai' Module," Trans. IEICE D-II, vol.J80-D-II, no.8, pp. 2254-2258, Aug. 1997 (in Japanese).
[110] http: //www.society.omron.co.jp/faceid/2/f2.html
[111] E.Osuna, R.Freund, and F.Girosi: "Training support vector machines: an application to face detection," Proc. of CVPR'97, 17-19, June 1997.
[112] CMU, OZ Project: http: //www.cs.cmu.edu/afs/cs.cmu.edu/project/oz/web/oz.html
[113] M.C.Park, T.Naemura, M.Kaneko, and H.Harashima: "Recognition of facial gestures using the GRIP method for head motion estimation and hidden Markov models," IEEE International Conference on Image Processing, vol.4, no.861, Oct. 1997 (in CD-ROM).
[114] S.-C.Pei, C.-W.Ko, and M.-S.Su: "Global motion estimation in model-based image coding by tracking three-dimensional contour feature points," IEEE Trans. Circuits and Systems for Video Technology, vol.8, no.2, pp.181-190, April 1998.
[115] IEEE Trans. on Pattern Analysis and Machine Intelligence (PAMI), Theme Section, "Face and Gesture Recognition," vol.19, no.7, pp. 675-785, July 1997.
[116] M.Pantic and L.J.M.Rothkrantz: "Automatic analysis of facial expressions: The state of the art," IEEE Trans. Pattern Analysis and Machine Intelligence (PAMI), vol.22, no.12, pp.1424-1445, Dec. 2000.
[117] F.I.Parke and K.Waters: "Computer facial animation," A.K.Peters, Ltd., Boston, 1996.
[118] P.Penev and J.Atick: "Local feature analysis : a general statistical theory for object representation," Network: Computation in Neural Systems, pp.477-500, Mar. 1996.
[119] A.Pentland, B.Moghaddam, and T.Starner: "View-based and modular eigenspaces for recognition," Proc. of IEEE Conference on Computer Vision and Pattern Recognition, pp. 84-91, July 1994.
[120] A.Pentland and T.Choudhury: "Face recognition for smart environments," IEEE Computer, vol.33, no.2, pp.50-55, Feb. 2000.
[121] G.Ball, D.Ling, D.Kurlander, J.Miller, D.Pugh, T.Skelly, A.Stankosky, D.Thiel, M.van Dantzich, and T.Wax: "Lifelike computer characters: the Persona project at Microsoft Research," *Software Agents*, J.Bradshaw Ed., Cambridge, MA: MIT Press, 1997.
[122] P.Phillips, H.Wechsler, J.Huang, and P.Rauss: "The FERET database and evaluation procedure for face-recognition algorithm," Image and Vision Computing, vol.16, pp. 295-306, 1998.
[123] S.Pigeon and L.Vandendrope: "The M2VTS multimodal face database," Proceedings of First International Conference on Audio and Video-Based Biometric Person Authentication, 1997.
[124] F.Pighin, J.Hecker, D.Lischinski, R.Szeliski, and D.H.Salesin: "Synthesizing realistic facial expressions from photographs," SIGGRAPH98 Conference Proc., pp.75-84, July 1998.

[125] F.Pighin, R.Szeliski, and D. H.Salesin: "Modeling and animating realistic faces from images," International Journal of Computer Vision, vol.50, no.2, pp.143-169, 2002.
[126] Proceedings of the International Workshop on "Lifelike animated agents --- Tools, affective functions, and applications," The Seventh Pacific Rim International Conference on Artificial Intelligence, Tokyo, Japan, Aug. 2002.
[127] Proc. of the Workshop on Virtual Conversational Characters: Applications, Methods, and Research Challenges, Melbourne, Australia, Nov. 2002. (http://www.vhml.org/workshops/HF2002)
[128] J.Rickel and W.L.Johnson: " Steve: an animated pedagogical agent for procedural training in virtual environments (extended abstract)," Proc. of the Workshop on Animated Interface Agents: Making them Intelligent, in conjunction with IJCAI-97, pp.71-76, Aug. 1997.
[129] T.Rist and J.Muler: " WebPersona: a life-like presentation agent for the World-Wide Web," Proc. of the Workshop on Animated Interface Agents: Making them Intelligent, in conjunction with IJCAI-97, pp.53-60, Aug. 1997.
[130] S.A.Rizvi, P.Phillips, and H.Moon: "The FERET verification testing protocol for face recognition algorithms," Proc. of IEEE Conference on Automatic Face and Gesture Recognition, pp. 48-53, April 1998.
[131] D.A.Rowland and D.I.Perrett: "Manipulating facial appearance through shape and color," IEEE Computer Graphics and Applications, vol.15, no.5, pp.70-76, Sep. 1995.
[132] http: //www.cs.rug.nl/~peterkr/FACE/face.html
[133] RWC Multimodal Database: http: //rwc-tyo.noc.rwcp.or.jp/people/toyoura/rwcdb/home.html
[134] T.Sakaguchi, H.Yamada, and S.Morishima: "Construction and evaluation of 3-D emotion space based on facial image analysis," Trans. IEICE A, vol.J80-A, no.8. pp.1279-1284, Aug. 1997 (in Japanese).
[135] A.A.Salah, E.Alpaydin, and L.Akarun: "A selective attention-based method for visual pattern recognition with application to handwritten digit recognition and face recognition," IEEE Trans. Pattern Analysis and Machine Intelligence (PAMI), vol.24, no.3, pp.420-425, Mar. 2002.
[136] A.Samal and P.A.Iyengar: "Automatic recognition and analysis of human faces and facial expressions: A survey," Pattern Recognition, vol.25, no.1, pp.65-77, Jan. 1992.
[137] N.Sarris, N.Grammalidis, and M.G.Strintzis: "Building three dimensional head models," Graphical Models, vol.63, no.5, pp.333-368, 2001.
[138] N.Sarris, N.Grammalidis, and M.G.Strintzis: "FAP extraction using three-dimensional motion estimation," IEEE Trans. Circuits and Systems for Video Technology, vol.12, no.10, pp.865-876, Oct. 2002.
[139] S.Sato and T.Kanade: "Name-It : Naming and detecting faces in video by the integration of image and natural language processing," Proc. of International Conference on Artificial Intelligence (IJCAI97), pp.1488-1493, Aug. 1997.

[140] H.Schneiderman: "A statistical approach to 3D object detection applied to faces and cars," Ph.D Thesis, Tech. Report 00-06, Robotics Institute, Carnegie Mellon University, Pittsburgh, May 2000.
[141] R.E.Shapire and Y.Singer: "Improving boosting algorithms using confidence-rated predictions," Machine Learning 37(3), pp.297-336, 1999.
[142] A.Shashua and T.Riklin-Raviv: "The quotient image: Class-based rerendering and recognition with varying illuminations", IEEE Trans. Pattern Analysis and Machine Intelligence (PAMI), Vol.. 23, No.2, 2001.
[143] T.Sim, S.Baker, and M.Bsat: "The CMU pose, illumination, and expression (PIE) database," Proc. of 5th International Conference on Automatic Face and Gesture Recognition, pp.53-58, Washington, D.C., USA, May 2002.
[144] A.Smolic, B.Makai, and T.Sikora: "Real-time estimation of long-term 3-D motion parameters for SNHC face animation and model-based coding applications," IEEE Trans. Circuits and Systems for Video Technology, vol.9, no.2, pp.255-263, Mar. 1999.
[145] http: //www.hoip.jp/web_catalog/top.html
[146] D.Stork and M.Hennecke: "Speechreading: an overview of image processing, feature extraction, sensory integration and pattern recognition techniques," Proc. of International Conference on Automatic Face and Gesture Recognition, pp. xvi-xxvi, Oct. 1996.
[147] A.Sugiura and M.Ueda: "An assessment of prosopagnosia by using face image synthesis," Trans. IEICE A, vol.J80-A, no.1. pp. 294-297, Jan. 1997 (in Japanese).
[148] K.-K.Sung and T.Poggio: "Example-based learning for view-based human face detection," IEEE Trans. Pattern Analysis and Machine Intelligence (PAMI), vol.20, no.1, pp.39-51, Jan. 1998.
[149] D.L.Swets and J.J.Weng: "Using discriminant eigenfeatures for image retrieval," IEEE Trans. Pattern Analysis and Machine Intelligence (PAMI), vol.18, no.8, pp.831-836 , Aug. 1996.
[150] H.Miwa, A.Takanishi, and H.Takanobu: "Development of a human-like head robot WE-3RV with various robot personalities," Proceedings of IEEE-RAS International Conference on Humanoids Robots, pp.117-124, 2001.
[151] H.Miwa, T.Okuchi, H.Takanobu, and A.Takanishi: "Development of a new human-like head robot WE-4," Proceedings of the 2002 IEEE/RSJ International Conference on Intelligent Robots and Systems, pp.2443-2448, 2002.
[152] H.Takanobu, T.Akizuki, A.Takanishi, K.Ohtsuki, D.Ozawa, M.Ohnishi, and A.Okino: "Jaw training robot that manipulates patient's jaw to sideway", Proc. IEEE International Conference on Intelligent Robots and Systems (IROS), Volume 1, pp1463-1468, 2002.
[153] Y.Takebayashi: "Spontaneous speech dialogue system TOSBERG II - towards the user-centered multimodal interface," Trans. IEICE D-II, vol.J77-D-II, no.8, pp.1417-1428, Aug. 1994 (in Japanese).
[154] A.Tefas, C.Kotropoulos, and I.Pitas: "Using support vector machines to enhance the performance of elastic graph matching for frontal face authentica

tion," IEEE Trans. Pattern Analysis and Machine Intelligence (PAMI), vol.23, no.7, pp.735-746, July 2001.

[155] D.Terzopoulos and K.Waters: "Analysis and synthesis of facial image sequences using physical and anatomical models," IEEE Trans. Pattern Analysis and Machine Intelligence (PAMI), vol.15, no.6, pp.569-579, June 1993.

[156] Y.-I.Tian, T.Kanade, and J.F.Cohn: "Recognizing action units for facial expression analysis," IEEE Trans. Pattern Analysis and Machine Intelligence (PAMI), vol. 23, no.2, pp.97-115, Feb. 2001.

[157] B.Tiddeman and D.Perrett: "Transformation of dynamic facial image sequences using static 2D prototypes," Visual Computer, vol.18, no.4, pp. 218-225, 2002.

[158] M.Turk and A.Pentland: "Eigenfaces for recognition," Journal of Cognitive Neuroscience, vol.3, no.1, pp.71-86, 1991.

[159] O.de Vel and S.Aeberhard: "Line-based face recognition under varying pose," IEEE Trans. Pattern Analysis and Machine Intelligence (PAMI), vol.21, no.10, pp.1081-1088, Oct. 1999.

[160] D.Valentin, H.Abdi, A.J.O'Toole, and G.W.Cottrell: "Connectionist models of face processing, a survey," Pattern Recognition, vol.27, no.9, pp. 1209-1230, Sept. 1994.

[161] T.Vetter: "Synthesis of novel views from a single face image," International Journal of Computer Vision, vol.28, no.2, pp.103-116, 1998.

[162] http: //www.viisage.com

[163] P.Viola and M.Jones: "Rapid object detection using a boosted cascade of simple features", Proc. IEEE Computer Vision and Pattern Recognition, volume 1, pp.511-518, 2001.

[164] H.Wang and S.-F.Chang: "A highly efficient system for automatic face region detection in MPEG video," IEEE Trans. Circuits and Systems for Video Technology, vol.7, no.4, pp.615-628, Aug. 1997.

[165] L.Wiskott, J.Fellous, N.Kruger, and C. von der Malsburg: "Face recognition by elastic bunch graph matching," IEEE Trans. Pattern Analysis and Machine Intelligence (PAMI), vol.19, no.7, pp.775-779, July 1997.

[166] H.Wu, Q.Chen, and M.Yachida: "Face detection from color images using a fuzzy pattern matching method," IEEE Trans. Pattern Analysis and Machine Intelligence (PAMI), vol.21, no.6, pp.557-563, June 1999.

[167] T.Yokoyama, Y.Yagi, M.Yachida, and H.Wu: "Face contour extraction by using active contour model based on axis symmetry," Trans. IEICE D-II, vol.J80-D-II, no.8, pp.2178-2185, Aug. 1997 (in Japanese).

[168] P.Belhumeur, J.Hespanha, and D.Kriegman: "Eigenfaces vs. fisherfaces: recognition using class specific linear projection," IEEE Trans. Pattern Analysis and Machine Intelligence (PAMI), vol.19, no.7, pp.711-720, July 1997.

[169] M.-H.Yang, D.J.Kriegman, and N.Ahuja: "Detecting faces in images: A s urvey," IEEE Trans. Pattern Analysis and Machine Intelligence (PAMI), vol.24, no.1, pp.34-58, Jan. 2002.

[170] L.Zhang: "Automatic adaptation of a face model using action units for semantic coding of videophone sequences," IEEE Trans. Circuits and Systems for Video Technology, vol.8, no.6, pp.781-795, Oct. 1998.

[171] A.Sato, A.Inoue, T.Suzuki, and T.Hosoi: "NeoFace --- development of face detection and recognition engine," NEC Res. & Develop., vol.44, no.3, pp.302-306, July 2003.

8

Document Analysis and Recognition

Toyohide Watanabe
Nagoya University, Nagoya, Japan

SUMMARY

Document image understanding aims to extract and classify meaningful data individually from paper-formed documents. Until recently, many methods and approaches have been proposed with regard to structure recognition for various kinds of documents, technical enhancement problems for OCR, and requirements for practical usage. Of course, though the technical research issues in the early stage are looked upon as complementary attacks for the traditional OCR, which is dependent on character recognition techniques, the application ranges or related issues were widely investigated or should be researched progressively. This chapter addresses current topics about document image understanding from a technical point of view as a survey.

1. INTRODUCTION

Document analysis and recognition focuses on analyzing document structures physically or logically and distinguishes individual items organically. The techniques in many current investigations are to extract and classify meaningful information from paper-formed documents automatically so as to complementarily support the traditional OCR techniques (1,2). Roughly speaking, the techniques for document analysis and recognition mainly focus on 2-dimensional/1-dimensional recognition (e.g., for page structures or constructive relationships among items), while the traditional OCR techniques chiefly focus on 0-

dimensional recognition (e.g., for characters or items) (3-5). Of course, the system for document analysis and recognition includes the ability of character recognition which enables to manipulate easily the extracted data for the applicable processing.

In comparison with traditional OCR, the techniques of document analysis and recognition deal with documents totally in point of constructive relationships among items, composition constraints among neighboring or related items, description rules for connection of items, etc. Until recently, many methods and approaches have been used in various types and kinds of documents such as tables, banking checks, journal pages, newspaper pages, application forms, business cards, official letters, official materials, technical reports (6-13), as well as in various diagrams, drawings, maps, etc. in the graphics recognition (14-18), and these systems have been implemented. However, currently usable products are not provided as their own complete systems, but are available as parts of functional OCR systems. Techniques combined with OCR are desirably applicable to the work in big companies like banks though the practical uses are too limited for consumers. This is because the document form is specified uniformly in the regularly organized process or it is required that many documents of the same type should be speedily managed at once in terms of the performance. Namely, in the current situation it is important for data management in big companies that many documents with the same form should be manipulated effectively and speedily. This situation is not always desirable in the research and development of document analysis and recognition. The functionality and effectiveness attended inherently with document analysis and recognition must be applied to a wide range of utilizations for end-users by means of successful products.

In this chapter we focus on the technical topics for document analysis and recognition. First, we discuss the document structure from a viewpoint of logical and physical structures. The documents, which have been investigated under the research and development of document analysis and recognition, can be evaluated with respect to the logical and physical features of document composition structure. In general, the physical features are called layout structures on the basis of the geometric and spatial relationships among composite elements. Second, we discuss the methods that have been proposed for individual documents until recently, in accordance with the features of approaches. The representative means of document structures are roughly classified into frame-based and rule-based in the traditionally developed methods and approaches. Also, the analysis and recognition methods based on these representations are systematically organized by means of the interpretation mechanism of individual knowledge representations.

2. DOCUMENT STRUCTURE

Generally, documents are organized systematically more or less under some configuration rules, description notations, and so on. Individual components which organize each page or whole pages of documents are associated mutually with constructive relationships. Usually, the constructive relationships are assigned to make individual components meaningful on a 2-dimensional space of page or 3-dimensional space of volume from a viewpoint of logical and physical structures. These constructive relationships are specified as a logical structure for the inherent property of an item and a layout structure for the location property of the item.

2.1 Logical Structure and Layout Structure

The logical structure and layout structure are the basic views for analyzing document configurations (19,20). The logical structure defines the content-based properties of individual document components with respect to their relationships among dependent and independent components: usually, the inclusive relationship is effective. The layout structure specifies the geometric and spatial positions of individual document components with respect to their physical page structures. In the layout analysis or structure recognition, the knowledge about layout structure is mainly used because this information can represent the structural positions of individual components explicitly on a 2-dimensional space and it is easy to analyze and recognize document components constructively.

Additionally, the layout structure may be classified into the logical and physical layout structures in accordance with the allocation means of individual document components (21,22). In some documents such as tables, individual composite elements are allocated into predefined fields with respect to the structural attributes systematically. In other documents such as article pages, each composite element is located relatively with respect to the physical features of other composite elements (e.g., the sizes, volumes, etc.). Of course, in the case of the article page, the approximate structure is physically specified in advance. Namely, the logical and physical layout structures are roles to distinguish the features of individual documents. Figure 1 shows a classification of typical documents with respect to the relationship between the logical and physical layout structures.

Of course, every existing document is not always distinguished well under such relationships. In Figure 1, the documents are classified into four types (23). In many cases, we often find out that currently used documents are edited as compound collections of different layouts. Such a complicatedly organized document is called the complex document (24). However, even if documents are complex, the composite sub-documents can be identified as one of four types of documents when they are analyzed on the basis of characteristics of their components and their structural relationships.

Document type 1

This type is too strongly dependent on the physical layout structure. The positions, lengths and so on of individual items are always fixed in advance. Examples include application forms, banking checks, and questionnaires.

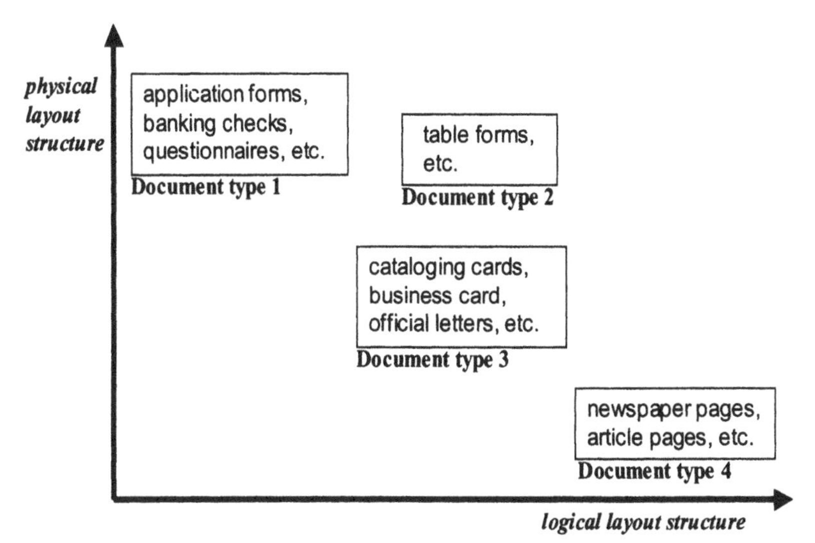

Figure 1: Document types

Document type 2
This type is specified by the logical layout structure more effectively than by the physical layout structure. Namely, the position of each item may be moved up/down or left/right from the normal location according to the interrelation of mutually related items or among previously allocated items. For example, cataloging cards, official letters and business cards are typical.

Document type 3
This type is dependent on the physical layout structure, as well as the document type 1. In comparison with document type 1, this type may be complex in structure (including hierarchical or repeated items) or the layout structure may be guided by other elements (including line segments, blank areas, etc.). The position, length, and so on, of each item is always fixed. For example, table forms are typical.

Document type 4
In this type, the positions, lengths, and so on, of individual items are ordinarily dependent on those of related items or other items. Generally, the global document structure is predefined by the physical layout structure in this type of document, and the allocation strategy for the practical locations of individual items is wholly specified by the constructive relationship among individual items. Namely, this type is related to the physical layout structure in terms of the whole output form, and is also arranged by the logical layout structure with respect to the locations

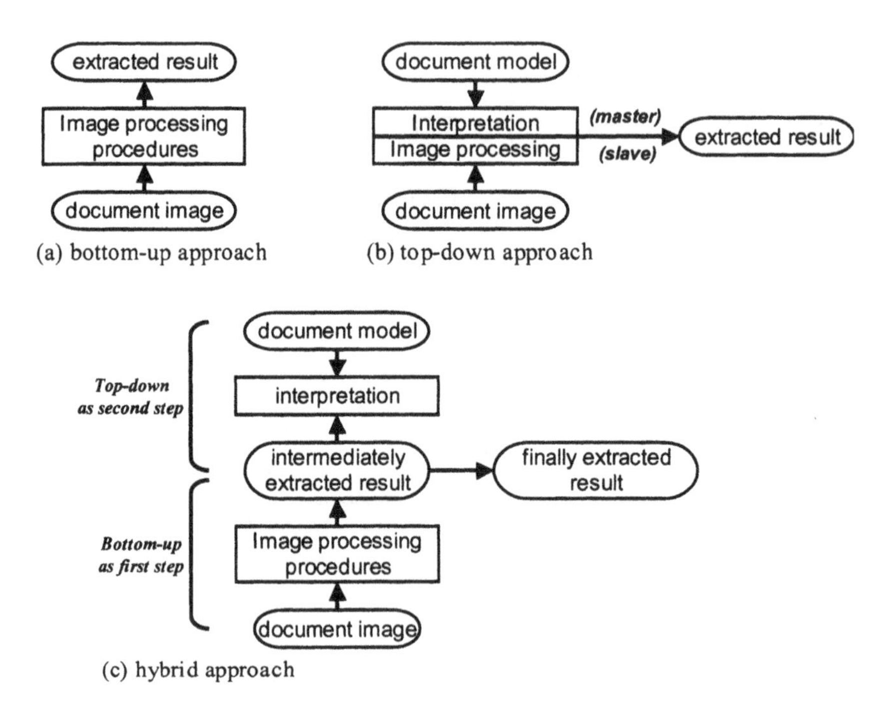

Figure 2: Top-down and bottom-up approaches

and shapes of individual items. For example, newspaper pages and article pages are typical.

Thus, we observe that different types of documents should be better processed so as to be consistent to the characteristics among items with respect to the logical layout structure and physical layout structure. At the very least, the optimal recognition methods ensure processing efficiency and recognition ratio as well as the flexibility, adaptability, and applicability of the processing mechanism.

2.2 Approach

Many approaches and methods have been proposed for various kinds of documents. Of course, these approaches and methods were more or less investigated by depending on application-specific document structures. The approach is categorized into the bottom-up and top-down methods (19,20,23). The bottom-up approach has been traditionally applied to an image processing techniques, while about 10 years ago the top-down approach was addressed interestingly to the knowledge-based processing. In the top-down approach, it is important to specify the knowledge about document features declaratively, which helps the analysis of structure and the interpretation of individually extracted items. This knowledge is called the document model.

It is not easy, however, to define this document model directly for various types of documents or complex documents. Thus, the hybrid approach which is composed complementarily of top-down and bottom-up approaches, is effective in many cases. In the first step the bottom-up approach is applied to the document images in order to approximately analyze and extract the composite elements (as the transformation from pixel-based data to vector-based and symbolic-based data). In the second step, the top-down approach is applicable to the interpretation of transformed vector-based and symbolic-based data with the document model. Of course, as yet another framework, the bottom-up and top-down approaches may be cooperatively integrated. Figure 2 shows such an individual approach conceptually.

2.3 Document Model

The document model, which is important in the top-down approach as shown in Figure 2, specifies the constructive features about document configuration, composition rule, description rule, data domain and so on. Namely, the document model is a kind of knowledge about document structure (19-23). Although it is better to define this document model more generally in order to apply many kinds of documents effectively, the currently proposed frameworks for specifying document models depend on application-specific or similar document classes. This is because it is difficult to interpret document structures successfully by using the more abstract document model. Additionally, the description information in document model is distinguished from the logical information and physical information (23,25,26).

2.3.1 Physical or logical representation

The layout structure of documents is dependent on application-specific usage. The difference between the physical representation and logical representation for knowledge specification is dominated with respect to 2-dimensional layout structures of documents. The physical representation is defined by knowledge specification means, which make use of coordinate data of individual items such as positions, sizes, lengths, etc. Logical representation is constructively specified by means of interrelated and interdependent relationships among individual items.

For example, consider the two table forms in Figure 3. These documents are different in their physical representation but may be the same in their logical representation. Of course, these representations are dependent on various specification views for document structures. However, the logical representation is abstracted more than the physical representation. The logical representation of document structure is very applicable to various documents of the same or similar types (or classes) if the inference mechanism, based on defined knowledge, is effe-

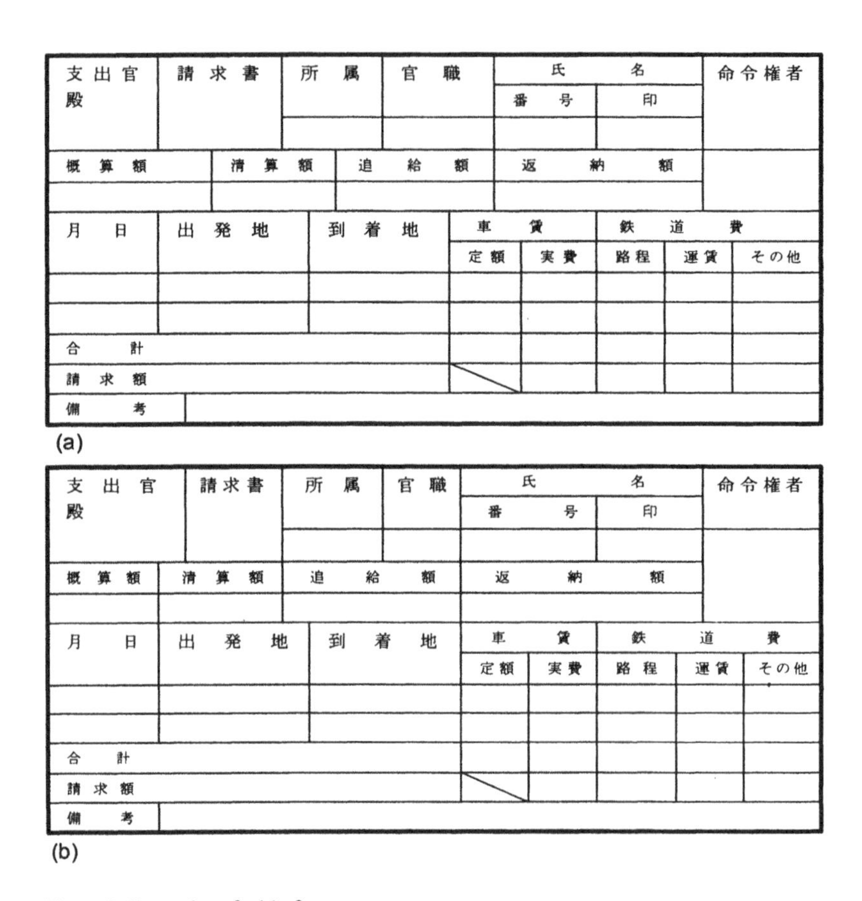

支出官 殿	請求書	所属	官職	氏名		命令権者
				番号	印	

概算額	清算額	追給額	返納額

月日	出発地	到着地	車賃		鉄道費		
			定額	実費	路程	運賃	その他
合計							
請求額							
備考							

(a)

支出官 殿	請求書	所属	官職	氏名		命令権者
				番号	印	

概算額	清算額	追給額	返納額

月日	出発地	到着地	車賃		鉄道費		
			定額	実費	路程	運賃	その他
合計							
請求額							
備考							

(b)

Figure 3: Examples of table forms.

ctively provided. For example, consider two document fragments in Figure 4. In the logical representation, the predicate "neighbor(A,B)" directly shows that item blocks (or fields) "A" and "B" are adjacent, and is applicable to (a) and (b). In the physical representation, the coordinate data of (a) and (b) must be checked interpretatively whether they are the same positions for x and y-axes. Of course, the meanings of neighboring relationships may be defined in accordance with individual processing and interpretation schema.

2.3.2 Abstraction level

In specifying the knowledge about documents, we can concentrate on various characteristics of documents and represent them as usable knowledge (25-28). In this case, the representation method depends on the use of knowledge and the interpretation of document images. We call the representation range of knowledge

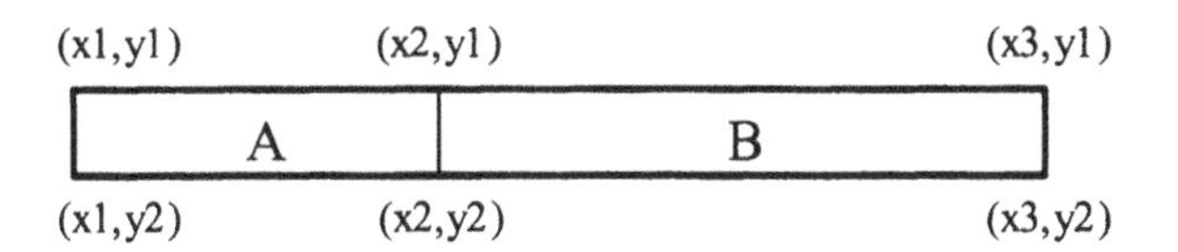

physical representation: ((x1,y1),(x2,y2)),((x2,y1),(x3,y2))

logical representation: **neighbor(A,B)**

(a)

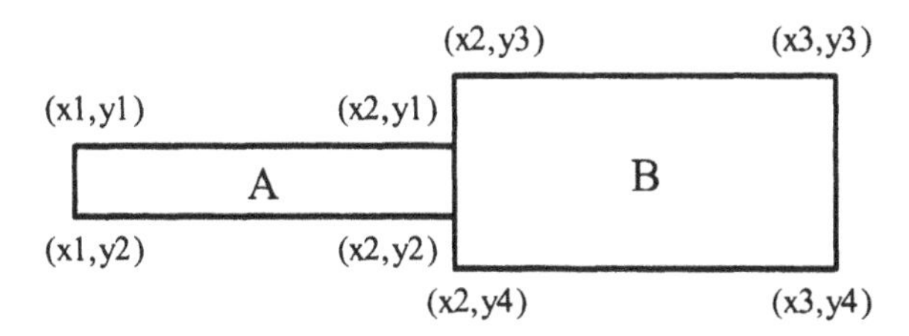

physical representation: ((x1,y1),(x2,y2)),((x2,y3),(x3,y4))

logical representation: **neighbor(A,B)**

(b)

Figure 4: Logical and physical representations.

the abstraction level. The higher the abstraction level is, the stronger the applicability of knowledge becomes. For example, consider the fragments of table forms shown in Figure 5.

These fragments are specified as the neighboring relationships among rectangular item blocks (or fields): "A", "B", "C" and "D". In Figure 5, three different fragment structures are illustrated. If we represent the adjacency relationships with commonly shared line segments among two rectangular item blocks (or fields), we can specify the relationships as illustrated in Figure 6. The arrow indicates the relationship, and the symbols "h" and "v", which are attended with arrows, show whether these neighboring blocks (or fields) are connected horizontally or vertically. Individual document fragments are represented by different neighboring relationships, respectively.

If we make use of the upper-left corners of individual rectangular item blocks (or fields), the resulting adjacency relationships are illustrated in Figure 7.

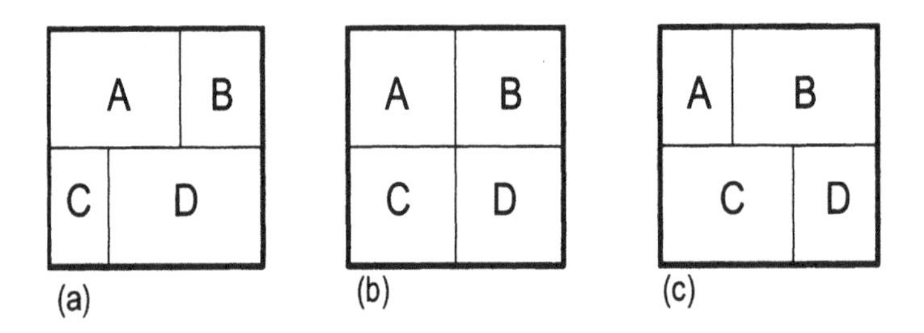

Figure 5: Fragments of table forms

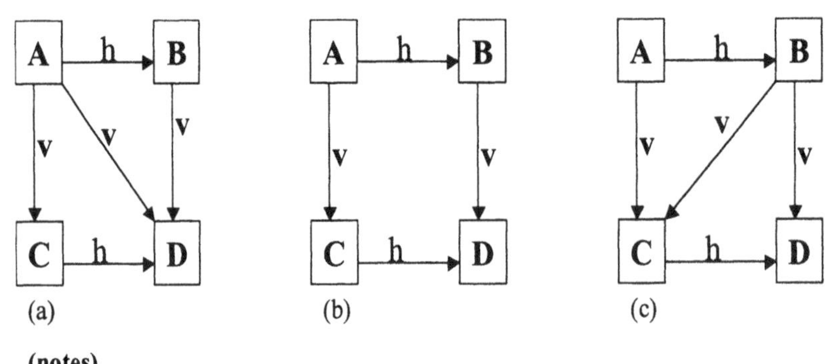

(notes) A,B,C,D: item blocks or fields
v: vertical adjacent relationship
h: horizontal adjacent relationship

Figure 6: Representation by commonly shared line segments.

In this figure, these different document fragments are specified by the same notations as those in Figure 6. The distinction between the representation in Figure 6 and that in Figure 7 is clear. Although the representation for adjacency relationships in Figure 6 is different in accordance with their geometric structures, those in Figure 7 are the same even if their geometric relationships are different. Of course, Figure 5(b) may be transformed into another adjacency relationship. Item block "B" is a neighbor of item block "D" by means of the arrow "v" in place of the arrow "h" between item blocks "C" and "D". Such a difference is derived from specification views of individual knowledge designers. Thus, both specifications are true for the layout analyzer: in our case, it is only one processing purpose that the layout analyzer judges whether the currently parsing layout structure is consistent with the predefined knowledge of this table form. At least, we can conclude that the representation means in Figure 7 are more strongly abstracted than those in Figure 6.

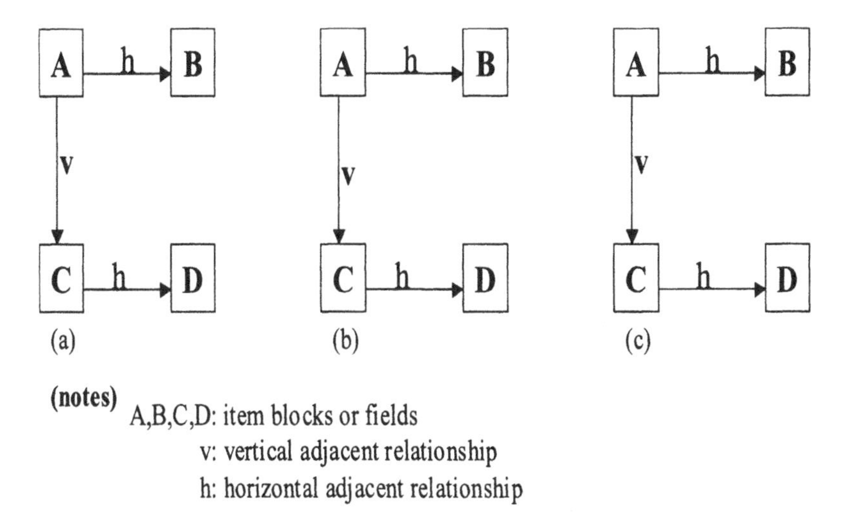

Figure 7: Representation by upper-left corners.

2.3.3 Semantic or syntactic information

The knowledge that is usable for analyzing and recognizing document images is mostly composed on the basis of the layout structure of documents. This is because the currently proposed methods focus mainly on geometric characteristics of document structures and distinguish individual items with the assistance of structure analysis one by one. However, these methods based on the models of layout structures are not always applicable to various documents. For example, consider the case where the layout structure is irregularly transformed from the original form or where the structural characteristics are not extracted sufficiently.

In general, the recognition results, which were distinguished only by means of applications of syntactic knowledge, may not always be correct. This is because processing based on the syntactic knowledge is heavily dependent on the representation means of objects. Thus, we must investigate other knowledge with a view to resolving such a problem. We call it semantic knowledge (25,26). Semantic knowledge is information that defines the domain of individual items and specifies the interdependent and interrelated relationships among items. Semantic information is complementary to syntactic information with a view to understanding document images. The issue about document image understanding is that it not only focuses on the development of an effective method for extracting and classifying individual items automatically, but also must concentrate on making sure that the identified results are valid.

Today, the phrase "ground truth" is regarded as an important concept in analyzing and recognizing documents correctly. The ground truth is related to the validation of recognized results. Namely, the semantic information is a basis of ground truth and takes a useful role to ensure ground truth (29).

3. DOCUMENT MODEL AND ANALYSIS METHOD

Generally, the recognition of document structure can be organized under the paradigm as illustrated in Figure 8. The document class recognition may not be addressed under the framework of document analysis and recognition explicitly (30). Much research and development has excluded document class recognition as the direct subject because the documents to be analyzed are too strongly dependent on the application and also the main features about layout structures of documents are often explicitly pre-specified. The layout recognition, item recognition, and character recognition are important modules to extract and classify the meaningful information from the document images, and document class recognition has an important role in expanding to a wide range of applications progressively (3-5,31).

Layout recognition separates individual item groups under a constructive relationship and is the most critical procedure in a document analysis and recognition. This procedure divides 2-dimensional document data into groups of 1-dimensional item sequence data. Item recognition identifies individual items as meaningful composite elements of documents under the descriptive relationship. This procedure transforms 1-dimensional item sequence data into ordered (or unordered) collections of 0-dimensional character data. Finally, character recognition extracts individual characters/symbols as meaningful words/notations under the data format or domain value. Of course, these procedures may not always be organized systematically or the objects to be recognized may not always be determined uniformly. This is because the recognition procedure is dependent on the specification of the document model. Some methods do not separate these three layer procedures independently and other methods organize three-layer recognition procedures explicitly. Layout recognition and item recognition are composed under complementary organization and the recognition level between these procedures is not explicit: in one method item recognition is not explicitly provided when the item sequence rule is very simple or when items are easily classified in the layout recognition.

The document model is classified into frame-based means and rule-based means. Also, the frame-based means are divided into list-based means, tree-based means, and graph-based means (19,23). Generally, the list-based means are adaptable directly to analyze the structure, and also are more effective because the knowledge representation is very simple. On the other hand, the rule-based means are applicable to analyze more complicated document structure because individual items as composite elements of documents are inferred through the constructive relationships among individually adjacent items(32-34). Namely, in the document types shown in Figure 1 the documents formed deterministically by the physical layout structure are specified by means of list-based knowledge representation. On the contrary, the documents specified irregularly by the logical layout structure should be analyzed through rule-based knowledge representation. On the contrary, the documents specified irregularly by the logical layout structure should be analyzed through rule-based knowledge representation.

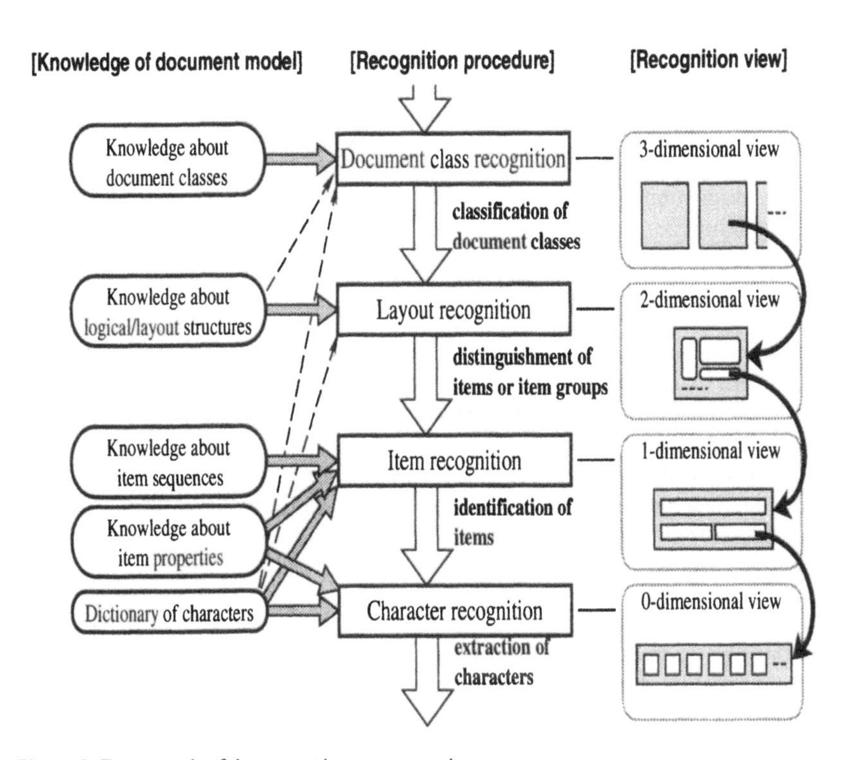

Figure 8: Framework of document image processing.

Concerning the layout analysis methods, the recognition paradigm is shown with respect to individual document types, hereafter. Layout recognition based on layout knowledge must analyzed and interpreted to the document structure to identify individual item data with the applicable document model, which specifies not only layout knowledge but also other various information about the documents. Namely, the knowledge-based layout recognition process is organized systematically as the model-driven approach. Here, we address different approaches for various types of documents and arrange individual frameworks from the viewpoint of knowledge representation and a processing mechanism.

Document type 1

All item data are always located at the predefined positions, often associated with some leading words (or key terms) and so on. Basically, individual data items can be assigned line by line. Thus, the layout knowledge is very simply and certainly specified on the basis of the locations of individual data items. The structure of layout knowledge is represented by the list (or frame). In this case, the layout

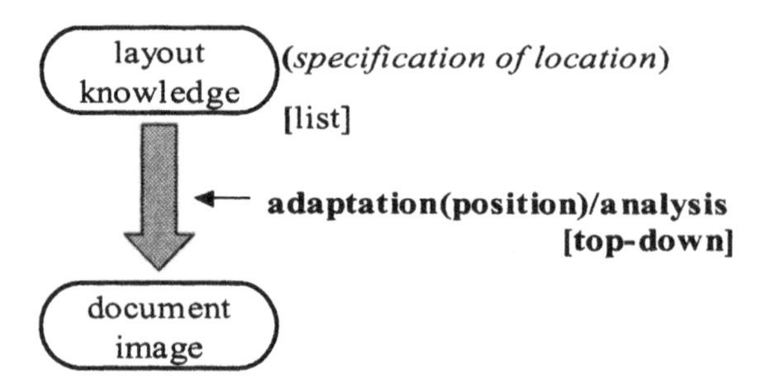

Figure 9: Framework of document type 1.

recognition process can be organized conceptually as shown in Figure 9. The document images are analyzed directly by appropriate image processing routines under the adaptation of layout knowledge. Of course, this process is mainly employed in the top-down approach.

For example, the method proposed in (35) is typical for application forms. And, many traditionally proposed knowledge-based (or model-driven) approaches are categorized into processing this type of document with the knowledge representation means.

Document type 2

In this type of document, individual data items are controlled mainly by the logical layout structure with respect to the location: the position of data items is variously alterable by other related data items. Thus, the layout knowledge cannot accommodate the location information effectively. Although some methods that use coordinate data to assign the positions of individual data items have been reported, these methods are too strongly limited concerning adaptability. Of course, these methods introduced the reasonable matching ranges for the pre-assigned coordinate data in order to eliminate disadvantages of adaptability, but they were unsuccessful in applying well to variously organized instances. Namely, in comparison with document type 1, it is better that the layout knowledge in document type 2 excludes coordinate data (as physical information) such as the position, size, length, etc. (25,26).

For example, the method proposed in (4,31) is very applicable to various documents of this type: business cards, cataloging cards, reference lists, etc. This layout knowledge is defined by the tree as a sequence set of partition operations, based on the neighboring relationships among individual data items (as logical information) (36), though many conventional methods adapt ad hoc means that specify the constructive structure, coordinate data, and so on, about individual data items (or physical properties of data items) directly. Figure 10 shows the layout recognition process based on such operations, conceptually. Of course, this process is composed in the top-down approach, as well as that of document type 1.

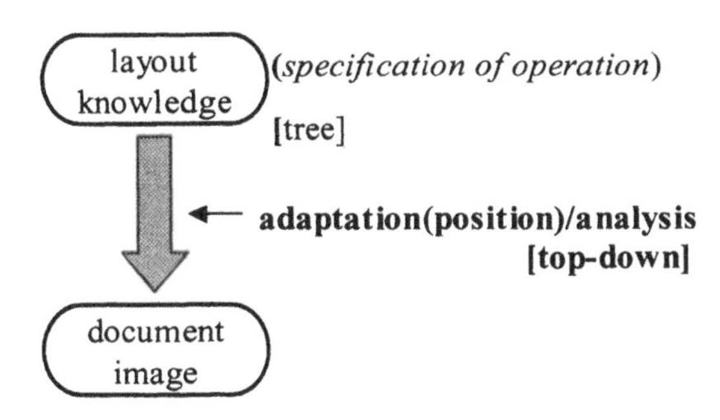

Figure 10: Framework of document type 2.

Document type 3

This type of document is commonly organized as a table form and individual data items may be allocated to the predefined positions as well as the document type 1. Moreover, individual data items may relate to other data items hierarchically or repeatedly (37). In some cases these data item's are surrounded with vertical or horizontal line segments (38) or in other cases they are separated by blanks (39). Of course, this type of document is constrained strongly by the physical layout structure. The layout knowledge can be represented by the tree because the hierarchical structure among data items can be regarded as an upper-lower relationship among nodes in the tree and the repeating structures can also be defined as the attribute of the upper node (40,41).

For example, the method proposed in (25-27,30,38,39), specifies a complex structure, using two different binary trees: a global structure tree and a local structure tree. The global structure tree specifies the characteristic structure such as hierarchy and repeating in addition to the constructive relationships among data items and groups. Also, the local structure tree defines in detail the connectivity among data items individually. Figure 11 shows a layout recognition process conceptually. In this figure, two different processing phases are illustrated: first, the analysis phase extracts the characteristic points (as upper-left corners of data item fields, generated from vertical and horizontal line segments) from document images by means of the bottom-up approach, and then the interpretation phase distinguishes the extracted characteristic points with the layout knowledge in the top-down approach.

Document type 4

In this type of document, individual data items and their composite elements are more complicatedly interrelated than the previously addressed document types. The positions of individual data items or composite elements are too strongly dependent on the previously allocated data items or other related composite elements under the physical layout structure. The layout knowledge can be defined by the rules.

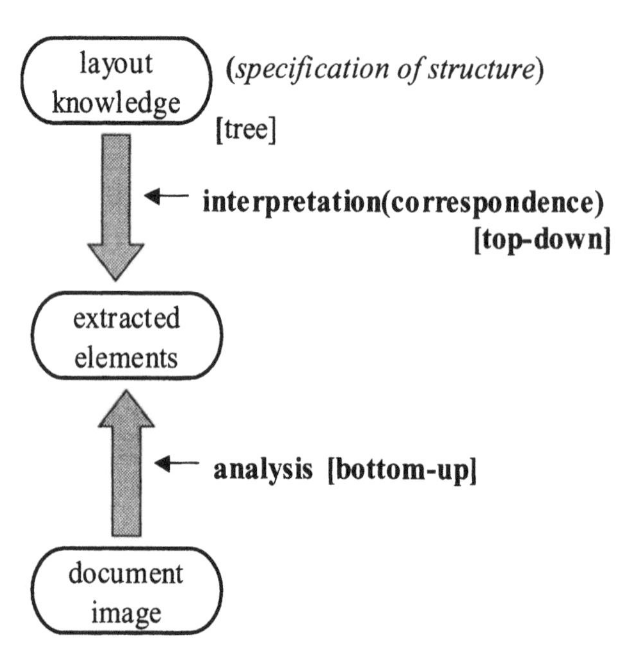

Figure 11: Framework of document type 3.

For example, the method of layout recognition of Japanese newspapers, proposed in (42), or the method of layout recognition of English newspapers, proposed in (43), is composed of a production system in order to establish the correct correspondence among interrelated data items and elements. The framework of this layout recognition is shown in Figure 12. The approximate processing mechanism is the same as that in document type 3: first, the analysis and labeling phase extracts candidates for composite elements or data items from document images and organizes the extracted candidates as a graph or list to determine the constructive relationship in the bottom-up manner. Second, in the top-down manner the interpretation phase interprets the constructive links among the extracted candidates (as a graph or list) with the layout knowledge (as a set of rules) and distinguishes the structure by clearly assigning connectivity among data items and/or composite elements. In this case, the layout meta-knowledge was used in (42) to make the control for rule interpretation easy. This is because the layout meta-knowledge can make the interpretation efficiency effective by classifying different rules into similarly related sets. Of course, this layout meta-knowledge itself is also represented as a set of rules. The method for the layout recognition of article pages (44) is more simple than those of newspapers because the physical layout structures of articles are more explicit than those of newspapers and also the logical layout structures are more simple than those of newspapers.

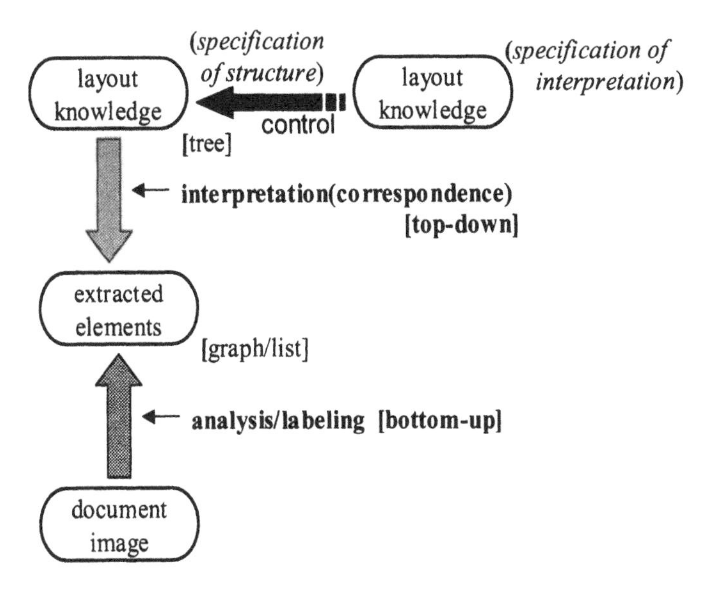

Figure 12: Framework of document type 4.

4. DOCUMENT TYPE AND LAYOUT KNOWLEDGE

If an appropriate representation means of layout knowledge and processing mechanisms can be well applied to various types of documents correspondingly, it is explicit for the recognition subject to make the recognition ratio and processing efficiency high. Of course, the recognition process is more or less dependent on the representation means of layout knowledge. In this case, we must pay attention that the representation means of layout knowledge, addressed in Section 3, are not always applicable to the corresponding document types, but other means may be usable to some document types. However, it is very important for us to focus on the relationship between document types and knowledge representation means. This is because the more strongly documents are dependent on the logical layout structure, the more complex the knowledge representation means become.

Figure 13 shows that individual document types are organized under the relationship between the physical layout structure and logical layout structure approximately. Additionally, in this figure these document types are arranged in accordance with the representation means of layout knowledge, especially, in the document type 3 (such as business cards (4), cataloging cards (37), business and official letters (45), etc.), the layout knowledge can be defined as a set of operations for partitioning regions hierarchically though the layout knowledge in the other document types specifies the properties of individual data items and/or relationships among data items directly. This is because the documents of document

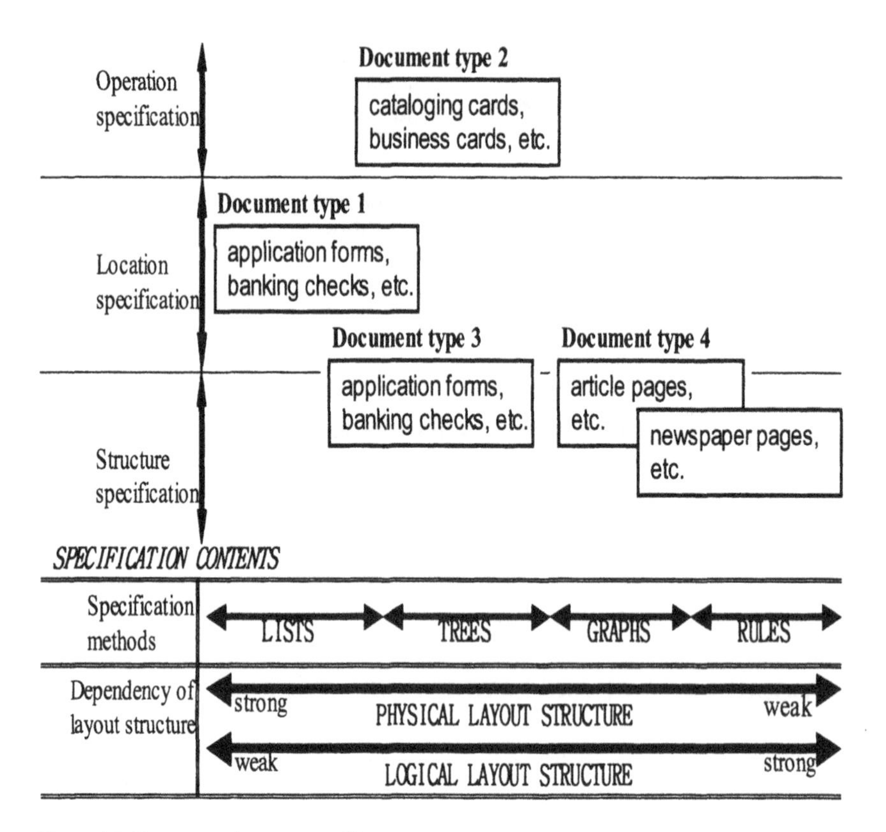

Figure 13: Representation means of layout structures.

type 3 are almost independent of the physical layout structures and are possibly derived from the logical layout structures.

We often observe, however, that many existing documents are composed of collections of various document types (24). Although individual composite documents are able to be successfully analyzed by means of the application of their own corresponding layout knowledge, the whole document structure is not uniformly specified. The document mixed complicatedly with different types of documents is called a complex document, and the new representation method is investigated on the basis of previously discussed viewpoints and/or successive research and development effects, so as to be applicable to various complex documents effectually. Watanabe and Sobue discussed the transformation method among different layout knowledge: transformation between document type 2 and document type 3 (24). Document type 2 is uniquely different from other knowledge representation means because the knowledge represents a sequence of operators, but does not do the physical structure information, related to layout structure, as illustrated in Figure 13. Thus, (24) showed that even if documents were complex the layout structure can be well-specified by the structure specification means such as the layout knowledge in the document type 3.

Furthermore, we discuss the relationship between the logical layout structure and physical layout structure from the viewpoint of knowledge representation means. In the four types of documents, various kinds of knowledge representation means are used: lists, trees, graphs (or networks), and rules. The list, tree, and graph are fundamentally frame-based representation means, and these are smartly applicable when the document structure is explicitly definable under the physical layout structure, except for document type 2. In document type 2 the knowledge representation means are not to specify the locations and/or structures, but to specify the operations.

If the document structure can be defined strongly by the physical layout structure, it is better to make use of more simple frame-based representation means: the list is more successful than the tree or graph in document type 1 because the processing efficiency is superior and the knowledge definition and management is easy. The tree is more effective than the graph in document type 3, similarly. Of course, in case of document type 3 the graph is adaptable but such selection is not smart. In comparison with frame-based layout knowledge representations, the rule-based knowledge representation is very powerful. However, this selection often generates drawbacks in processing efficiency because the knowledge representation becomes complex and also the interpretation phase exhausts much time, navigating extensive possibilities.

When we try to recognize or analyze the layout structure of some documents, it is very important to make use of the most simple knowledge representation means as possible, (for most basic and characteristic properties of document structures), and then specify other properties effectively that are inherently dependent on individual documents.

5. OTHER ISSUES

In this chapter, we mainly addressed model-based document analysis and recognition approaches from the viewpoint of knowledge representation and analysis methods. However, the research and development issues, attended with document analysis and recognition, are not limited to the discussion points of this chapter, but extend to various kinds of documents (e.g., maps, diagrams, music books, engineering drawings, business charts, etc.) (14-18,46-48), data-driven approaches for technical problems (e.g., segmentation of complicated composite document elements, separation among touched characters or line segments, extraction of character strings from attended backgrounds, etc.), recognition or analysis of complex documents or mixed types of documents, recognition of handwriting characters from filled-in types of documents, document processing for applications or on applications (e.g., OCR, information retrieval, WWW server, etc.) and so on (6-18).

Of course, these topics relate to the subject of document analysis and recognition must be investigated as a fundamental means for the construction and utilization of usable information bases. Currently, the results derived from research and development in document analysis and recognition are not always usable or applicable as their own complete systems or packages. Document analysis and recognition is looked upon as a functional module of OCR, and takes an important role in data input tasks.

6. CONCLUSION

In this chapter, we addressed the subject of layout recognition. Many documents have their own layout structures, more or less, and the layout recognition methods work well, using the layout knowledge that must be established effectively from such inherent layout structures under the model-driven (or top-down) approach. Such a framework is very smart with regard to the flexibility, applicability, and adaptability of processing abilities, and also makes the processing efficiency and recognition ratio high.

The research and development history for document analysis and recognition is about 20 years. Also, many useful products have been proposed: in particular, character recognition and drawing interpretation. Additionally, since 1990 the knowledge-based approach using a model-oriented understanding paradigm has eagerly been investigated, and currently presents the most basic framework. Of course, the approaches and methods related closely to or complementarily to this approach also were investigated in a wide range of applications to recognize and analyze the various kinds of documents, attaining a high recognition ratio, making the processing efficiency sufficient, and extracting meaningful data from complicated documents effectively. Future topics for document analysis and recognition, should include that these subjects are expanded to a wide range of applications.

REFERENCES

(1) H.S.Baird, "Anatomy of a page reader", *Proc.of MVA'90*, pp.483-486, 1990.

(2) M.Ejiri, "Knowledge-based approaches to practical image processing", *Proc.of MIV'89*, pp.1-8, 1989.

(3) T.Watanabe, and T.Fukumura, "An architectural framework in document image understanding", *Proc.of ICARCV'94*, pp.1431-1435, 1994.

(4) J.Higashino, H.Fujisawa, Y.Nakano, and M.Ejiri, "A knowledge-based segmentation method for document understanding", *Proc.of ICPR'86*, pp.745-748, 1986.

(5) T.Watanabe, Q.Luo, Y.Yoshida, and Y.Inagaki, "A stepwise recognition method of library cataloging cards on the basis of various kinds of knowledge", *Proc.of IPCCC'90*, pp.821-827, 1990.

(6) L.O'Gorman, and R.Kasturi, "Document Image Analysis Systems", *IEEE Computer*, vol.25, no.7, pp.5-8, 1992.
(7) L.O'Gorman, and R.Kasturi, "*Document Image Analysis*", IEEE Computer Society Press, 1995.
(8) IAPR TC-10/11, "*Proc.of ICDAR'91*, vol.1/vol.2", AFCET-IRISA/INRIA, 1991.
(9) IAPR TC-10/11, "*Proc.of ICDAR'93*", IEEE Computer Society Press, 1993.
(10) IAPR TC-10/11, "*Proc.of ICDAR'95*, vol.1/vol.2", IEEE Computer Society Press, 1995.
(11) IAPR TC-10/11, "*Proc.of ICDAR'97*, vol.1/vol.2", IEEE Computer Society Press, 1997.
(12) IAPR TC-10/11, "*Proc.of ICDAR'99*, vol.1/vol.2", IEEE Computer Society Press, 1999.
(13) IAPR TC-10/11, "*Proc.of ICDAR'01*", IEEE Computer Society Press, 2001.
(14) R.Kasturi, and K.Tombre (eds.), "Graphics recognition: methods and applications", *Lecture Notes in Computer Science 1072*, 1996.
(15) K.Tombre, and A.K.Chhabra (eds.), "Graphics recognition: algorithms and systems", *Lecture Notes in Computer Science 1389*, 1998.
(16) A.K.Chhabra and D.Dori (eds.), "Graphics recognition: recent advances", *Lecture Notes in Computer Science 1941*, 1999.
(17) IAPR TC-10, "Graphics recognition", *Proc.of GREC'01*, 2001.
(18) T.Watanabe: "Recognition in maps and geographic documents: features and approach", *Proc.of GREC'99 (Lecture Notes in Computer Science 1941)*, pp.39-49, 2000.
(19) T.Watanabe, Q.Luo, and N.Sugie, "Structure recognition method for various types of documents", *Int'l J.of Machine Vision and Applications*, vol.6, pp.163-176, 1993.
(20) T.Watanabe, Q.Luo, and T.Fukumura, "A framework of layout recognition of document understanding", *Proc.of SDAIR'92*, pp.77-95, 1992.
(21) H.Masai, and T.Watanabe, "Identification of document types from various kinds of document images based on physical and layout features", *Proc.of MVA'96*, pp.369-372, 1996.
(22) H.Masai, and T.Watanabe, "Document categorization for document image understanding", *Proc.of ACCV'98(Lecture Notes in Computer Science 1352)*, pp.105-112, 1998.
(23) T.Watanabe, "A guide-line for specifying layout knowledge", *Proc.of Document Recognition and Retrieval VI, in SPIE/EI99*, vol.3651, pp.162-172, 1999.
(24) T.Watanabe, and T.Sobue: "Layout analysis of complex documents", *Proc.of ICPR2000*, vol.3, pp.447-450, 2000.
(25) T.Watanabe, and Q.Luo: "A multi-layer recognition method for understanding table-form documents", *Int'l J.of Imaging Systems and Technology*, vol.7, pp.279-288, 1996.
(26) T.Watanabe, and T.Fukumura, "A framework for validating recognized results in understanding table-form document images", *Proc.of ICDAR'95*, pp.536-539, 1995.
(27) T.Watanabe, Q.Luo, and N.Sugie, "Knowledge for understanding table-form documents", *IEICE Trans.on Inf. and Syst.*, vol.E77-D, no.7, pp.761-769, 1994.
(28) T.Watanabe, Q.Luo, and N.Sugie, "Toward a practical document understanding of table-form documents: its framework and knowledge representation", *Proc.of ICDAR'93*, pp.510-515, 1993.
(29) Y.Wang, I.T.Phillips and R.Haralick: "Automatic table ground truth generation and a background-analysis-based table structure extraction method", *Proc.of ICDAR'01*, pp.528-532, 2001.
(30) T.Watanabe, Q.Luo, and N.Sugie, "Layout recognition of multi-kinds of table-form documents", *IEEE Trans.on PAMI*, vol.17, no.4, pp.432-445, 1995.

(31) Q.Luo, T.Watanabe, Y.Yoshida, and Y.Inagaki, "Recognition of document structure on the basis of spatial and geometric relationships between document items", *Proc.of MVA'90*, pp.461-464, 1990.
(32) D.Niyogi, and S.Srihari, "A rule-based system for document understanding", *Proc.of AAAI'86*, pp.789-793, 1986.
(33) F.Esposito, D.Malerba, G.Semeraro, E.Annese, and G.Scafaro, "An experimental page layout recognition system for office document automatic classification: an integrated approach for inductive generalization", *Proc.of ICPR'90*, pp.557-562, 1990.
(34) J.L.Fisher, S.C.Hinds, and D.P.D'amatoi, "A rule-based system for document image segmentation", *Proc.of ICPR'90*, pp.567-572, 1990.
(35) J.Higashino, H.Fujisawa, Y.Nakano, and M.Ejiri, "A knowledge-based segmentation method for document understanding", *Proc.of ICPR'86*, pp.745-748, 1986.
(36) G.Nagy, "Hierarchical representation of optical scanned documents", *Proc.of ICPR'86*, pp.347-349, 1986.
(37)C.F.Lin, and C.-Y.Hsiao, "Structural recognition for table-form documents using relaxation techniques", *Int'l J.of Pattern Recognition and Artificial Intelligence*, vol.12, no.7, pp.985-1005, 1998.
(38) T.Watanabe, H.Naruse, Q.Luo, and N.Sugie, "Structure analysis of table-form documents on the basis of the recognition of vertical and horizontal line segments", *Proc.of ICDAR'91*, pp.638-646, 1991.
(39) T.Sobue, and T.Watanabe, "Identification of item fields in table-form documents with/without line segments", *Proc.of MVA'96*, pp.522-525, 1996.
(40) Q.Luo, T.Watanabe, and N.Sugie, "Structure recognition of table-form documents on the basis of the automatic acquisition of layout knowledge", *Proc.of MVA'92*, pp.79-82, 1992.
(41) H.Kojima, and T.Akiyama, "Table recognition for automatic document entry system", *Proc.of SPIE'90*, pp.285-292, 1990.
(42) Q.Luo, T.Watanabe, and N.Sugie, "A structure recognition method for Japanese newspapers", *Proc.of SDAIR'92*, pp.217-234, 1992.
(43) S.Tsujimoto, and H.Asada, "Understanding multi-articled documents", *Proc.of ICPR'90*, pp.551-556, 1990.
(44) K.Kise, K.Momota, M.Yamaoka, J.Sugiyama, N.Babaguchi, and Y.Tezuka, "Model based understanding of document images", *Proc.of MVA'90*, pp.471-474, 1990.
(45) A.Dengel, and G.Barth, "High level document analysis guided by geometric aspects", *Int'l J.of Pattern Recognition and Artificial Intelligence*, vol.2, no.4, pp.641-655, 1988.
(46) N.Yokokura, and T.Watanabe, "Recognition of composite elements in bar graphs", *Proc.of MVA'96*, pp.348-351, 1996.
(47) N.Yokokura, and T.Watanabe, "Recognition of various bar-graph structures based on layout model", *Proc.of ACCV'98(Lecture Notes in Computer Science 1352)*, pp.113-120, 1998.
(48) N.Yokokura, and T.Watanabe, "Layout-based approach for extracting constructive elements of bar-charts2, *Proc.of GREC'97(Lecture Notes in Computer Science 1389)*, pp.163-174, 1998.

9

Recent Progress in Medical Image Processing

Junichiro Toriwaki and Kensaku Mori
Nagoya University, Nagoya, Japan

Jun-ichi Hasegawa
Chukyo University, Toyota, Japan

SUMMARY

This chapter surveys the recent study and development of medical image processing with the focus on applications to computer-aided diagnosis (CAD) and computer-aided surgery (CAS). From the engineering viewpoint, CAD is the application of image pattern recognition and understanding, and CAS is more closely related to visualization and virtual object manipulation. In this chapter, we present mainly the processing of X-ray images including conventional X-ray images (two dimensional) and CT images (three dimensional) intended to realize CAD and CAS. Topics include a few examples of CAD of 2D chest X-ray images and stomach X-ray images and chest CT images, visualization of 3D CT images, navigation diagnosis of the virtual human body, virtual and augmented endoscopy, and virtual resection. Concepts of extended CAD and the virtual human body are also briefly discussed.

1. INTRODUCTION

Use of images of the human body started in 1885, when the X-ray was discovered by Roentgen. It brought about the innovation in medicine by providing a tool to observe the inside of the body without surgical invasion.

Digital computer usage appeared about 1945 as an extremely powerful tool of numerical computation, and rapidly spread to a wide field of applications during the following ten years. Introduction of computers in medicine first began in the management and retrieval of medical records. Computer processing of sensed data acquired from the human body was first tried in the diagnosis of waveforms such as ECG and EEG.

Utilizing the computer for diagnosing medical images had also been studied actively in various aspects including generation, enhancement, measurement, analysis, and retrieval. In fact, medical image processing has been one of the most important areas of image processing since the beginning of digital image processing (1) (2). Deriving diagnostic decisions from X-ray images by a computer was expected in the early stages and was often called automated diagnosis (3).

Because really automated diagnosis has been known to be too difficult, the primary target of studies has recently shifted to making full use of computer processing for parts of the procedure to reach a final diagnosis of medical images. Today the term "computer-aided diagnosis (CAD)" has become popular for these types of medical image processing.

The invention of CT in the early 1970's was the greatest innovation after the discovery of the X-ray. It affected much of image processing as well as medicine. From a technological viewpoint, the significance of digital image processing was established by CT as well as by the appearance of satellite images, both around the beginning of the 1970's. The CT drastically changed X-ray diagnosis in every field of medicine including CAD.

In this chapter we present a survey of medical image processing with the focus on applications of image pattern recognition and understanding to CAD. Preceding to this chapter, one of the authors published a few survey articles about medical image processing (4 - 9). Ref. (5) covers topics in computer-aided surgery (CAS), and Ref. (6) written in Japanese treats mainly studies of CAD. This survey is based on both Refs. (5) and (6), being revised by selecting topics, summarizing them, and adding more recent work. The reference list was also largely updated.

2. BIOMEDICAL IMAGE PROCESSING AND CAD

2.1 BIOMEDICAL IMAGE PROCESSING

Biomedical (medical) image processing include almost every field of image processing applicable to medicine such as those listed in Table 1 (although by no means exhaustive). Progress is so rapid in all fields that it is impossible to give surveys of all areas in one chapter. Thus we limit the scope of this chapter to the subareas of CAD, which were studied actively in the last few years.

Table 1 Fields and Topics of Medical Image Processing

1. Imaging
   Image acquisition: X-ray CT, MRI, US, PET, Microscope images, etc.
   Reconstruction algorithm
2. Image Enhancement and Restoration
   Image transformation: FFT, geometrical,
   Fourier domain processing: filtering
   Noise reduction
   Grey value manipulation: γ-correction, histogram transformation
3. Visualization and Image Generation
   Computer graphics techniques: volume rendering, surface rendering,
   Moving eye views rendering
   Virtual human body manipulation
4. Image Retrieval
   Image database: development, retrieval
5. Image Transmission
   Coding and decoding
   Telemedicine applications: aids for tele-diagnosis and telesurgery
6. Image Analysis and Recognition
   Image measurement
   Computer-aided diagnosis applications
   marking, segmentation, feature extraction, classification
   Computer-aided intervention applications
   image guided surgery, registration, deformation
   Computer-aided surgery applications
   registration, deformation, simulation

2.2 COMPUTER-AIDED DIAGNOSIS

We define CAD in this chapter as the method or the procedure for diagnosis that makes full use of a computer. In particular, a computer is utilized for acquiring information useful as the basis of diagnosis and for assisting doctors to derive di-

agnosis from images of the patient's body. The way to implement CAD and functions realized as CAD differs greatly, depending on the ability of the computer that can be available for CAD. It should be noted that the computer in CAD executes only limited parts (or functions) of the whole procedure to reach the diagnosis. Various kinds of CAD in this sense are shown in Fig.1. Inputs to a CAD system (or a CAD procedure) of images are two dimensional images or three dimensional images. The outputs may be measurement values, images with marks given by the computer, three dimensional images generated by the computer, or diagnostic recommendation concerning a patient's health. Let us enumerate some examples.
(i) Imaging: Most of recent imaging in medicine is realized by using a computer. reconstruction of a cross-section image from projections (= computed tomography) is a typical example.
(ii) Marking: To detect shadows that are suspected to be abnormal from X-ray images and present images with marks to indicate them.
(iii) Feature measurement: Measuring image features that may be useful for diagnosis such as the size of nodules and the cardio-thoracic ratio.
(iv) Classification: Deriving a diagnostic decision concerning a given image. Classifying a chest X-ray image into "normal" and "suspected to be abnormal" is a well-known example.
(v) Visualization: To visualize a 3D CT image of a patient by surface rendering or volume rendering.

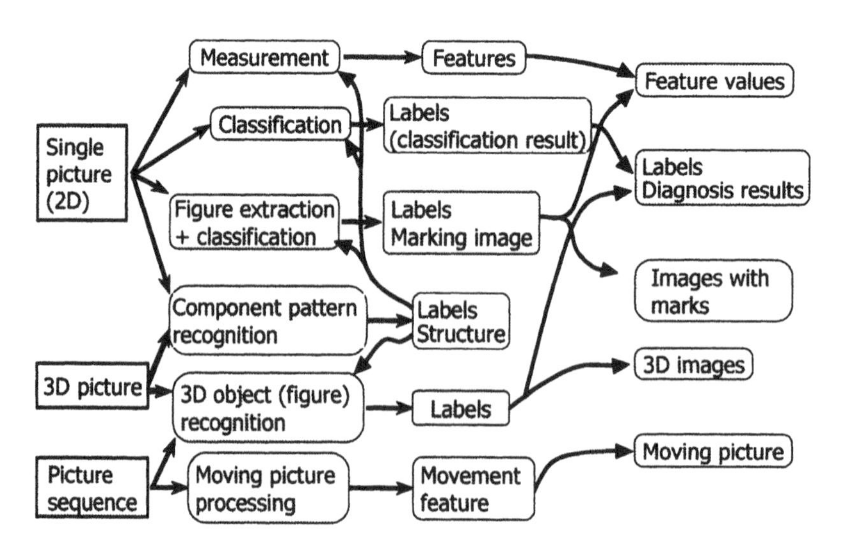

Fig.1 Types of computer-aided diagnosis (CAD).

(vi) Segmentation: To recognize organs existing in input 3D images is the inevitable processing for many CAD systems such as cancer detection and virtual endoscopy, as will be discussed later. This corresponds to the segmentation of objects from given 3D images.

3. CAD OF 2D X-RAY IMAGES

In this section we present the CAD of 2D (two dimensional) X-ray images.

3.1 MARKING AND AID FOR SCREENING

3.1.1 Outlines

Given a conventional X-ray image, a CAD system may be considered which detects shadows suspected to be abnormal and marks them on an input image. The system output is a 2D image typically shown in graphic display with marks on nodule shadows. This function is called marking. Marking has been a useful aid to screening. Now in Japan several kinds of screening are performed for breast, lung, and stomach cancer using conventional X-ray images. CAD systems for them have been studied actively for about forty years. Let us introduce some of them briefly. The goal common to these studies are to develop a system that detects candidates of abnormal shadows to arouse doctor's attention and to eliminate normal images.

Two kinds of decision errors occur, the false positive (FP) and the false negative (FN)* The trade-offs between them are always a critically important problem. Probabilities of both types of errors are evaluated experimentally using sample images and used as a popular index of the system performance. They are calculated in either individual nodules base, film base, or case base. It is said that a CAD system is clinically useful if the number of FP marks is one or two per one film and the FN rate is about 0.02 or less. It should be noted, however, that such performance indices evaluated experimentally strongly depend on a set of sample images used in the experiment. A database of images including various types of typical abnormal shadows have been developed recently for both designing and evaluating CAD systems by a common criteria (for example, see (10) for CADM database), but the absolute values of error probabilities are still significantly dependent on the quality of images contained in such database.

3.1.2 CAD of mammograms

CAD of mammograms used for detection of women's breast cancer has been stud-

---

*False Positive (FP) = the error that a normal case is decided as abnormal
False Negative (FN) = the error that an abnormal case is decided as normal
FP rate = The number of cases decided as abnormal / the number of normal cases
FN rate = The number of cases decided as normal / the number of abnormal cases
Average recognition rate = The number of correctly classified cases / the number of the whole cases

ied extensively, and now at least one CAD system is commercially available in the U.S.A. ((11) and [Example 2]).

Targets to be detected are massive shadows with vague irregular edges (tumors) and micro calcification, small points of high brightness value with a diameter of less than about 1 mm.

Example 1 According to (12), for 4184 mammograms from 1037 cases (including 241 cases of cancer) taken at the National Cancer Center Hospital East, the detection probability of tumors was 0.9, false negative shadows were 1.35/one image, the detection probability of micro calcification was 0.94, and the false positive was 0.4/one image on the average. Image processing algorithms employed here were developed by H. Kobatake et al. (12). Computation time for one image was about 3 minutes by SUN SPARC STATION. Almost the same level of results were reported by several other groups (6) (13).

Example 2 The first commercially available CAD system of mammograms was provided in 1998 from a company in the U.S.A. This was the first commercially available CAD system of any kind of X-ray images. According to its catalogue, the detection probability of micro calcification clusters was 0.98 and that of tumors was 0.75 for 1083 cases (11 - 12).

Algorithms are basically common for these examples as shown below. First, the enhancement of massive shadows and calcifications is performed by various local operators (spatial filtering), and then theresholding is applied to the enhanced input image to extract candidate shadows. Then detected figures are classified into "normal" and "suspected to be abnormal" or "benign" and "malignant" by shape features such as the irregularity of the border.

Recognition of component patterns such as the rib cage, chest wall, and breast area are also done. Sometimes the classification of density of the mammary gland is also used. Prior to the appearance of the commercial CAD system above, extensive study by K.Doi et al. showed the effectiveness of CAD in the diagnosis of mammograms by using several thousand X-ray images (14). Recently CAD for 3D images is beginning using ultra sound images and CT images.

3.1.3 CAD for conventional chest X-ray images

Study of CAD for conventional chest X-ray images has begun earliest among various X-ray images. For example the author's group in Japan started in 1964 the study of the CAD of chest photofluorograms (or indirect chest X-ray film) (3) (8). In those days, we had established in Japan the mass screening system to detect tuberclosis. Since all people working in companies and all students and pupils in universities and schools took X-ray photofluorograms once or twice a year, a large number of films needed to be diagnosed in one or two weeks*. Therefore we

---

*Numbers of chest photofluorograms taken in one year in Japan were 25, 528, 720 in 1983 and 24, 488, 666 in 1988(15).

aimed that any shadows in the lung suspected to be abnormal were marked on a film. As a result we expected to eliminate all normal case films beforehand.

Example 3 The CAD system for chest photofluorograms called AISCR-V3 was developed by the authors' group in 1973 (16 - 19) (Fig.2). For 134 chest photo-

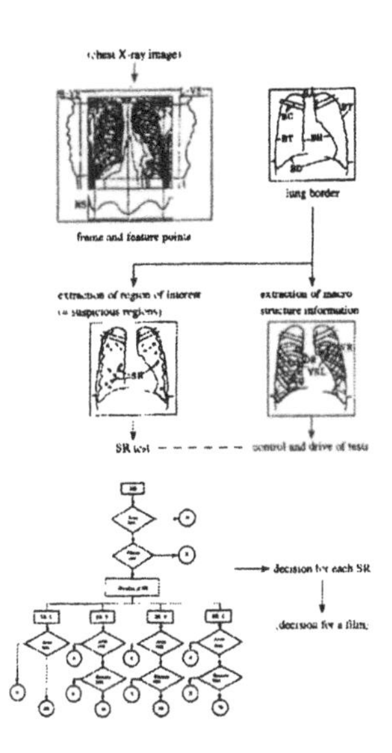

(a) Flow of the system AISCR-V3 (19) (see Example 3 in the text for details).

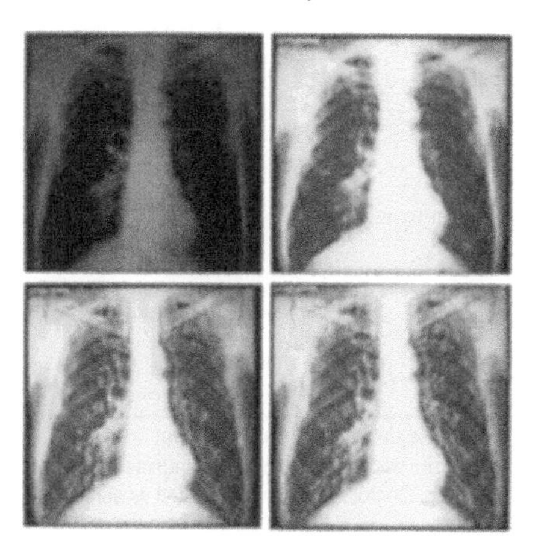

(b) Examples of results

fluorogramrs, the performance of doctors and the CAD system was compared. Doctors decided 76 films were normal (Group I), and 58 were abnormal (Group II). For Group I, results were later evaluated and confirmed by using direct chest X-ray films and other available diagnostic records. The results are summarized in Table 2. Furthermore this table also shows a decision table achieved by using both the doctor's decision and results by CAD. This example suggests the possibility that a doctor's diagnosis can be improved significantly by appropriately integrating decision by CAD.

The same type of CAD system for direct (standard) chest X-ray films was developed in Japan by H. Natori's group for diagnosing lung cancer (20 - 21). Its performance had an FN ratio of 0.078 for 192 standard chest X-ray images with lung tumors and an FP ratio of 0.28 for 74 normal X-ray images.

Nodule detection in the lung using standard chest X-ray images had been studied in several other groups (22 - 23) in the last few years. A systematic survey with a comprehensive list of papers was given with a brief introduction of results in (24).

Table 2 Results of Classification by the CAD System AISCR-V3 described in Example 3*.

(1) Sample Set A (Group I)

True class	Result by AISCR-V3		
	Normal	Abnormal	Total
Normal	40	27	67
Abnormal	4	5	9
Total	44	32	76

(2) Sample Set B(Group II)

True class	Result by AISCR-V3		
	Normal	Abnormal	Total
Normal	10	13	23
Abnormal	13	22	35
Total	23	35	58

(3) Sample Set A + B

True class	Result by AISCR-V3 + Doctor		
	Normal	Abnormal	Total
Normal	40	36	76
Abnormal	10	48	58
Total	50	84	134

Note Sample set A : Films the doctor classified as Normal.
Sample set B : Films the doctor classified as Abnormal.
Result by AISCR-V3 + Doctor: Films are classified as abnormal if either of CAD and doctor classified it as abnormal.

* Input images are chest phorofluorograms taken in actual screening.

3.1.4 CAD for stomach X-ray images

Double contrast stomach X-ray images (denoted simply by stomach X-ray images below) have long been used for screening stomach cancer in Japan. In this case CAD is also desired to detect shadows suspected to be abnormal and by doing this to reduce the number of images to be examined by doctor in detail.

From the technical viewpoint this problem is characterized by shape features of cancer lesions in X-ray images. That is, a key to suspecting the existence of cancer is the convergence of line patterns on X-ray images of the stomach, which correspond to fold patterns on the inside wall of the stomach. Then we need to recognize the characteristic spatial arrangement of line patterns that are different from nodules or massive circular shadows. Hasegawa et al. developed a novel nonlinear filter calculating an index of the degree of concentration of line patterns to each point in an image (25 - 26). This index was called the concentration index (25) and later extended to a 3D image of the lung (26).

The CAD of stomach X-ray images was studied by Hasegawa et al. and Fukushima et al. (25) (27-28).

Example 6 For 77 stomach X-ray images (all including cancer), the detection probability was 0.92 with 9.3 FP shadows per one image in average by the CAD system (29) (Fig.3).

3.1.5 CAD for CT images

Use of CT images for the detection of lung cancer in its early stage has recently been expected as a possible way to overcome the limitation of standard (conventional) chest X-ray images and chest photofluorograms. One problem in the use of CT images for screening is the number of images to be diagnosed. Typically more than 30 slices (cross-section images) will be taken from one patient. Thus CAD of chest CT images is expected to reduce the number of images to be diagnosed by doctors.

Many papers have been published by Niki et al. (30), and Yamamoto et al. (31 - 33) in Japan concerning CAD of chest CT images. Let us introduce some

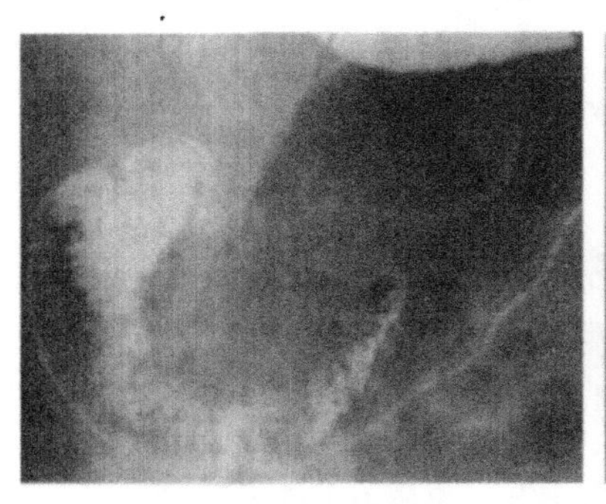

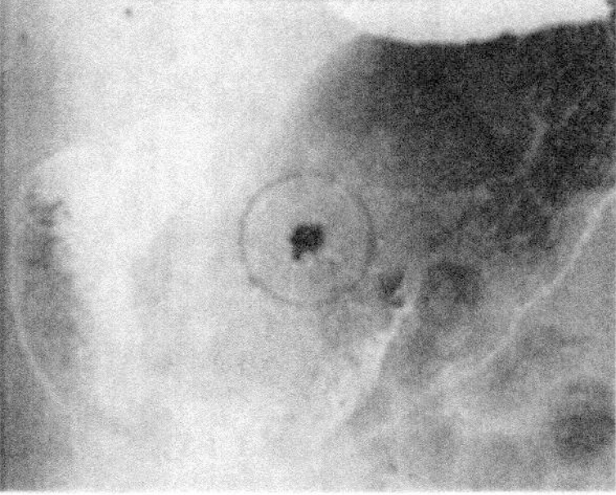

Fig.3 CAD for stomach X-ray image (result of marking) (25).

pertinent examples.

Example 4 For chest CT images obtained from 460 cases (1600 slices) taken by helical CT, all of the rank E nodules were detected by the CAD system with 5~7 FP nodules per one case on the average (34). The rank E means the case that all of three medical experts coincidently claimed as lung cancer. Thin system has been tested for more than three years at the National Cancer Center Hospital Central and showed almost the same results (34) (30).

Example 5 A CAD system for CT images by Yamamoto et al. was applied to 176 cases of CT images taken by the helical CT for screening. The performance was 5.1 FP nodules per one case on the average with no FN error. They expect that the ratio of images doctors had to examine could be reduced to about 15% of all images by using this CAD system as the second reader (32) (Fig.4).

The screening of lung cancer by using chest CT images was first proposed in Japan in 1994 (31), and has been studied actively (33 - 34). Papers concerning this topics have been published in the annual meeting of the Society of Thoracic CT Screening in Japan (35) since 1994 (35). Effectiveness of CT in lung cancer detection was also reported in several papers in radiology (36) (37). Progress in CAD of CT images closely related these studies in medicine. In the design and development of medical systems using CT images, the trade-off between cost (or risk) and benefit is critically important. Ref. (38) discussed this problem systematically. Imaging parameters are selected considering the screening use, and this severely affects the image processing methods in CAD. Presently the slice thickness and the reconstruction pitch are both set to be 10mm for screening use in Japan.

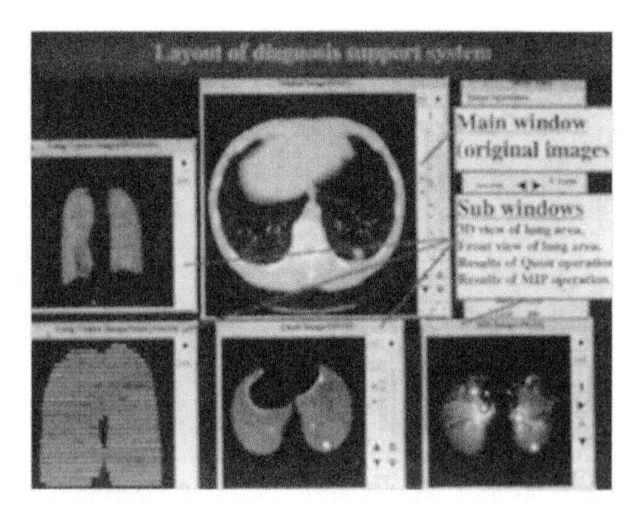

Fig.4 CAD for X-ray CT images. (Figure by S. Yamamoto (31).)

Two types of basic strategies are possible for processing of CT images. One is the combination of 2D image processing for each slice (slice-base approach) and the other is to employ 3D image processing algorithms that directly access 3D images consisting of volume data obtained by helical CT (39 - 40). CAD systems described in two examples above employ both of these two types by integrating them appropriately.

3.2 CAD FOR MEASUREMENT

One of important functions of CAD systems is to measure and calculate features useful for diagnosis and treatment. Let us introduce here a few examples.

3.2.1 Degree of advance

It is very significant to give quantitative features to indicate how serious the state of diseases is or in which stage of the disease is. One of examples is the CAD of pneumoconiosis from standard chest X-ray images. It was required to establish the standard measure of the advance of the disease, because pneumoconiosis is one of typical occupational diseases and governments or companies had to cover the insurance for damage of patients. The CAD for this was studied actively in the 1970's (3) (12) (24) (41 - 43).

3.2.2 Feature measurement

The Extraction of quantitative features from medical images has been studied since the beginning of CAD by computer. For example, one of the earliest paper concerning CAD treated measuring the size of the heart and calculation of cardio-thoracic ratio from a chest X-ray image (2). Feature measurement itself is not a difficult problem for computer, and was reported in many articles (3) (42 - 44). The most important issue to be solved for automatic feature measurement is how to recognize the target organ or figures to be measured in a given medical image. Thus the automated measurement from chest X-ray images, CT images, and others have only recently been put into practice in spite of very early starting of research. Today various feature measurement functions are implemented in commercial medical imaging systems.

Change detection among X-ray images taken at different times from the same patient is also regarded as a kind of feature measurement. In this case the registration between two images and nonlinear deformation of images are required to eliminate insignificant differences among images. Recent development of new techniques such as warping contributes much to this purpose. One recent report shows that ROC (Receiver Operator Characteristic) of doctors is improved significantly by presenting doctors signs of differences between the left and right lung of one image detected by the computer (45).

4. CAD OF 3D CT IMAGES

4.1 3D IMAGES AND VIRTUAL HUMAN BODY

A three-dimensional digital image (3D image) is a set of density values stored on a three dimensional array. Typical examples are obtained by scanning the human body by helical multidetector CT scanner. Such images are regarded as a reconstruction of the human body or its parts on computer memory, therefore, we call them the virtual human body (VHB) (Fig.5).

New methodology has been used for new applications of CAD to 3D images and new types of CAD have become possible by the use of 3D images and VHB. First, high-level visualization techniques for 3D data were developed for diagnosing 3D images. Second, new ways of diagnosis and treatment were developed by

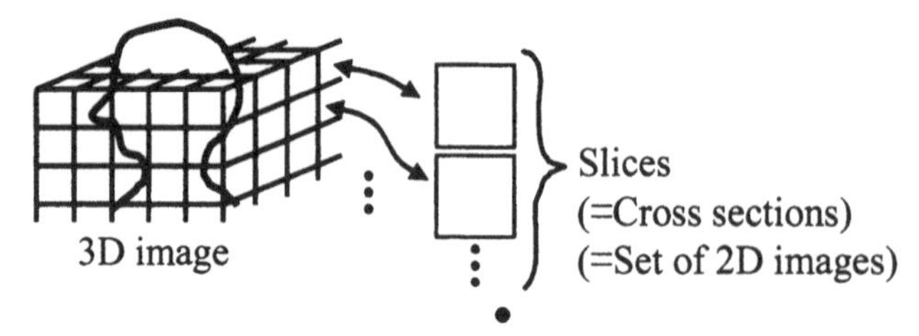

- (a) A 3D image of the human body

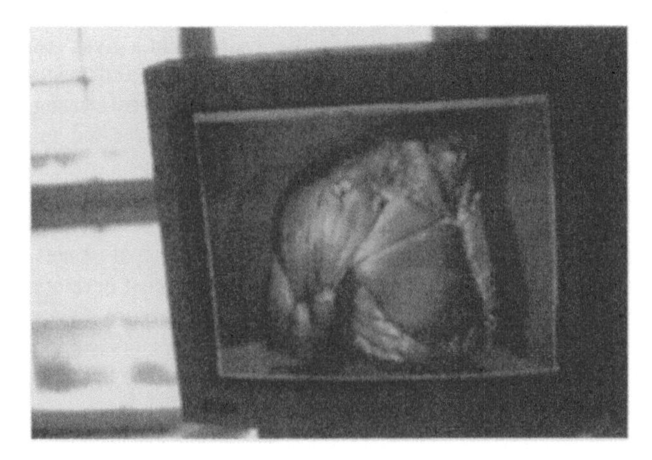

- (b) VHB visualized on computer display (by K. H. Hohne) (52) (79)

Fig.5 Virtual(ized) human body (VHB).

using VHB, such as virtual endoscopy and image-guided surgery. Next we describe some of them briefly in the following subsections.

Concerning image processing algorithms, much work has been done because properties of 3D digital images are remarkably different from those of 2D images. Details are omitted here due to space limitation. A few monographs and a collection of survey articles were published (46 - 51). Ref. (51) is the only systematic textbook for 3D image processing presently available.

4.2 USE OF VIRTUAL HUMAN BODY

The virtual human body (VHB) is considered as a kind of a virtual environment or a virtual space. We classify its applications from two viewpoints: (a) the combination of real and virtual space and (b) search or modification (Table 3) (5) (52).

4.2.1 Viewpoint 1 Real and Virtual body

Here we can consider three cases: using only the real body, and only the virtual body separately, and using both simultaneously or by fusing them.

4.2.2 Viewpoint 2 Search and modification

Given a new virtual environment, the first step is usually to examine or search it carefully without changing it. Next we will try to modify it, if necessary. Examining the real human body corresponds to the traditional diagnosis. Searching the VHB will be a new type of CAD tool called the navigation diagnosis (53). It includes virtual endoscopy as a special case. This will be described in Section 4.3. Fusing images of both the real and the virtual human body will provide a novel type of CAD tool, such as the fusion endoscope (or the augmented endoscope). Modification is utilized in the form of surgical simulation (both preoperative and intraoperative). This was described in (5) and will be omitted here (5) (54) (55).

Table 3 Use of Virtual Human Body

	Virtual human body	Real + virtual human body	Real human body
Search	Navigation diagnosis Virtual endoscopy Training system	Fusion endoscopy (augmented)	Diagnosis
Modification	Preoperative planning Surgical simulation Virtual resection	Intra- operative aid Surgical navigator	Treatment

4.3 VISUALIZATION

4.3.1 Outline
Since the whole of a 3D image with gray values at each point can not be seen by human eyes, methods and tools for rendering a 3D image (= volume data) are extremely important. This is one of most important characteristics in diagnosing 3D medical images. We will introduce here important methods developed for rendering medical images (56 - 57). First we should denote a simple but basic fact that an original image is three dimensional and that a picture plane for rendering is always two dimensional. Thus rendering always accompanies the loss of information due to this dimensionality gap.
4.3.2 Cross Sections
Display of cross sections of a 3D image is trivial as a rendering method in principle, but still very useful practically. If a 3D image of very high spatial resolution is available, it is not difficult to generate a cross section in the arbitrary direction. Using a cross section in the direction of the sagittal cut is advantageous for reducing the number of cross sections to be read carefully.

One complicated way to calculate a cross section is to generate a cross section that contains the center line of thin long organs such as the urethra and pancreatic duct. This type of cutting plane is often called the curved planar reformation (CPR), and several algorithms have been reported (56) (88), but algorithms to calculate it have not always been well-established.
4.3.3 Projection\
Cross sections carry information along a plane defined in the 3D space. It cannot provide information of points not contained in the rendered plane. The most frequently used method to draw the structure of the 3D space on a 2D plane is the mapping of each point in the 3D space onto a 2D plane according to mathematically defined equations. Two kinds of mappings (or projections) - the orthogonal

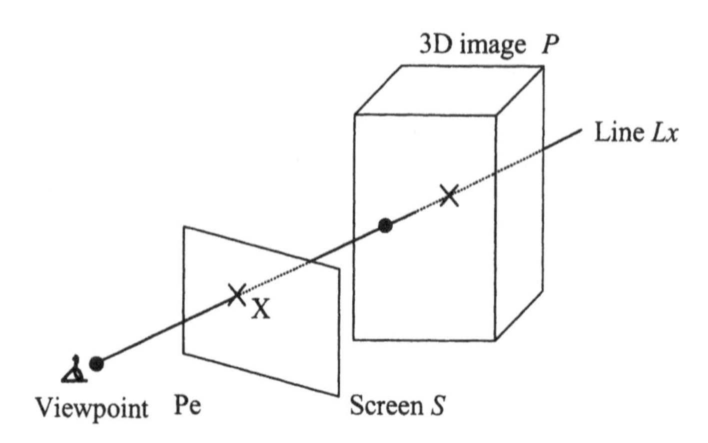

• **Fig.6** Visualization of a 3D image based on mapping of points and density values.

and the perspective transformation - are widely used (58).

This type of geometrical transformation (mapping) determines the point-to-point correspondence between each point in the 3D space and that on the 2D picture plane. This correspondence is not one-to-one because the 3D space degenerates to a 2D plane. This causes a difficult problem: how the density value of a point on a 2D plane is calculated.

Let us assume, for example, the virtual view point Pe in the 3D space containing a 3D picture *P* (Fig.6). Assume next an arbitrary point X on a 2D display screen *S*. If we employ the perspective transformation for the geometrical mapping from the 3D space *P* to the plane *S*, all points in the 3D space on the line *Lx* connecting points X and Pe will be mapped onto the same point X. Thus how to determine the density value of X becomes a serious problem. Principles now adopted widely are (a) integration of density values along the line *Lx*, and (b) selection of the specific density value along the line *Lx*. Well known examples are volume rendering for the type (a), and the maximum intensity projection for the type (b). The idea of the maximum intensity projection is self-evident and the explanation will not be necessary. The realization of the volume rendering algorithm is worth studying carefully, and many articles have been published, but details are skipped here. Related references include (56) (59 - 64).

4.3.4 Visualization with Moving Viewpoints

In rendering a 3D object or in visualizing a 3D image we assume a suitable viewpoint in the 3D space including objects. If we change the location of the viewpoint

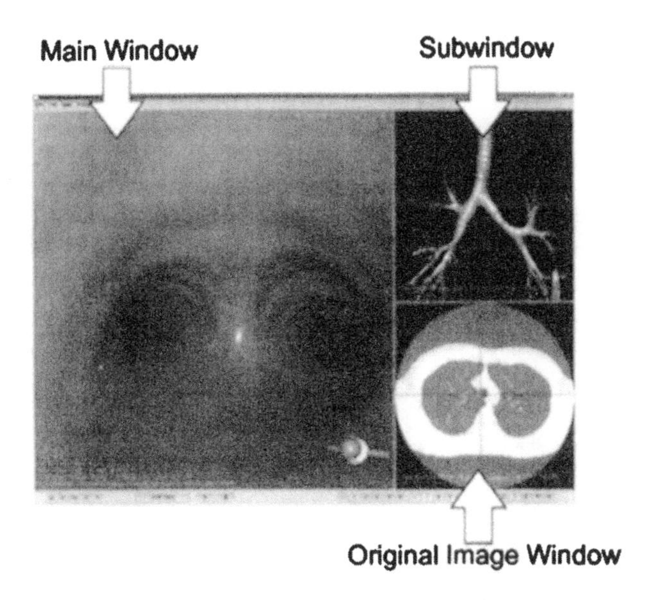

Fig.7 Example of virtual endoscopy (67) (68).

continuously and generate 2D pictures corresponding to each viewpoint rapidly enough (more than five frames per second, for example), then we can obtain picture sequences that look like moving pictures taken when walking around or navigating through the 3D space by the vehicle. By controlling the location and the direction of the viewpoint interactively we can feel as if we were operating a vehicle or an airplane and flying through the virtual space (= the virtual human body). Examples will be introduced in the next subsection.

4.3.5 Virtual Endoscopy and Navigation Diagnosis

The methodology described above was realized as the virtual endoscopy system (VES) and the navigation diagnosis. VES is a tool for diagnosis, which enables us to move (or "fly through") the inside of tubular organs (such as the bronchus) and organs with cavity (such us the stomach), and examine the inside walls of the organs (62) (65-69) (Fig.7). Now research of VES is very active and has been extended to various techniques such as virtual laryngoscopy, colonoscopy, bronchoscopy, cystoscopy, and other vessels (62). A numbers of papers were also published in conference proceedings such as SPIE, RSNA, and CARS. Use of VES for screening colon cancer has also begun (70 - 71).

4.3.6 Virtual Resection

An alternative way to visualize the inside wall of organs is to extend the whole surface of 3D objects onto a 2D plane. To do this the surface of a 3D organ is cut along an appropriate cutting line and extended over a 2D plane. To calculate the shape of the extended surface, two basic procedures are possible. One is to extract a 3D object beforehand and the border surface of the object is projected onto a 2D plane using mathematical transformation. The other is to map a suitable subset of

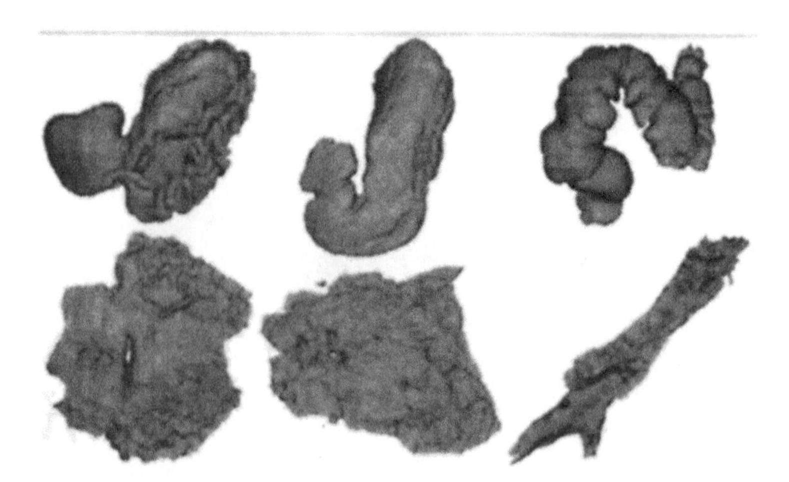

Fig.8 Virtual resection (unraveling). The outside views (top) and unraveled images (bottom) (73).

voxels in the neighborhood of the object surface in 3D space onto a 2D plane, and then the border of the organ is determined on a "extended" 3D picture. Detailed procedures for performing the 2D expansion will be different according to each problem. Applications to the colon (62) (72) and stomach (73) have been reported. In both, fine fold patterns existing on organ's walls are significant for diagnosis such as cancer detection. Therefore the smooth shape of the organ's surface is first determined, and by expanding it the transformation parameters for 2D expansion are calculated. The transformation used to map 3D surfaces onto a 2D plane is usually nonlinear and not expressed by a simple mathematical form, therefore all shapes in the 3D space will be distorted seriously. In our experiment we employed a mass-spring model to implement this transformation by computer (73) (Fig.8).

As a method of visualization, this virtual resection (or unraveling) is in clear contrast to virtual endoscopy or the view with a moving viewpoint (72). The virtual endoscopy image is more real and familiar because it is similar to an actual endoscopy image, although affected by the perspective transformation, but it is more convenient to observe tubular organs such as the bronchus and colon. However, undisplayed parts remain if many folds or convex parts exist on the organ wall such as colon CT (74). For a large cavity organ like the stomach, very careful repetition of the "fly-through" is required to prevent oversights. On the other hand the whole of an image of the organ extended on the 2D plane can be examined at one glance. Understanding original forms from presented images is not so easy for virtual resection images because of deformation, although oversights due to occlusion does not happen.

4.4 FUSION OF REAL AND VIRTUAL IMAGES

4.4.1 Outline

A virtual image produced from the virtual human body can present doctors pictures of the arbitrary parts of the body seen from any viewpoint. If a virtual version of the current patient is available, the inside of the body will be seen by pictures generated from the virtual human body. By superimposing such pictures on images of an actual patient's body, we can provide doctors images of patients which are augmented by images obtained from virtual body. This can be regarded as the fusion of real and virtual human bodies. The registration among the actual body and the virtual body images is the key technology to achieve this.

4.4.2 Augmented (Fusion) Endoscopy

In an actual endoscopy, doctors examine the state of the surface of the organ wall through images seen in the visual field of the endoscope. By superimposing images of virtual organs produced from the virtual human body of the same patient, doctors can acquire information concerning invisible parts of organs. For example, vessels and bones beyond the bronchus wall will be observed in bronchoscopy by presenting images of those organs generated from the virtual human body. Automated tracking of the position and the direction of the endoscope camera was reported in (89) to obtain camera parameters for generating images of virtual organs at the same position (Fig.9).

4.4.3 Image Guided Intervention

In order to execute the surgical planning accurately images such as CT and MRI which are taken prior to operation can be used effectively. Positioning and movement of surgical instruments are controlled based on those preoperative images. Registration of the patient's organ, preoperative CT images, and surgical instruments is critically important here.

Image-guided intervention was first developed in neurosurgery in the 1970's and has been extended in orthopedics and traumatology (75).

The use of preoperative images was supported by the assumption that the tissue to be treated was rigid and would not move with respect to a reference employed in each case. However this assumption is not correct for most of soft tissue. One solution to this was the use of intraoperative images taken during the operation in the operating room. In this case matching among preoperative images, intraoperative images, operation instruments and the actual patient's body are important problems and various algorithms have been developed (76). Sometimes particular equipment such as those in stereotactic surgery were developed for setting reference and matching images. The method of stereo vision based upon the geometrical arrangement of two cameras may be applicable in some cases. Presentation of the parts of the actual human body to be manipulated in the operation augmented with the virtual image inside the organ (i.e. the fusion of virtual and real images) may also be used here very effectively (55) (75).

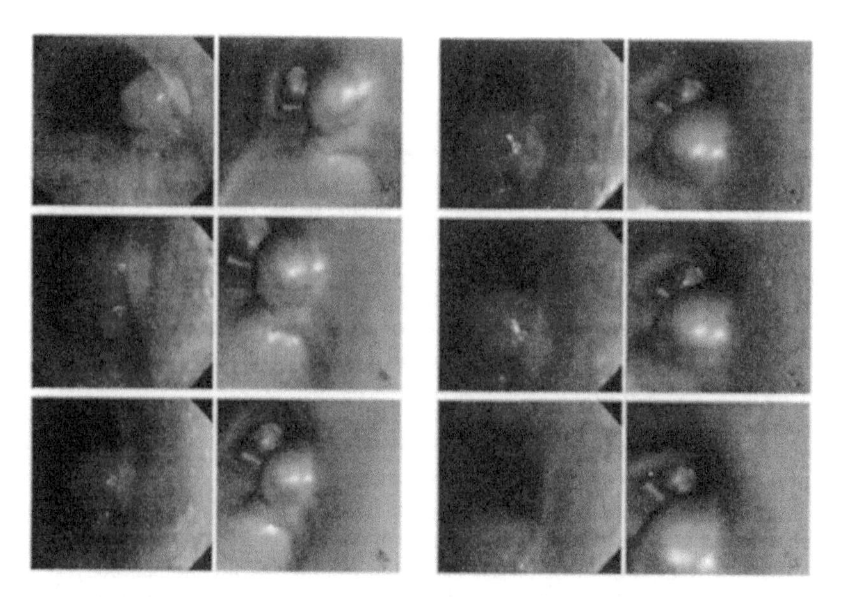

Fig.9 Automated tracking of an endoscope camera for augmented endoscopy. Frames of real endoscope images (left) and virtual endoscope images at the corresponding position and direction of camera (right) (89) .

5. CONCLUSIONS

We presented a brief introduction of the state of the art of medical image processing. This field is often called biomedical computing, and covers the wide range of image processing used in medicine including image acquisitions and visualization to image understanding. This is a sequel to several survey articles the author had published since 1999 (5 - 9) in academic journals both in Japanese and English. In this chapter the stress was put on diagnostic aid of X-ray images, computer diagnosis, and visualization. New types of aid by computer such as virtual endoscopy, c the use of the virtual human body, and image-guided intervention were also introduced briefly. Also a very short historical perspective since the beginning of CAD were inserted in several places. Finally, Fig.10 showed the whole of image information which is now becoming available in diagnosis and treatment. This is too diverse to be presented in the limited pages of this chapter. The authors expect the reader to refer to other articles in the series to supplement the limitation of those contents.

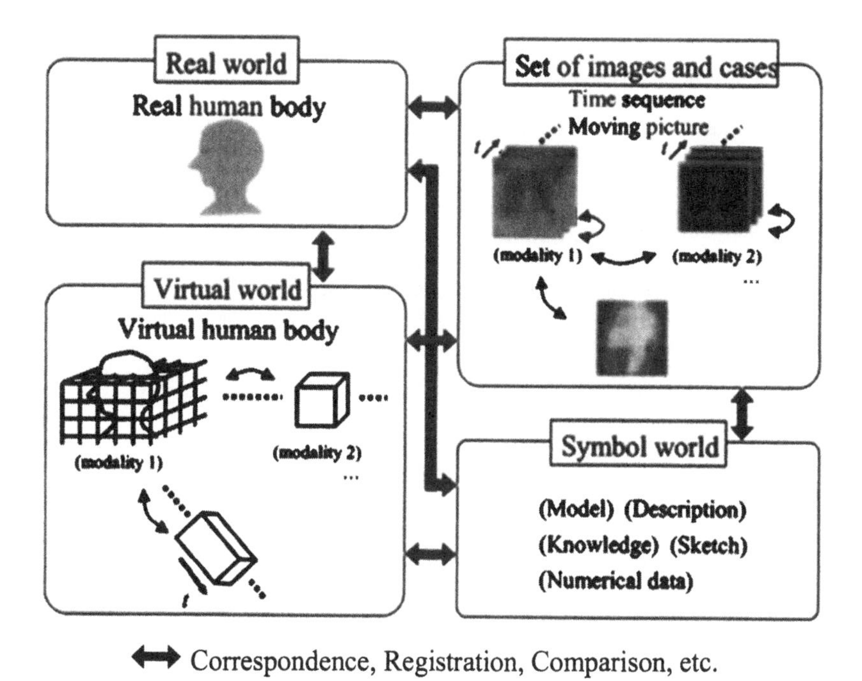

Fig.10 World of images and information available in modern diagnosis and treatment.

ACKNOWLEDGEMENTS

The authors thank Mr. Yuichiro Hayashi of Nagoya University. for preparing an electronic version of the manuscript and figures.

Supplementary comment on references

A series of editorials have been published in IEEE Trans. on MI, which were written by the leaders of major medical imaging research and development groups in the world. They include each groups' facilities, publication records, and philosophy, and will be extremely useful for the readers. See (8) (77-79).

Several well organized surveys are also available on CAD (4 - 6) (24) (57) and on CAS (80). Conference Proceedings closely relating to CAD and CAS are CARS, MICCAI (55), SPIE (81), and RSNA (5). Special issues of journals relating to the topics treated here are in (82 - 85). Monographs, textbooks and collection of articles for medical image processing were published in the last few years (51) (56) (62) (75) (86) (87).

REFERENCES

(1) GC Cheng, RS Ledley, DK Pollock A Rosenfeld eds., Pictorial Pattern Recognition. Thompson Books Co., Washington D.C., 1968.
(2) H Becker, W Nettleton, P Meyers, J Sweeney, C Nice, Jr., Digital computer determination of a medical diagnostic index directly from chest X-ray images. IEEE Trans. Biomed. Eng. BME-11:67-72, 1964.
(3) J Toriwaki, Y Tateno, T Iinuma eds., Computer Diagnosis of Medical X-ray Images. Springer-Verlag Tokyo, 1994 (Japanese).
(4) J Toriwaki, A Shimizu, K Mori, Computerized analysis of chest CT images. K Doi, H MacMabon, ML Giger, KR Hoffmann, eds. Computer-aided Diagnosis in Medical Imaging (Proc. of the 1st International Workshop on Computer Aided Diagnosis 1998.9), Elsevier, Amsterdam, 35-44, 1999.
(5) J Toriwaki K Mori, Recent progress in biomedical image processing - Virtualized human body and computer-aided surgery. Trans.on Information and Systems of Japan, E-82D(3): 611-628, 1999.
(6) J Toriwaki, Trends and future problems in computer aided diagnosis of X-ray images. Trans. on IEICE, Japan, J83-D-II(1): 3-26, 2000.
(7) J Toriwaki, Forty years progress in computer aided diagnosis of X-ray images. Journal of the Society of the Computer Aided Diagnosis of Medical Images, Japan, 5, 6:1-12, 2001. (http://www.toriwaki.nuie.nagoya-u.ac.jp/)
(8) J Toriwaki, Editorials Forty Years of CAD for X-ray Images. IEEE Trans. on MI (in press)
(9) J Toriwaki, Progress in image processing technologies for computer aided diagnosis of medical images - a historical survey and future development. Paper of the Technical Group of Medical Imaging, IEICE, Japan, 102(56):27-34 , No.MI2002-21,2002
(10) http://www.toriwaki.nuie.nagoya-u.ac.jp/~cadm/japanese/index.html
(11) ImageCheker™, Computer Aided Detection for Mammography, R2 Technologies Inc., 1998

(12) H Kobatake, K Murakami, H Takeo, S Nawano, Computerrized detection of malignant tumors on digital mammogrtams. IEEE Trans. on Medical Imaging, 18(5):369-378, 1999.
(13) Y Hatanaka, T Hara, H Fujita, S Kasai, T Endo, T Iwase, Development of an automated method for detecting mammographic masses with a partial loss of region. IEEE Trans. on Med. Image., 20(12):1209-1214, 2001.
(14) K Doi, H MacMahon, S Katsuragawa, RM Nishikawa, Computer-aided diagnosis in radiology : potential and pitfalls, European J. of Radiology, 31:97-109,1999.
(15) Jornal of Health and Welfare Statistics, Health and Welfare Statistics Association, Japan, 41 (9), 1994
(16) J Toriwaki, Y Suenaga, T Negoro, J Toriwaki, T Fukumura, Pattern Recognition of Chest X-ray Images. Computer Graphics and Image Processing, 2(3/4):252-271, 1973.
(17) J Toriwaki, J Hasegawa, T Fukumura, Computer analysis of chest photofluorograms and its application to the automated screening. AUTOMEDICA, 3:63-81, 1980.
(18) Y Suenaga, T Negoro, T Fukumura, Pattern reognition of chest X-ray images. Proc. of the First International Conf. on Pattern Recogition:125-137, 1973.
(19) J Hasegawa, J Toriwaki, T Fukumura, Software system AISCR-V3 for automated screeing of chest photofluorograms, Trans. of the Institute of Electornics and Communication Engineers of Japan, J66-D(10):1145-1152, 1983 (Japanese).
(20) H Suzuki, N Inaoka, H Takabatake, M Mori, H Natori, A Suzuki, An experimental system for detecting lung nodules by chest X-ray image processing. Proc.SPIE Conf. On Physiology and Function from Multidimensional Images, Biomedical Image Processing II, 1450:99-107, 1991
(21) H Suzuki, N Inaoka, Development of a computer-aided detection system for lung cancer diagnosis, Proc.SPIE Conf. on Medical Imaging VI; Image Processing, 1652:567-751, 1992.
(22) H Yoshimura, M Giger, K Doi, H MacMahon, S Montner, Computerized scheme for the detection of pulmonary nodules. A nonlinear filtering technique, Investigat. Radiol., 27:124-127, 1992.
(23) M Giger, K Doi, H MacMahon, Image feature analysis and computer-aided diagnosis in digital radiography. Automated detection of nodules in peripheral lung fields, Med Phys., 15(2):158-166, 1988.
(24) Bv Ginnneken, BM ter Haar Romeny, MA Viergever, Computer-Aided Diagnosis in Chest Radiography. A Survey, IEEE Trans. on Medical Imaging, 20(12):1228-1241, 2001.
(25) J Hasegawa, T Tsutsui, J Toriwaki, Automated extraction of cancer lesions with convergent fold patterns in double contrast X-ray images of stomach. Trans. of the Institute of Electornics, Information and Communication Engineers (IEICE) of Japan, J73-D-II(4):661-669, 1990.
(26) Y Mekada, Y Hirano, J Hasegawa, J Toriwaki, H Ohmatsu, K Eguchi, Three dimensional concentration index-a new feature to evaluate vessels convergence in nodules in lung of 3D X-ray CT images. CAR'96 Computer Assisted Radiology(Proc. of the International Sym. on Computer and Communication Systems for Image Guided Diagnosis and Therapy, Paris, June 1996):1013, 1996.
(27) S Fukushima, H Tsujinaka, Computer-based discrimination of abnormal stomachs using radiograms. IEEE Trans. Biomedical Engineering, BME 32:353-362.
(28) T Soma, S Fukumura, Feature extraction and quantitative diagnosis of gastric roentgenograms. InK.Preston, M.Onoe (eds) Digital Processing of Biomedical Images, Plenum, New York:323-334.

(29) Y Mekada, J Hasegawa, J Toriwaki, S Nawano, K.Miyagawa, Automated extraction of cancer lesions from double cotrast X-ray images of stomach. Proc. 1st International Workshop on Computer-Aided Diagnosis, in K.Doi et al. eds. : Workshop on Computer-Aided Diagnosis in Medical Imaging, Elsevier Science B.V.:407-412, 1999.
(30) N Niki, Y Kawata, M Kubo, H Ohmatsu, R Kakinuma, M Kaneko, M Kusumoto, N Moriyama, A CAD system for lung cancer based on 3D CT images. Proc. of 15th International Congress and Exhibition, Computer Assisted Radiology and Surgery:701-705 (CARS2002), 2002.
(31) S Yamamoto, I Tanaka, M Senda, Y Tateno, T IInuma, T Matsumoto, Image processing for computer-aided diagnosis of lung cancer by CT (LSCT). Systems and Computers in Japan, 25(2):250-260, 1994.
(32) S Yamamoto, H Takizawa, H Jiang, T Nakagawa, T Matsumoto, U Tateno, T Iinuma, M Matsumoto, A CAD system for lung cancer screening test by X-ray CT. Proc. of 15th International Congress and Exhibition, Computer Assisted Radiology and Surgery:605-610 (CARS2001), 2001.
(33) T Okumura, T Miwa, J Kako, S Yamamoto, M Matsumoto, Y Tateno, T IInuma, M Matusmoto, Image processing for computer-aided diagnosis of lung cancer screening system by CT (LSCT). Proc. of SPIE Medical Imaging 1998:1314-1322, 1998.
(34) N Niki, Y Kawata, M Kubo, A CAD system for lung cancer based on CT image. Proc. CARS2001:593-598, 2001.
(35) The Journal of Thoracic CT Screening, 1994 ~, Japan Society of Thoracic CT Screening, 2002.
(36) CI Henscke, DI McCauley, DF Yankelevitz, D Naidich, G McGuinness, OS Miettiinen, DM Libby, NK Altorki, JP Smith, Early lung cancer action project. overall desing and findings from baseline screening, THE LANCET, 354:99-105,1999.
(37) S Sone, S Takashima, F Li, Z Yang, T Honda, Y Maruyama, M Hasegaw, T Yamanda, K Kubo, K Hanamura, K Asakura, Mass screening for lung cancer with mobile spiral computed tomography scanner. THE LANCET, 351:1242-1245, 1998.
(38) T IInuma, Cancer screening and radiation. Japan Medical Association Journal, 44(6): 2283-289, 2001.
(39) K Mori, J Hasegawa, J Toriwaki, K Katada, Y Anno, Automated extraction of lung cancer lesions form multi-slice chest CT images by using three-dimensional image processing. Trans. of the Institute of Electronics, Information and Communication Engineers of Japan, J76D-II(8):1587-1594, 1993 (Japanese).
(40) K Mori, J Hasegawa, J Toriwaki, Y Anno, K Katada, A procedure with position-variant thresholding and distance transformation for automated detection of lung cancer lesions from 3-D chest X-ray CT images. Medical Imaging Technology, 12(3): 216-223, 1994.
(41) J Hasegawa, X Chen, J Toriwaki, Measurement of round opacities in the lung of X-ray images towards quantitative diagnosis of pneumoconiosis. Proceedings of the VIIth International Pneumoconioses Conference(1988.8):1358-1363, 1990.
(42) E Hall, R Kruger, S Dwyer III, D Hall, R Mclaren, G Lodwick, A survey of pre-processing and feature extraction techniques for radiographic images. IEEE Trans. Comput., C-20:1032-1044, 1971.
(43) R Kruger, J Townes, D Hall, S Dwyer III, G Lodwick, Automated radiographic diagnosis via feature extraction and classification of cardiac size and shape descriptors, IEEE Trans. Biomed. Eng., BME-19(3):174-186, 1972.
(44) N Sezaki and K Ukena, Automatic computation of the cardiothoracic ratio with application to mass screening. IEEE Trans. Biomed. Eng., BME-20:248-253, 1973.

(45) S Tsukuda, A Heshiki, S Katuragwa, Q Li, H MacMahon, K Doi, Detection of lung nodules on digital chest radiographs. potential usefulness of a new contralateral subtraction techniqaue, Radiology, 223(1):199-203, 2002.
(46) N Nikolaidis, I Pitas, 3-D Image Processing Algorithms. John Wiley and Sons, Inc., New York, 2001.
(47) G T.Herman, Geometry of Digital Spaces, Birkhauser, Boston,1998.
(48) G Bertrand, A Imiya, R Klette eds., Digital and Image Geometry, Advanced Lectures, LNCS 2243, Springer Verlag, Heidelberg, 2001.
(49) R Klette, A Rosenfeld, F Sloboda eds., Advances in Digital and Computational Geometry, Springer, 1998.
(50) D Caramella, C.Bartolozzi eds., 3D image processing Techniques and Clinical Applications, Springer, New York, 2002.
(51) J Toriwaki, Three Dimensional Digital Image Processing, Shokodo, Tokyo, 2002 (Japanese).
(52) Abstracts, International Symp. on the Virtual Human Body state of the arts and visions for medicine, Hamburg, 2002.7.1, 2002.
(53) J Toriwaki, Virtualized human body and navigation diagnosis, BME (Journal of Japan Society of Medical Electronics and Biological Engineering, 11(8):24-35, 1997 (Japanese).
(54) T Yasuda, Y Hashimoto, S Yokoi, J Toriwaki, Computer system for craniofacial surgical planning based on CT images, IEEE Trans. on Medical Imaging, 9(3):270-280, 1990.
(55) Proc. of MICCAI (Medical Image Computing and Computer Assisted Intervention) 2001, 2002.
(56) V Visualization, IN Bankman ed., Hndbook of Imaging, Academic Press, London, 2000.
(57) J Toriwaki, K Mori, Visualization of the human body toward the navigation diagnosis with the virtualized human body, Journal of Visualization, 1(1):111-124, 1998.
(58) A Watt, F Policarpo, The Computer Image, Addison-Wesley, U.S.A., 1998.
(59) RA Drebin, L Carpenter, P Hanrahan, Volume rendering. Computer Graphics. Proc. ACM SIGGRAPH'88, 22(3):65-74, 1988.
(60) M Levoy, Display of surfaces from volume data. IEEE Computer Graphics & Applications 8(3):29-37, 1988.
(61) A Kaufman, Volume Visualization, IEEE Computer Society Press, 1991.
(62) P Rogalla, JT van Scheltinga eds., Virtual endoscopy and related 3D techniques. Springer, New York, 2001.
(63) K Mori, Y Suenaga, J Toriwaki, Fast volune rendering based on software optimisation using multimedia instructions on PC platforms, Proc. of CARS2002 : 467-472, 2002.
(64) M Yoshioka, K Mori, Y Suenaga, J Toriwaki, Fast volume rendering by software: usefulness in a virtual endoscopy system. Medical Imaging Technology, 190(6):477-486, 2001.
(65) DJ Vining, AR Padhani, S Wood, EA Zerhouni, EK Fishman, JE Kuhman, Virtual bronchoscopy: a new perspective for viewing the tracheobronchal tree. Radiology, 189, 1993.
(66) Special issue: Virtual Endoscopy-Present Status and Perspective, Clinical Gastroenterology. (RINSYO SYOKAKI NAIKA), 17.6 NIPPON MEDICAL CENTER, TOKYO, 2002 (Japanese) .

(67) K Mori, J Hasegawa, J Toriwaki, S Yokoi, H Anno, K Katada, A method to extract pipe structured components in three dimensional medical images and simulation of bronchus endoscope images. Proc. of 3D Image Conf. 94:269-274, 1994 (Japanese).

(68) K Mori, J Hasegawa, J Toriwaki, Y Anno, K Katada, Extraction and visualization of bronchus from 3D CT images of lung. N Ayache ed.: Computer Vision, Virtual Reality and Robotics in Medicine (Proc. First International Conference, CVRMed'95, Nice, France, April 1995), Lecture Note in Computer Science, 905:542-548, Springer.

(69) J Toriwaki, Study of computer diagnosis of X-ray and CT images in Japan - a brief survey, Proc. of IEEE Workshop on Biomedical Image Analysis:155-164, 1994.

(70) 2002 Scientific Program, Radiological Society of North America, 2002.

(71) Y Ge, DJ Vining, DK Ahn, DR Stelts, Colorectal cancer screening with virtual colonoscope. Proc.SPIE Conf. on Physiology and Function from Multidimensional Images, 3660:94-105, 1999.

(72) G Wang, SB Dave, BP Brown, Z Zhang, EG McFarland, J Haller, MW Vannier, Colon unraveling based on electrical field - Recent progress and further work. Proc.SPIE Conf. on Physiology and Function from Multidimensional Images, 3660:125-132, 1999.

(73) K Mori, Y Hoshino, Y Suenaga, J Toriwaki, J Hasegawa, K Katada, An improved method for generating virtual stretched view of stomach based on shape deformation. Proc. of CARS 2001, :425-430, 2001.

(74) Y Hayashi, K Mori, J Hasegawa, Y Suenaga, J Toriwaki, A Method for detecting undisploayed regions in virtual colonoscopy and its application to quantitative evaluation of fly-through methods, Proc. of MICCAI 2002, pp.631-638, 2002

(75) P Suetens, Fundamentals of Medical Imaging. Cambridge University Press, Cambridge, 2002.

(76) JV Hajnal, DLG Hill, DJ Hawkes, Medical Image Registration. CRC Press, Boca Raton, 2001.

(77) RA Robb, Guest Editorial : The biomedical imaging resource at Mayo Clinic. IEEE Trans. on Medical Imaging, 20(9):854-867, 2001.

(78) JK Udupa, GT Herman, Guest Editorial : Medical image reconstruction. processing, visualization, and analysis : the MIPG perspective, IEEE Trans. on Medical Imaging, 21(4) :281-295, 2002.

(79) KH Hoehne, Guest Editorial : Medical Image Computing at the Institute of Mathematics and Computer Science in Medicine. University Hospital Hamburg-Eppendorf, IEEE Trans. on Medical Imaging, (In press)

(80) J Duncan, N Ayache, Medical image analysis : Progress over two decades and the challnges agead. IEEE Trans. on Pattern Analysis and Machine Intelligence, 22(1):85-106, 2000.

(81) Proceedings of SPIE, 4683, 2002.

(82) Special issue on Technologies for next generation medical image processing. Trans. IEICE, Japan, J83-D-II, 2000.

(83) Special issue on Virtual and augmented reality in medicine, Proc.IEEE, 86(3), 1998.

(84) Special issue on Computer-aided intervention, IEEE Trans. on Medical Imaging, 17(10), 1998.

(85) Special issue on computer-aided diagnosis, IEEE Trans. on Medical Imaging, 20(12),2001.

(86) JS Suri, SK Setarehdan, S Singh eds., Advanced Algorithimic Approaches to Medical Image Segmentation. Springer, 2002.

(87) RN Strickland, Image-processing techniques for tumor detection, Mercel Dekker, Inc., N.Y., 2002.
(88) S Achenbach, W Moshage, D Ropers, M Bachmann, Curved multiplannar reconstruction for the evaluation of contrrast-enhancced electorn beam CT of the coronary arteries. AJR, 170:895-899, 1998.
(88) K Mori, D Deguchi, J Sugiyama, Y Suenaga, J Toriwaki, CR Maurer Jr., H Takabatake, H Natori, A method for camera motion tracking of a bronchoscope using epipolar geometry analysis and intensity image registration of real and virtual endoscopic images, Medical Image Analysis, 6:321-336, 2002

10

Image Processing for Intelligent Transport Systems

Shinji Ozawa
Keio University, Tokyo, Japan

SUMMARY

Image processing of vehicle's is discussed in this chapter. When a vehicle is in a factory, image processing is applied for design and inspection, and when a vehicle is on the road, image processing is useful for Intelligent Transport Systems, which recently have been widely developed. There has been much research and implementation using image sensors to get information for traffic control and vehicle control. The image seen from a camera located beside or on the road can be used for vehicle detection, velocity of a car or car group measurement, parking car detection, etc. Moreover, the image seen from a camera located in the vehicle can be used for preceding car detection, measurement of distance to the preceding car, obstacle detection, lane detection, etc. In this chapter, studies of image processing for vehicles on the road are described.

1. INTRODUCTION

Recently, the ITS (intelligent transport systems) have been widely developed.
It consists of three sub-systems such as

1. ATMS (advanced traffic management systems),
2. ATIS (advanced traveler information systems), and
3. AVCS (advanced vehicles control systems).

ATMS observe the traffic flow, and controls the road and traffic. ATIS obtain road and vehicle condition widely and broadcast them to drivers. AVCS operate vehicles automatically according to measured data.

Technology for ITS include sensing for vehicle detection, vehicle-vehicle and vehicle-roadside communication, automatic vehicle control, and human interface to give information to drivers.

Originally, civil engineering measured traffic parameters. Instead of manual data collection, traditional sensors have been used to collect data. If the condition is controlled, for example if cars move just under the sensor with an assumed direction and assumed speed, a loop coil can accurately detect whether the car is under the sensor or not. But unfortunately sometimes a car passes through so slowly, or a car moves in a direction which is not assumed. In this case, the results of the sensor are not correct.

When the image sensor is used, many algorithms can be applied not only to detect the presence of cars(, the same function as loop coil), but also to detect the car motion, etc. The advantage of image processing is because that image includes information is spread along the road, called spatial information, and the image sequence includes information of the loci of the object, called temporal information. Image processing can achieve various requirements with high accuracy.

In this chapter, image processing for driving assistance , for traffic control, and for supervisory and road control facilities are discussed. Much research has been done, both in machine vision and its applications.

The early papers about ITS using image processing appeared around 1974, for traffic flow measuring by Takaba [1](1974), velocity and vehicle classification by Takaba [2](1977), incident detection by Takaba [3](1984), vehicle-license number recognition by Williams [4](1989), road signpost detection by Akatsuka [5](1987), and queue detection by Hoose [6](1989).

On the other hand, research about autonomous land vehicles was developed by Waxman [7](1987), Kanade [8](1988), Turk [9](1988), Dikninson [10](1990), and others. And they had been specified for roads following method such as by Tsugawa [11](1979), Brooks [12](1981), Ozawa [13](1986), Graefe [14](1986), Liou [15](1987), DeDementhon [16](1987), and Koyama [17](1988).

Application areas were organized in Japan, U.S.A., and Europe, as VERTIS (VEhicle and Road Transport Information Systems), ITS-America (Intelligent Transport Systems, America) and ERTICO (Intelligent Transport Systems, Europe), respectively. They have had many projects: ADVANCE (ATMS/ATIS), PATH (AVCS) in the united states, PROMETHEUS and DRIVE in Europe, AHS (Automated Highway Systems), TSCS (Traffic Signal Control System), VICS (Vehicle Information and Control System), and many projects organized by the government in Japan.

Proposals of new algorithms and reports of field tests have appeared in the journals or at conferences, such as the ITS AMERICA Annual Meeting, The International Conference on Road Traffic Monitoring (RTM), the International Symposium on Automotive Technology and Automation (ISATA), The Vehicle Navigation and Information Systems Conference (VNIS), and the ITS World Congress.

From a machine vision point of view, the technologies are specified as an object recognition problem. But there are still many difficulties. For example, it must work outdoors where illumination is not controlled and objects are hardly defined by the intensity of the pixels. In this chapter, first the requirements from an application are described and then image processing for this purpose is discussed.

2. INTELLIGENT TRANSPORT SYSTEMS (ITS)

2.1 Relation Between Road, Drivers, and Vehicles

Figure 1 shows the relation between driver, vehicle, and road. Originally, the vehicle and the road were separately developed, in the sense of technologies. However, by driving, the driver observes around his vehicle, recognizes road conditions, and then operates the vehicle.

Today, some advanced technologies are located in a vehicle, for example, power steering, automatic gear shift, anti-lock breaks, and so on. They help drivers to operate the vehicle. If a sensor can detect the preceding vehicle, measure the distance to that vehicle, detect a traffic accident or traffic congestion far from the vehicle, then it helps the driver's observation and recognition. Figure 1 shows such a relationship. If these function are fully automated, it is called "automatic driving", and is the aim of the AHS (Automated Highway System).

2.2 The Concept of ITS

ITS consists of 9 development areas for 20 users services. The main users are drivers, carriers, and management agencies. The 9 development areas are :

1. Advances in navigation system.
2. Electronic toll collection.

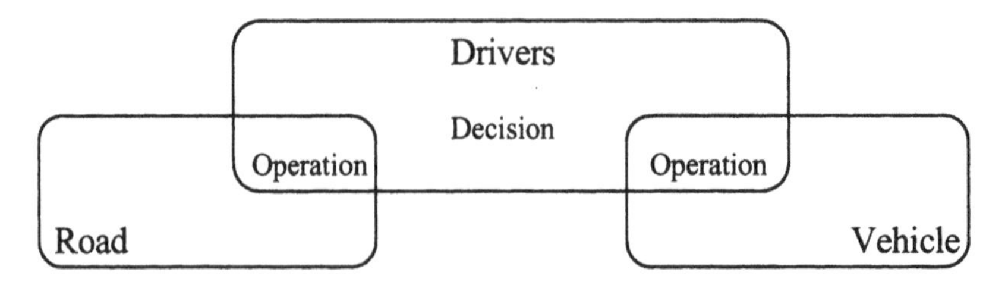

Figure 1 Relation between the driver, vehicle, and road.

3. Assistance for safe driving.
4. Optimization of traffic management.

5. Increasing efficiency in roads.
6. Support for public transportation.
7. Increasing efficiency in commercial vehicle operation.
8. Support for pedestrians.
9. Support for emergency vehicle operations.

3. MEASUREMENT IN ITS

3.1 Measurement System

Figure 2 shows a vehicle on the road and a display board, which are examples of the information collecting and display equipment.

Usually, as shown in Figure 3, traffic control, regulation, or caution action are done manually by a traffic officer at a control center while gathering the basic information. Some of the traffic caution is done automatically by a system such as to display a message on a presentation board for drivers. Figure 3 shows a graphic representation of a road and a vehicle on the road. The Φ mark shows equipment, Θ and O show a vehicle and the event to be detected. Sensing should be done at many fixed points on the road. Traffic flows and the presence of a car stopping may occur in the middle lane, in the branch or merge section, or in the intersection. That information is sent to the control center.

(a) Vehicle on the road

(b) Display board

Figure 2. Measurement system.

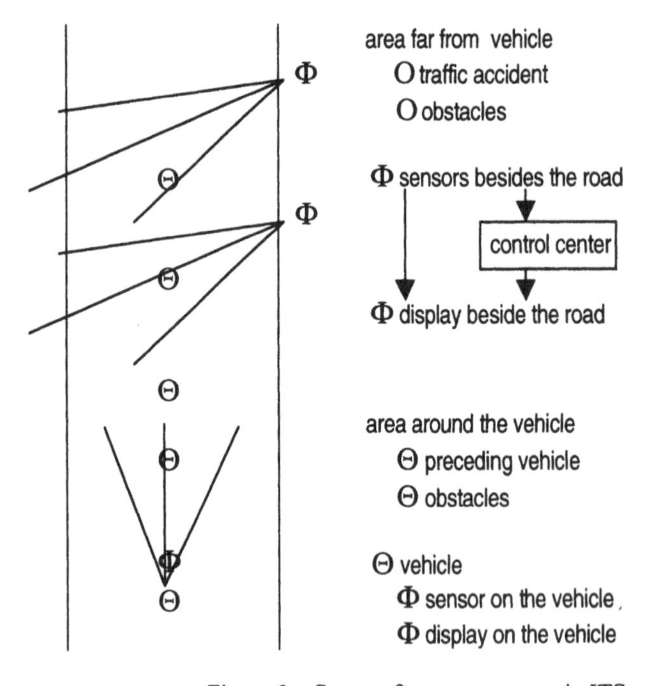

Figure 3. System for measurement in ITS.

In the end, two kinds of images are shown in Figure 4. They are input for image processing.

(a) Image seen from above the road (b) Image seen from the vehicle

Figure 4. Input images.

3.2 Item to Be Measured

Often the item to be measured can not be measured directly, because of the difference between the item which can be detected by physical sensors and that which is required by the system. The requirements are

1. Information about the lane.
2. Information of route.
3. Behavior of a driver.
4. Traffic conditions.

Information about the lane is needed for safe driving without going out of the lane. The items to be measured are shape of the lane mark, condition of the surface of the road, obstacles in front of the vehicle and so on. Usually drivers have many routes to reach a destination. The position of the vehicle and information of traffic congestion of the road or parking lot, can help drivers to select the route.

Drivers do not always drive carefully or sometimes doze off. If a system can detect poor behavior, it can signal a warning to the driver for the safety.

Recent progress of hardware and software includes analyzing an image and may detect traffic condition such as density of the vehicle, group velocity, and abnormal driving behavior.

3.3 Sensors and Detection

There are special purpose instrumentation and sensors to detect vehicles as is shown in Table 1. There are available for linear traffic flow and due existence/velocity of the vehicle.

Table 1 Sensors to detect vehicle

Sensors	Principle
air pipe, optical fiber (pressure)	Existing
supersonic wave (time flying)	Existing
microwave, Infrared	
supersonic wave, microwave (frequency shift) microwave, infrared	Moving
loop coil	Moving

Recently, communication using the radio, LCX or wireless IC card were applied to vehicle identification. Sensors to detect a driver's behavior (driving technique, fatigue, dozing) have been used to monitor brain waves, ECG, eye motion, face image and so on.

Sensors to detect the condition of the atmosphere are physical sensor (vibration, friction, slippage, water, temperature) and supersonic wave and laser radar, which determine the distance to the preceding vehicle. Image processing can not only be a substitute for special purpose instrumentation but also can process spatially and

(a) $f_t(x,y)$

(b) $grad \cdot f_t(x,y)$

Figure 5 The differential images.

temporally diverse information. Image processing for ITS is strongly needed for most of these requirements.

4. IMAGE SENSOR

The first step of traffic measurement is vehicle detection as shown in Figure 5(a). It can not be done using only a thresholding technique, because sometimes the background and the vehicle have the same intensity. Differential images are used for this purpose and there are three ways.

The first is spatial difference, which is shown in Figure 5(b). The edge of the lane separator appears in the differential image.

The second is a difference in the background image as $g_t(x,y) = f_t(x,y) - b_0(x,y)$ where $b_0(x,y)$ is the image without any cars. The differential image $g_t(x,y)$ should have a large value only if the pixels correspond to the vehicle. Figure 6(a) shows the differential image. But the background image changes because of a time-varying environment, so much noise remains on the vehicle as well as on other parts. There have been many studies to acquire an adaptive background image $b_t(x,y)$, such as by Kagehiro[18], Inoue[19], Long[20], Karman[21].

The last is a temporal difference as $g_t(x,y) = f_t(x,y) - f_{t-1}(x,y)$. Figure 6(b) shows the temporal difference image. But shadows of the vehicle appear. And background image without shadows is required, as researched by Fujimura[22], Kileger[23], and Wixson[24].

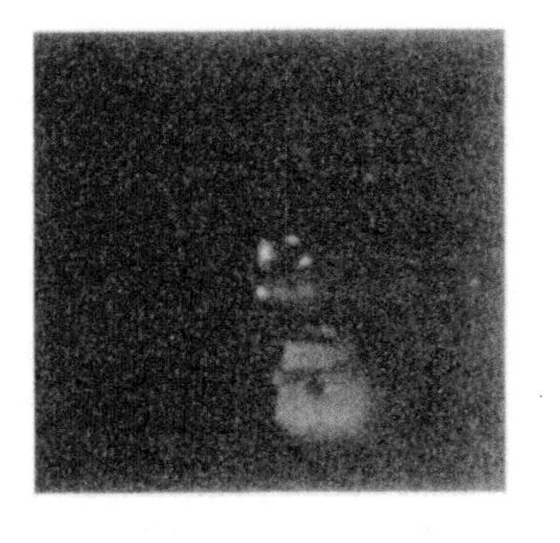

(a) $f_t(x,y)-b_t(x,y)$ (b) $f_t(x,y)-f_{t-1}(x,y)$

Figure 6. The differential images

5. IMAGE PROCESSING FOR DRIVING ASSISTANCE

5.1 Finding a Driving Lane Mark

Usually, the first step of vehicle control is to extract the lane mark from the road image. It is relatively easy to extract this from a continuous straight edge in fine conditions as is seen in Figure 7(a) Ishikawa[25], but is difficult to extract from an intersection or dotted lane mark Nohsoh [26], or in rainy conditions as is seen in Figure 7(b) and investigated by Tamai [27], Tamai[28], and Takahashi [29].

The second step is 3D reconstruction of the road, which is to transform pixels $p(x,y)$ in the image to 3D coordinates $P(X,Y,Z)$ The highway is constructed under regulations of road construction. Those rules are also used for reconstruction, and were investigated by Ozawa [13], Sakurai [30], Zen [31], and

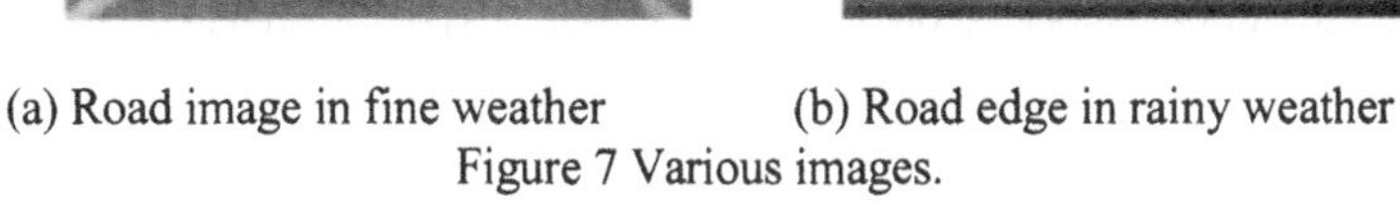

(a) Road image in fine weather (b) Road edge in rainy weather

Figure 7 Various images.

Yang [32]. A parallel assumption of both edges of the road allows the use of vanishing points to extract the slope of the road or tilt angle of the camera Kluge [33], Nohsoh [34]. The assumption that the road surface in front of vehicle is horizontal and the road segments are orthogonal allows one to extract corresponding points from pixels on the road edge. Edge point pairs $P_l(X_l, Y_l, Z_l)$ and $P_r(X_r, Y_r, Z_r)$ correspond to image point pairs $p_l(x_l, y_l)$ and $p_r(x_r, y_r)$, respectively as researched by Kobayashi [35], and Tanba [36]. And when the lane marks are approximated as a polynomial curve, the curve pair and some basic rules can reconstruct 3D information of the road (Nohsoh [37], Nohsoh [38]). In fact, this problem requires the freedom of six degrees from the vehicle position. More accurate measurements are required to separate road configuration and vehicle position (Nohsoh [37], Nohsoh [38]). Dynamics of the road and vehicle are applied to this problem by Tanba [39], Negishi [40], and Horie [41].

5.2 Detecting the Preceding Car

There have been many proposals to detect a preceding vehicle such as by Yoshizawa [42], using optical flow Chiba [43], Gicchatti [44], Ninomiya [45], using correlations of images Okuno [46], and using stereo matching Saneyoshi [47], Shimomura [48], Shimomura [49]. These algorithms can also be applied to detect a following vehicle (Tsunashima [50]). It works better than the case of preceding vehicle detection, because the direction of the background scene is contrary to that of a following vehicle. When a single camera is used the assumption that shadow of the vehicle lays just on the road, or the width between the tail lamps of the preceding vehicle is known, then the 3D distance to the preceding vehicle can be calculated.

5.3 Detecting Obstacles

The direction guiding equipment and vibration sensors on the driving lane are an examples of the obstacle detection. It can be assumed that dangerous material has height, and only if the object is located in the lane is it considered as obstacle. These assumptions are used for in the methods by Mallot [51], Ferrari [52], and Carlson [53].

6. IMAGE PROCESSING FOR TRAFFIC CONTROL

6.1 Finding a Moving Car

When the camera is located on or beside the road, a system can detect the

(a) $f_t(x, y)$ (b) $f_{t+1}(x, y)$

Figure 8 Traffic measurement in the tunnel.

existence of a car or a car passing through the region, by processing within a fixed window that corresponds to a fixed area on the road surface by spatial differentiation.

By the detection of shadows on the road surface just under the vehicle or the detection of the brightness of the tail lamp, the presence of a car is detected (Sakai [55], Sasaki [56], Kagesawa [57]). In a tunnel the light condition is good, so temporal difference works well to find the pixels corresponding to the moving car. Figure 8 shows the results of this system, The shadows under the vehicle are well detected. The existence and velocity of this vehicle is calculated and can be determined that the case of (a) and (b) is "normal."

6.2 Finding Accidental Stopping

Tracking a moving region extracted from the temporal difference may give loci of a moving object. Short loci can be neglected as noise, and a short gap can be connected easily using its motion history. Congestion and accidents are extracted by this process as investigated by Momozawa [58], Williams [60], Tsuge [61], Kimachi [62], and Iwata [63].

Instead of the complicated future matching method, projection of a moving region is provided (Furusawa [59], Furusawa [64]). It gives $proj_x[y]$ frame by frame and it can be considered a tempo spatial map. It is easier to analyze this than to analyze the moving region. By analysis, detection of congestion, traffic accidents, and avoidance can be done by Kamijo[65].

6.3 Counting Cars in Parking Lot

At a parking lot located in rest area facilities of the expressway, counting cars is required as a service to drivers. Many methods to count cars have been proposed. When parking areas are drawn, it is useful for an algorithm to have determined

(a) $D_t(x,y)$ (b) loci of the vehicle

Figure 9 Vehicle counting in a parking lot.

such as cars entering or exiting from the area, Ueda [66], occlusion of a separation line shows the existence of a vehicle Kawakami [67], and characteristic values inside the area that are changed Mizukoshi [70] and Maeda [71].
The method of tracking the moving objects for the whole area of the outdoor parking lot was proposed. Figure 9 shows that it is effective to use time differential images to extract moving objects from stationary objects (Hasegawa [68], Hasegawa [69]). However, a moving object can often be taken from many regions in the differential images. Therefore, the method to match two differential images and to link moving regions, allows a system to connect the reliable moving regions.

6.4 Estimating Traveling Time

Various sensors and algorithms are used to estimate traveling time. The future traveling time should be estimated from the history of past traveling time from here to there. Past traveling time is measured several ways. However, it can be measured from the length of a traffic jam and average velocity at a fixed point on the road, and it has become a popular measuring method to use car identifications at entrance and exit of the road by vehicle license plate number reading as researched by Jonson [72], Kato [73], Kanayama [74], Williams [4], Dunn [75], Takaba [76], Do [77], Fahmy [78], Comelli [79], Nijhuis [80], Cui [81], and Kim[82]. Next civil engineering predicts the future travel time $T(t+1)$. It is calculated using an auto regression model of $v(t+1)$ as follows.

$$v(t+1) = \alpha_0 \cdot v(t) + \alpha_1 \cdot v(t-1) + \alpha_2 \cdot v(t) + \alpha_3$$

$T(t+1)$ is easily calculated from $v(t+1)$.

6.5 Others

It was required to classify a vehicle as "large" or "small" by its size as investigated by Rourke [83], and Hsu [84].

It was also required to measure traffic flow at an intersection and was done after two dimensional analysis of vehicle movement by Aoki [85], Yamamoto [86], and Ichihara[87].

7. NEW SENSOR AND ITEM TO MEASURE

7.1 Enlarging the Dynamic Range of the Video Camera

Conventional TV cameras have a dynamic range of 0 to $5 \cdot 10^2 cd/m^2$ where as outdoor image processing requires a dynamic range of 0 to $10^4 cd/m^2$. Usually an automatic iris process is used, and it creates a partially saturated image, as shown in Figure 10. To evaluate the intensity of temporal average (pixel by pixel) gives a suitable condition of exposure or shutter speed. This idea and its realization were proposed by Yamada [88].

7.2 Enlarging the View Angle of Video Camera

Usually the entire view angle of a video camera is about 30°, and sometimes this is too narrow. When a fish eye lens is used, the view angle is extended to 180° or more, however it serves the non-linear image as well as the low resolution image. When the fish eye lens is used and the vehicle moves straight, image sequence including front view, side view, and rear view of the vehicle can be obtained with only one fixed camera. A 3D model of the vehicle can be constructed as shown by Kato [89]. Multi-cameras are also available to synthesize wide view as demonstraited by Ichihara [90], and Kanamoto [91].

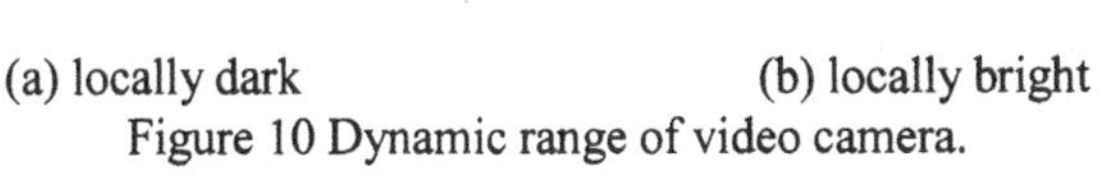

(a) locally dark (b) locally bright

Figure 10 Dynamic range of video camera.

7.3 Detecting an Occluded Vehicle

When traffic is congested, the regions of two vehicles are merged and thus difficulty in separation by algorithms was investigated by Koller [92], Beymer|cite [93], Ikeda [94], Kuboyama [95], and Gradner[96].

7.4 Detecting an Obstacle

A tidal wave, an avalanche of earth and rocks, falling rocks, shoulder erosion and land slides impede normal traffic flow. It is expected that a detection and warning system for these events can be developed.

7.5 Measuring Visibility

One primary factor of driving is the change of weather phenomena. For example, bad visibility and condition of the surface of the road, because of rain or snow, is dangerous. Under such conditions, traffic accidents and congestion increase. Weather information is related to the condition of the road surface, visibility and traffic flow, and it is expected that integration of such information gives an accurate understanding about the road conditions.

There have been many method investigated but one uses polarized light and is available.

7.6 Detecting Fire

To maintain safe driving, it is important to detect fire immediately, so optical fiber, infrared cameras, and pyro-electric sensors are used.

8. CONCLUSIONS

Image sensing techniques in vehicle and traffic control were surveyed in this chapter. First, a sensor in ITS was described. Next, research about practical applications using image processing were introduced with several techniques: 1) to find a moving car, 2) to find accidental stopping, 3) to count cars in a parking lot by tracking, and 4) to measure the travel time. Finally, a new sensor and item to measure was discussed.

However some problems still remain, but it is expected to be solved with the development of the contributions of computer vision fields.

REFERENCES

[1] S.Takaba, T.Taniguchi, and F.Tashiro, "The traffic flow measuring system based on extraction of image information," Proc. IEICE Conf. no.2396, 1974. (in Japanese).

[2] T.Kaneko and S.Tanaka, "A method to measure velocity and class of vehicle using image based traffic flow measuring system," Proc. IEICE Conf., no.1017, 1979 (in Japanese).

[3] S.Takaba, "Incident Detection in Street Network Using Real-Time Simulation," Proc. of SCSC, no.1, pp.348-353, 1984.

[4] P.G.Williams, "Evaluation of Video-Recognition Equipment for Number-Plate Matching," RTM'89 no.7, pp.89, 1989.

[5] H.Akatsuka and S.Imai, "Road signposts recognition system," Proc. of S.A.E., pp.189--196, Feb.1987.

[6] N.Hoose, "Queue Detection Using Computer Image Processing," RTM'89, no.8, pp.94, Feb.1987.

[7] A.M.Waxman, J.LeMoigne, L.S.Davis, B.Srinivasan, T.R.Kushner, E.Liang, and T Siddalingaiah, "A visual navigation system for autonomous land vehicles," IEEE Journal of Robotics and Automation, vol.3, no.2, pp.124--141, 1987.

[8] C.Thorpe, M.H.Hebert, T.Kanade and S.A.Shafer, "Vision and Navigation for The Carnegie-Mellon NAVLAB," IEEE Trans. PAMI, vol.10, no.3, pp.362-373, 1988.

[9] M.A.Turk, D.G.Morgenthaler, K.D.Gremban, and M.Marra, "VITS-A Vision system for autonomous land vehicle navigation," IEEE Trans.PAMI, vol,10, no.3, pp.342--361, 1988.

[10] S.J.Dickinson and L.S.Davis, "A flexible tool for prototyping ALV road following algorithms," IEEE Tran. Robotics and Automation, vol.6, no.2, pp.232--242, 1990.

[11] S.Tsugawa, T.Yatabe, T.Hirose, and S.Matsumoto, "An automobile with Artificial intelligence," Proc.IJCAI, no.6, pp.893--895, 1979.

[12] R.A.Brooks, "Symbolic reasoning among 3D models and 2D images," Artificial Intelligence, vol.17, no.1, pp.285--348, 1981.

[13] S.Ozawa, A.Rosenfeld, "Synthesis of a Road Image as Seen from a Vehicle," Pattern Recognition, Vol.19, No.2, Feb.1986.

[14] E.D.Dickmanns and V.Graefe, "Dynamic Monocular Machine Vision," MVA'88, pp.223--240, 1988.

[15] S.Liou, "Road following using vanishing points," Computer Vision, Graphics and Image processing, vol.39, pp.116, 1987.

[16] D.DeMenthon, "A zero-bank algorithm for inverse perspective of a road form a single image," Proc.IEEE International Conference on Robotics and Automation, pp.1444--1449, 1987.

[17] C.Koyama, K.Watanabe and K.Kanatani, "Reconstruction of 3D road geometry from images for ALV," IIPJ Tech.Rep., 88-CV55, no.1, 1988. (in Japanese).

[18] T.Kagehiro, and Y.Ohta: "Adaptive Background Reconstruction from Road image sequence", Sensing Symposium Technology at Industry pp.15-20, 1994 (in Japanese).

[19] K.Inoue and W.Seo, "Background image generation by cooperative parallel processing under severe outdoor condition," MVA'92 pp.215--218. 1992.
[20] W.Long and Y.Yang, "Stationary background generation: an alternative to the difference of two images," Pattern Recognition vol.23, no.12, pp.1351--1359. 1990.
[21] K.P.Karman and A.Brandt, "Moving object recognition using an adaptive background memory," Proc. Time varying Image Processing, 1990.
[22] K.Fujimura, T.Hasegawa, S.Ozawa, "A Method for Generating Adaptive Background to Detect Moving Object," Trans.SICE., vol.33, no.9, pp.963--968, Sept.1997 (in Japanese).
[23] M.Kilger, "A shadow handler in a video-based real-time traffic monitoring system," Proc. IEEE workshop on Applications of computer vision pp.1060--1066. 1992.
[24] L. Wixson, "Illumination assessment for vision-based traffic monitoring," IEEE Proc. ICPR, vol.2, pp.56--62, 1996.
[25] S.Ishikawa, H.Kuwamoto, and S.Ozawa, "Visual Navigation for Autonomous Vehicle Using White Line Recognition," IEEE Trans. PAMI, no.11, Nov 1987.
[26] K.Nohsoh and S.Ozawa, "A Real Time Lane Marker Detection Method for Highway Driving Road Images," IEICE Trans.Inf.&Syst., vol.J74-D-II, no.5 pp.662--666, May 1991(in Japanese).
[27] Y.Tamai, T.Hasegawa and S.Ozawa, "The Lane Detection under Rainy Condition," IEJ Tech.Repo., RTA-95-16, June1995 (in Japanese).
[28] Y.Tamai, T.Hasegawa and S.Ozawa, "The Ego-Lane Detection under Rainy Condition," Third Annual World Congress on ITS, no.3258, 1996.
[29] A.Takahashi, Y.Ninomiya, M.Ohta, and K.Tange, "Lane detection method to improve robustness in various environments," IEICE Tech.Rep.,PRMU98-93, pp.9--14, 1998 (in Japanese).
[30] K.Sakurai, H.Zen, H.Ohta, Y.Ushioda, and S.Ozawa, "Analysis of a Road Image as Seen from a Vehicle," Proc. of IEEE First International Conference on Computer Vision, June 1987.
[31] H.Zen, K.Sakurai, T.Kobayashi, and S.Ozawa, "Analysis of a Road Image as Seen from a Vehicle," IEICE Trans.Inf.&Syst., vol.71-D, no.9 pp.1709--1717 , Sept.1988 (in Japanese).
[32] J.Yang and S.Ozawa, "Correspondence in Road Image Sequence," IEICE Trans. vol.E79-A, no10, pp.1664--1669, OCT.1996.
[33] K.Kluge, "Extracting road curvature and orientation from image edge points without perceptual grouping into features," Proc. of IV'94, pp.109, 1994.
[34] K.Nohsoh and S.Ozawa, "A Real Time Highway Image White Line Detection System Based on Vanishing Point Estimation," Trans. IEEJ, vol.113-C, no.2, pp.139--148 Feb.1993 (in Japanese).
[35] T.Kobayashi and S.Ozawa, "A Method to Extract Corresponding Points in Road Image Analysis," IEICE Trans.Inf.&Syst., vol.J71-D-II, no.5 pp.827--830 March 1989 (in Japanese).

[36] N.Tanba, T.Kobayashi and S.Ozawa, "Analysis of a Road Image Seen from a Vechicle --A Study of the Superelevation--," IEICE Trans.Inf.&Syst.,Vol.J73-D-II,No.1 pp.125-129 January 1990 (in Japanese).
[37] K.Nohsoh and S.Ozawa, "A Simultaneous Estimation of Road Structure and Camera Position from Continuous Road Images," IEICE Trans.Inf.&Syst., vol.J76-D-II, no.3 pp.514--523 March 1993 (in Japanese).
[38] K.Nohso and S.Ozawa, "A Simultaneous Estimation of Vehicle Location and CameraMotion Deduced from Road Configuration and Continuous Road Images," IEICE Trans.INF.&SYST., vol.77-D-II, no.4 April 1994 (in Japanese).
[39] N.Tanba, M.Chiba and S.Ozawa, "Identification of System Dynamics for the Vehicle from Road Image Sequences," IEICE Trans.Inf.&Syst., vol.J75-D-II, no.3 pp.490--499 March 1992 (in Japanese).
[40] S.Negishi, M.Chiba and S.Ozawa, "Automatic Tracking of Highway Road Edge Based on Vehicle Dynamics," IEICE Trans.Inf.&Syst., vol.77-D-II, no.5, May 1994 (in Japanese).
[41] H.Horie and S.Ozawa, "Recognition of Forward Environment on Highway from Monocular Road Image Sequences," IEEJ Trans., vol.117-C, no.5,pp.648--657, May.1997 (in Japanese).
[42] A.Yoshizawa, "A Model Based Preceding Vehicle Detection," Proc. ISATA'92, no.3, pp.537, 1992.
[43] M.Chiba and S.Ozawa, "Detection of Optical Flow by Mode of Intersections of Constraint Equations," ITEJ Trans. vol.45, no.10, pp.1199--1206, Oct.1991 (in Japanese).
[44] A.Giachettei, et.al., "The recovery of optical flow for intelligent cruise control," Proc. of IV'95, pp.91, 1995.
[45] Y.Ninomiya and M.Ohta, "Moving objects detection from optical flow," IEICE Tech.Rep.,PRMU97-28, pp.25--31, 1997 (in Japanese).
[46] A.Okuno, K.Fujita and A.Kutami, "Visual navigation of autonomous on road vehicle," Roundtable discusion on vision based vehicle guidance, 1990.
[47] K.Saneyoshi, K.Hanawa, Y.Sogawa, and K.Arai, "Stereo image recognition system for drive assist," IEICE Tech.Rep.,PRMU97-30, pp.39--46, 1997 (in Japanese).
[48] N.Shimomura, "A study on preceding vehicle tracking on curved roads using stereo vision," IEICE Tech.Rep.,PRMU97-27, pp.15--23, 1997 (in Japanese).
[49] N.Shimomura, K.Nohsoh, and H.Takahashi, "A study on measuring headway distance using stereo disparity and the size of a preceding vehicle on image," IEICE Tech.Rep.,PRMU98-95, pp.21--28, 1998 (in Japanese).
[50] N.Tsunashima and M.Nakajima, "Extraction of the front vehicle using projected disparity map," IEICE Tech.Rep.,PRMU98-94, pp.15--20, 1998 (in Japanese).
[51] H.A.Mallot, H.H.Bulthoff, J.J.Little, and S.Bohrer, "Inverse perspective mapping simplifiers optical computation and obstacle detection," Biological Cybernetics, vol.64, no.3, pp.177--185, 1991.

[52] F.Ferrari, E.Grosso, G.Sandini, and M.Magrassi, "A Stereo Vision System for Real Time Obstacle Avoidance in unknown Environment," IEEE Int. Workshop on Intell,Robots and System,IROS'90.

[53] S.Carlsson and J.O.Eklundh, "Object Detection Using Model Based Prediction and Motion Parallax," Proc. Roundable Discussion on Vision-Based Vehicle Guidance '90no.5, pp.1--2, 1990.

[54] T. Sakai et al., "A Development of Traffic Measuring System in Tunnel using ITV Camera", Sumitomo Denki, vol.134, pp.86-92, 1989 (in Japanese).

[55] K.Sakai, "In Tunnel Traffic Flow Measuring and Monitoring System Using ITV Cameras," Proc. ISATA'92, pp.581. 1992

[56] K.Sasaki, "In-Tunnel Traffic Flow Measuring and Monitoring System Using ITV Cameras," Proc. ISATA'93, pp.639. 1993.

[57] M.Kagesawa, S.Ueno, K.Ikeuchi and H.Kashiwagi, "Recognizing vehicles in infrared images using IMAP parallel vision board," IEEE Trans. ITS, vol.2, no.1, pp.10--17, 2001.

[58] M. Momozawa et. al., "Accident Vehicle Automatic Recognition System by Image Processing Technology," IEEE International conference on Image Processing, pp.566-570, 1992.

[59] H.Furusawa et. al., "A Proposal of Traffic Accident Detection System using Image Processing, "SICE Trans. Pattern Measuring, vol.16, pp.9, 1991 (in Japanese).

[60] B.Williams, "CCATS-An Evaluation of Speed and Count Accuracy," Proc. ISATA'94, pp.405. 1994.

[61] A.Tsuge, H.Takigawa, H.Osuga, H.Soma, and K.Morisaki, "Accident Vehicle Automatic Detection System by Image Processing Technology," Proc. VNIS'94, pp.45, 1994.

[62] M.Kimachi, K.Kanayama, and H.Teramoto, "Incident Prediction by Fuzzy Images Sequence Analysis," Proc. VNIS'94, pp.51, 1994.

[63] T.Iwata, "Implementation and Evaluation of Incident Detection Equipment in Tunnel," Proc. ISATA'94, pp.441, 1994.

[64] H.Furusawa, "Accident Vehicle Image Detection System Using DTT Method," Proc. ISATA'92, pp.537, 1992.

[65] S.Kamijo, Y.Matsushita, K.Ikeuchi and M.Sakauchi, "Incident detection at Intersections utilizing hidden Markov model," 6th World Congress on Intelligent Transport Systems, 1999.

[66] K.Ueda, I.Horiba, K.Ikegaya, and H.Onodera, "An Algorithm for Detecting Parking Cars by the Use of Picture Processing" IEICE Trans. INF.&SYST., vol.J74-D-II, no.10, pp.1379--1389, 1992 (in Japanese).

[67] H.Kawakami, M.Tsutiya, T.Ikeda, and J.Tajima, "Detection of parking condition from image," Proc. SICE Tech. Rep. of Pattern measurment no.3, 1991 (in Japanese).

[68] T. Hasegawa, and S. Ozawa, "Counting cars by tracking of Moving Object in the Outdoor Parking Lot", IEICE Trans. INF.&SYST., vol.J76-D-II, no.7, pp.1390--1398, 1993 (in Japanese).

[69] T.Hasegawa, K.Nohsoh, and S.Ozawa, "Counting Cars by Tracking Moving Objects in the Outdoor Parking Lot," Proc. VNIS'94, pp.63, 1994.

[70] N.Mizukoshi, I.Horiba, K.Ueda, and H.Onodera, "Parking Condition Discrimination System of Image Processing Type Using a Neural Network Model(on Tokyo Ring Expressway Niikura Parking Area," Proc. VNIS'94, pp.69, 1994.

[71] E.Maeda, A.Shio and K.Ishii, "Robast object extraction using normalized principal component features," IEICE Trans., vol.J75-D-II, no.10, pp.1660--1672, 1992 (in Japanese).

[72] A.S.Johnson, "Evaluation of an Automatic Number-Plate Recognition System for Real-time Automatic Vehicle Identification," Proc. ISATA'91, pp.443, 1991.

[73] K.Kato, "Automatic License Plate Reader Utilizing Image Processing," Proc. ISATA'91, pp.435, 1991.

[74] K.Kanayama, "Development of Vehicle-License Number Recognition Apparatus Using Real Time Image Processing and its Application to Travel-Time Measurement," Proc. ISATA'92, pp.5229 1992.

[75] D.Dunn, "Compatibility Specifications for AVI Equipment in California," Proc. ISATA'92, pp.575, 1992.

[76] S.Takaba, "Estimation and Measurement of Travel Time by Vehicle Detectors and License Plate Readers," Proc. VNIS'91, pp.247, 1991.

[77] M.A.Do, "Integration of AVI System and the Enforcement Camera System in a Multi-lane Road Environment," Proc. VNIS'94, pp.75, 1994.

[78] M.M.Fahmy, "Automatic Number-Plate Recognition:Neural Network Approach," Proc. VNIS'94, pp.99, 1994.

[79] P.Comelli, P.Ferragina, M.N.Granieri and F.Stabilr, "Optical recognition of motor vehicle license plates," Trans. on IEEE Vehicular Technology, vol.44, no.4, pp.790--799, 1995.

[80] J.A.G.Nijhuis, etal., "Car license plate recognition by neural networks and fuzzy logic," Proc. of IEEE International networks and fuzzy logic. pp.2185--2903, 1995.

[81] Yuntao Cui and Qian Huang, "Character extraction of license plates from video," IEEE,CVPR'97, pp.502--507, 1997

[82] S. K. Kim. D. W. Kim and H. J. Kim, "A recognition of vehicle plate using a genetic algorithm based segmentation," Proc. IEEE ICIP, vol.2, pp.761--766, 1996.

[83] A.Rourke, "Automatic Road Vehicle Classification from Video Images," Proc. ISATA'92, pp.567, 1992.

[84] T.P.Hsn, "Model of Real-Time Automatic Vehicle Classification Using Image Processing," Proc. ISATA'94, pp.429, 1994.

[85] M.Aoki, "Road Traffic Measurement at an Intersection by an Image Sequence Analysis," Proc. ISATA'91, pp.427, 1991.

[86] T.Yamamoto, "Two-Dimensional Vehicle Tracking Using Video Image Processing," Proc. VNIS'92, pp.93, 1992.

[87] E. Ichihara, H. Takao and Y. Ohta, "Visual assistance for drivers using road-side cameras," Proc. IEEE Int. Conf. On ITS, pp.170-175, 1999.

[88] K.Yamada, M.Nakano, and S.Yamamoto, "Wide Dynamic Range Image Sensor for Vehicles",Sensing Symposium Technology at Industry, pp.1--4, 1994 (in Japanese).

[89] K.kato, T.Nakanishi, A.Shio, and K.Ishii, "Automobile's shape extraction from image sequence captured through a monocular extra-wide viewing angle," Sensing Symposium Technology at Industry, pp.27—32, 1994 (in Japanese).

[90] E.ichihara, H.Takao and Y.Ohta, "Bird's-eye view for highway drivers using roadside cameras," IEICE Tech.Rep.,PRMU98-98, pp.45--52, 1998.

[91] M.Kanamoto, I.horiba, and N.Sugie, "Wide-angle image in the parking situation image," IEICE Tech.Rep.,PRMU98-99, pp.53--58, 1998 (in Japanese).

[92] D.Koller, J.Weber and J.Malik, "Robust Multiple car tracking with occlusion reasoning," Proc. ECCV pp.189--196. 1994.

[93] D. Beymer and J. Malik, "Tracking vehicles in congested traffic," Proc. ITS pp.8--18. 1996.

[94] T. Ikeda, S. Ohnaka and M. Mizoguchi, "Traffic measurement with a road-side vision system -indevidual tracking of overlapped vehicles," IEEE Proc. of ICPR'96, no.5, pp.859--864, 1996.

[95] H. Kuboyama and S. Ozawa, "Measurement of heavy traffic in a tunnel from image sequences," IEICE Tech.Rep.,PRMU98-100, pp.59--66, 1998 (in Japanese).

[96] W. F. Gradner and D. T. Lawton, "Interactive model-based vehicle tracking," IEEE Trans. PAMI, vol.18, no.11, pp.1115--1121, 1996

Index